Placing Critical Geographies

This book explores the multiple histories of critical geography as it developed in 14 different locations around the globe, whilst bringing together a range of approaches in critical geography.

It is the first attempt to provide a comprehensive account of a wide variety of historical geographies of critical geography from around the world. Accordingly, the chapters provide accounts of the development of critical approaches in geography from beyond the hegemonic Anglo-American metropoles. Bringing together geographers from a wide range of regional and intellectual milieus, this volume provides a critical overview that is international and illustrates the interactions (or lack thereof) between different critical geographers, working across a range of spaces. The chapters provide a more nuanced history of critical geography, suggesting that while there were sometimes strong connections with Anglo-American critical geography, there were also deeply independent developments that were part of the construction of very different kinds of critical geography in different parts of the world.

Placing Critical Geographies provides an excellent companion to existing histories of critical geography and will be important reading for researchers as well as undergraduate and graduate students of the history and philosophy of geography.

Lawrence D. Berg is Professor of Critical Geography at the University of British Columbia in Canada. His research focusses on neoliberalism and the cultural politics of academic knowledge production. He is co-founder of *ACME: An International Journal for Critical Geographies.*

Ulrich Best is based at a federal research institute in Germany, where he works on labour conditions and workers' health and safety. He has published on borders, migration, urban exclusion and racism and on the history of critical geography in Germany.

Mary Gilmartin is a Professor in the Department of Geography at Maynooth University, Ireland. Her research focusses on contemporary migration and mobilities.

Henrik Gutzon Larsen teaches Human Geography at Lund University. His research addresses urban geography and housing, political geography and history of geographical thought.

Placing Critical Geographies

Historical Geographies of
Critical Geography

Edited by
Lawrence D. Berg, Ulrich Best,
Mary Gilmartin, and Henrik Gutzon
Larsen

Routledge
Taylor & Francis Group

LONDON AND NEW YORK

First published 2022
by Routledge
2 Park Square, Milton Park, Abingdon, Oxon OX14 4RN

and by Routledge
605 Third Avenue, New York, NY 10158

Routledge is an imprint of the Taylor & Francis Group, an informa business

British Library Cataloguing-in-Publication Data
A catalogue record for this book is available from the British Library

Library of Congress Cataloging-in-Publication Data
A catalog record has been requested for this book

ISBN: 978-1-4094-3141-1 (hbk)
ISBN: 978-1-4094-3142-8 (pbk)
ISBN: 978-1-315-60063-5 (ebk)

DOI: 10.4324/9781315600635

Typeset in Times NR MT Pro
by KnowledgeWorks Global Ltd.

Contents

List of illustrations

Figures

Tables

Boxes

Contributors

Nadia Abu-Zahra, University of Ottawa, International Development and Global Studies

Abel Albet, Universitat Autònoma de Barcelona, Departament de Geografia

Kye Askins, Everyday Activist

Bernd Belina, Goethe University, Department of Human Geography

Lawrence D. Berg, University of British Columbia, Institute for Community Engaged Research

Ulrich Best, Federal Institute for Occupational Safety and Health (BAuA), Germany

Michel Bruneau, Director of Research emeritus, CNRS-University of Bordeaux-Montaigne

Kerry Burton, University of Exeter, Department of Geography

Rodolphe De Koninck, Université de Montréal, Department of Geography

Elena dell'Agnese, Università di Milano-Bicocca, Dipartimento Di Sociologica e Ricerca Sociale

Robyn Dowling, University of Sydney, School of Architecture, Design and Planning

Ghazi-Walid Falah, American University in the Emirates (AUE)

Tamami Fukuda, Osaka Prefecture University, Graduate School of Humanities and Sustainable System Sciences

Maria-Dolors García-Ramon, Universitat Autònoma de Barcelona, Departament de Geografia

Mary Gilmartin, Maynooth University, Department of Geography

Takeshi Haraguchi, Kobe University, Graduate School of Humanities and Faculty of Letters

Richard Howitt, Macquarie University, Department of Geography and Planning

Henrik Gutzon Larsen, Lund University, Department of Human Geography

Ari Lehtinen, University of Eastern Finland, Department of Geographical and Historical Studies

Robyn Longhurst, University of Waikato, Department of Geography

Brij Maharaj, University of KwaZulu-Natal, Environmental Sciences

Claudio Minca, Università di Bologna, Dipartimento di Storia Culture Civiltà

Gustavo Montañez, Universidad de Externado, Bogotá

Koji Nakashima, Kanazawa University, Department of Geography

Matthias Naumann, Alpen-Adria-Universität Klagenfurt, Department of Geography and Regional Studies

Jo Norcup, University of Nottingham, School of Geography & University of Warwick, Centre for Caribbean Studies

Joe Painter, Durham University, Department of Geography

Linda Peake, York University, The City Institute at York University

Blanca Ramírez, Universidad Autónoma Metropolitana-Xochimilco, Departamento de Teoría y Análisis

Maano Ramutsindela, University of Cape Town, Department of Environmental & Geographical Science

Marcella Schmidt di Friedberg, Università di Milano-Bicocca, Dipartimento di Scienze Umane per la Formazione "Riccardo Massa"

Eric Sheppard, University of California, Los Angeles, Department of Geography

James D. Sidaway, National University of Singapore, Department of Geography

Kirsten Simonsen, Roskilde University, Department of People and Technology

Anke Strüver, University of Graz, Institut für Geographie und Raumforschung

Wing-Shing Tang, Formerly Hong Kong Baptist University, Department of Geography

Perla Zusman, Universidad de Buenos Aires, Facultad de Filosofía y Letras

1 Introduction

Placing critical geographies

Lawrence D. Berg, Ulrich Best,
Mary Gilmartin, and Henrik Gutzon Larsen

The primary objective of this book is to extend our understanding of the development of critical geographies (plural) across a wide range of academic settings. In doing so, we also see this volume as contesting what we term the hegemonic history of critical geography in contemporary geographical scholarship. This hegemonic history is part of a wider Anglo-American hegemony in geography (see, e.g., Berg and Kearns, 1998; Gregson et al., 2003; Paasi, 2005; Timar, 2004). It is thus no surprise that the hegemonic history of critical geography reduces the multiple and complex histories of critical geographies around the world to a singular story that reinforces Anglo-American hegemony, where critical geography is understood to have originated in the United Kingdom and the United States and 'diffused' outward to the peripheries of academic knowledge production. These hegemonic stories tend to reproduce liberal epistemologies that construct all differences as the same kind of difference whilst at the same time reducing space to a simple, flat plane free of politics. The ironies of a critical geography that fails to critically interrogate its own stories should not be lost on us (Berg, 2004).

It has been well over three decades since geographers started to think seriously about the way that space is not merely an empty container through which people move and in which things happen (Gregory and Urry, 1985), and that 'geography matters' (Massey and Allen, 1984). Thus, it is now, ostensibly, commonplace for critical geographers that space has become a category for analysis that allows us 'to understand the multiple, and often contradictory ways in which it is recursively constitutive of power relations of domination and subordination, and in turn the ways that such relations (re)constitute human experience of and in place' (Berg, 2004, 554). The upshot is that there has been a plethora of geographical studies that analyse the way that geography actually matters in the constitution of social (and spatial) relations.

Notwithstanding the more than 30 years in which geographers and others have taken geography seriously and started to analyse the difference that space makes, there still exist topics that seem to escape such critical geographical analysis. Given the power of liberalism as a technology for

DOI: 10.4324/9781315600635-1

understanding the Self in the present neoliberal epoch, it is not surprising that (with few exceptions, see, e.g., feminist geographies, Black Geographies) the socio-spatial politics of academic knowledge production tend to be overlooked or avoided in critical geographic scholarship. There thus exists a tendency to draw on a liberal understanding of the Self as a primary guiding factor in understanding our own role in reproducing the uneven geographies of academic knowledge production. In this regard – and like the flat-earth politics of liberalism elsewhere – our 'good intentions' are seen as sufficient to inoculate us against complicity in reproducing the spatial marginalization of colleagues outside the academic 'centres' via the hegemonic socio-spatial politics of academic knowledge production. Indeed, it is only recently that we have started to see more self-reflexive scholarship as the basis for the retelling of histories of critical geography. Trevor Barnes and Eric Sheppard (2019), for example, recently edited a volume that seeks to map the variegated spatial histories of radical geography in 'North America and Beyond'. This is an important step, not least because their book traces some histories 'beyond' North America (namely in Mexico, Japan, South Africa and the Francophone countries). At the same time, it is only a start to the necessary task of charting the wider terrains of those historical geographies of knowledge production that exist 'beyond' North America (and the United Kingdom, we would add).

With the above in mind, this book is intended to provide a more thoroughgoing engagement with different spaces and histories of what we have tended to call critical geography. Moreover, and given that the names we choose for ourselves matters, it is important to acknowledge that 'critical geography' and 'radical geography' are not necessarily names for the same objects. The term critical geography originated much later than radical geography: the former coming into wide use in the United Kingdom, the United States and Canada during the mid-1990s with the latter coming into use in the late 1970s and early 1980s in these same spaces. We use critical geography specifically to denote a wider *coalition* of geographers drawing on a significant range of critical social theories. Critical geography is thus more diverse than radical geography and, hopefully, more open to different approaches than the latter, which has tended to draw more narrowly on Marxist and socialist theory. Indeed, it is possible that critical geographers can draw on theoretical perspectives that are often found in opposition to each other in specific use but who can see that it might be politically strategic to work in a larger coalition as a more general political strategy. A good example might be Marxists and Anarchists who would disagree vehemently over the role of the State in emancipatory movements specifically, but who might also find it useful to join together to contest other forms of socio-spatial marginalization arising within capitalism and liberalism. This book is built on the assumption that there are a wide variety of 'critical geographies', and we resist the temptation to define them, in part at least, because lists always have the potential for exclusion. We want 'critical geography' to be defined, in a contingent fashion then, by the authors included in this volume who write about specific histories of critical geography in particular places.

What kind of space (in what kind of critical geography)?

We believe strongly in the truism that 'geography matters'. But just as importantly (and just as names matter), so does the kind of social theory used to both produce and index geographical difference. Here we draw on the works of one of the authors in this volume, Kirsten Simonsen (1996), in order to provide a sense of how different conceptualizations of spatial difference might help us to understand the variegated landscapes of and differences within and between a range of contexts for the production of critical geographies. Simonsen (1996) talks about three different conceptualizations of 'space' that have been used by geographers to understand 'spatial difference': 1. space as material environment; 2. space as difference; and 3. space as social spatiality. In what follows, we focus on the latter two approaches to think about how spatial difference matters in the production of critical geographies 'in place'. Conceptualizations of space as material environment, we suggest, can lead to the kind of thinking that ends in notions of such problematic models as spatial diffusion theory, whereby geographical theory is produced in the centres of production (read: United Kingdom and United States) and this knowledge unproblematically diffuses out to the spatial margins. With this in mind, we believe that thinking about space as difference and as social spatiality provide much richer and more nuanced ways of thinking about the ways that critical geography developed differently (and similarly) in various locations around the globe. Accordingly, we feel it will help to discuss space as difference and space as social spatiality in order to help us to both better understand and better contextualize the historical geographies of critical geography presented in this volume. We hope also to outline some of the key strengths of the book as well as some of its key weaknesses along the way. Of course, there is a good deal of overlap between the different spatial conceptualizations presented by Simonsen and we hope that this messiness will also come through in the discussion that follows.

Space as difference

Simonsen (1996, 499) argues that 'stated very simply, the essence of this conception [of space as difference] is that different places, regions or localities are substantially different – in a material as well as an immaterial sense – and that this difference influences social processes and social life'. This accords significantly with our arguments in the book: critical geography developed differently in different places because those places are substantially different from each other, and this kind of difference makes a difference in the scholarly development of critical geographies in those places. In some ways, we have taken such difference for granted and somewhat uncritically mapped this difference onto nation states or linguistic groupings. The irony here will not escape the critical thinking of critical geographers. Moreover, some of the practical problems of this approach should be obvious with a glance at the table of contents for this book. Yet, even with the limitations of state- and language-centric spatial differentiation, we feel this is at least a good starting point for developing more nuanced historical geographies of critical geography beyond the metropoles. Put another way – we have to start somewhere. But additionally, we should understand the kind of differences that space makes in poststructuralist terms as relational (Simonsen, 1996, 501–502).

Space as social spatiality

The problems that arise because of our use of state boundaries and language differences for the starting points for thinking about the geographies of critical geography might be made slightly less problematic if we think about the difference that space makes in terms of social spatiality. In this regard, Simonsen (1996, 502–503) notes that 'the spatial forms an integrated part of social practices and/or social processes – and that such practices and processes are all situated in space (and time) and all inherently involve a spatial dimension. This is so in all scales of social life'. In this sense, then, spatial difference needs to be understood as a fundamental social category. Space isn't merely a thing through which people move and objects exist, but instead, space must be seen as generative. Drawing on Henri Lefebvre in *La production de l'espace* (Lefebvre, 1986), Simonsen (1996, 503) thus argues that we should 'shift our attention from "things in space" to the actual 'production of space'. If space is produced, then it follows that space must also be productive. This highlights the important role of social relations in the production of space, and spatial relations in the production of the social.

So, in our conceptualization of this project, we wanted to capture, through the work of our contributors, the social production of space and the spatial production of the social. Geography matters, in myriad ways. The chapters you are about to read attempt, with varying degrees of success, to capture the difference that space makes and the different social relations that lead to different conceptualizations and understandings of the spatial.

Strengths and challenges

In soliciting chapters for this volume, we asked contributors to reflect on the histories of critical geography in specific places. In most cases, the contributors live and work – now or previously – in the places they have written about. We have thus prioritized 'insider' accounts, and we discuss the implications of this later. Here, it is important to highlight that we were not prescriptive, initially, about the focus of each chapter. Instead, we left it to the contributors to define critical geographies and to recount what those geographies looked like. We believe strongly that such geographic accounts need to come from those who are enmeshed in those geographies. As a consequence, this edited volume contains 14 grounded accounts that articulate the range of meanings of critical geographies in particular spaces and across specific geographical networks. The lack of a coherent and overarching definition of critical geographies is a strength of this collection. If we are to interrogate how critical geography is made, it is imperative to recognize and validate the meaning of critical geography in different places. The resulting chapters thus vary in substance and style, but provide a detailed patchwork of the histories of critical geographies. However, the chapters are also subjective interpretations of what critical geography is, and has been, in particular places. In drawing attention to these subjectivities, we also acknowledge one of the difficulties with compiling a collection of this type, since the acts of compilation and cataloging – even when motivated by a desire for greater inclusivity – are themselves acts of exclusion.

We encountered many challenges in developing this edited volume, and how we addressed these challenges has shaped the final form of the book. The contributors to this collection are drawn from our international networks – many developed through international conferences such as the International Conference of Critical Geography (ICCG), the Nordic Geographers Meeting (NGM) and the Annual Meeting of the American Association of Geographers. Networks that draw from encounters at international conferences are limited because of how participation in international conferences is restricted to people with the necessary economic, social and mobility capital. In other words, our networks reflect our own subjectivities and positionalities; different editors, with different networks of authors, would have produced a different outcome.

Our focus on elucidating the critical geographies extant in particular places around the globe means that we didn't specify that authors for each contribution had to cover specific epistemological or ontological approaches. Instead, the contributors were free to shape the story of critical geography in the way they deemed most appropriate, and that reflected their subjective assessments of historical developments. Accordingly, there are gaps and tensions within and between the chapters of the book. An epistemological or ontological approach would have been more prescriptive, with instructions to address the presence, or absence, of specific critical engagements, such as radical, feminist, queer, postcolonial or Black geographies; or of specific approaches to understand the world. Instead, the collection contains different articulations of critical: some of which align with the hegemonic understanding of the term; others of which provide alternative perspectives on how the term is understood. As a whole, then, the collection both reflects space as difference, and shows how space in turn influences the social relations of knowledge production.

As we discussed earlier, our focus on covering particular material spaces in the book, and the resultant focus on states or linguistic groups, produced their own exclusions. While the range of states included in the book deliberately extends beyond the Anglosphere, it is still limited. The framing of chapters by linguistic groups or regions serves to enforce the marginalization of those outside the 'core': for example, there is no Anglophone chapter, but there is a Francophone chapter; the United Kingdom has a separate chapter, while Brazil is covered in a chapter on Latin America. The emphasis on states masks the significant spatial variations within states, and the implications of these variations for understanding grounded critical geographies. Equally, the bounding of states and regions places limits on the extent to which connections between local areas are identified and understood. As an example, the chapters as a whole pay little attention to the critical geographies of rural space, with the resultant association of the histories of critical geographies with the urban. The chapters thus reinforce a clear spatial hierarchy that has intensified in recent years with the prioritization of the urban through discourses of planetary urbanism. It is also important to acknowledge the issue of the spatio-temporal positioning of the contributors. The histories of critical geographies are written from vantage points that, often inadvertently, privilege the preoccupations of the time. Livingstone, in his broader disciplinary history, described this as a 'situated geography', concluding that the 'geographical tradition … can only be articulated in the midst of the particularities of time and place' (Livingstone, 1992, 28, 358). This also holds true for the histories of critical geographies, regardless of the broader failure to interrogate these particularities in the search for a shared and liberatory project.

While our guidance to contributors about framing their chapters was minimal, this was not the case when we prepared the chapters for publication. To start, the chapters are written in English, which is not the first, or even second, language for many contributors. This places additional demands on contributors: as Fregonese (2017, 195) points out, writing in a second or third language involves 'sourcing thoughts and terms from two, or often more, languages', leading to 'an in-between linguistic realm in which one can very easily become stranded'. Our preparation and editing of contributions served to shape the linguistic expression into a more standardized form and format, thus somewhat erasing the diversity of modes of storytelling inspired by different places and languages. Müller (2021, 21) warned of this when he wrote that '[i]n the pressure to conform to anglocentric expectations of scholarship, we may lose the diversity of concepts, themes, styles and epistemic locations that should be the hallmark of any discipline'. Our editorial interventions, particularly by the editors who have English as a first language, serve to make the patchwork of chapters more monochrome.

Just as there are geographies to the production of this volume, so are there histories. Often frustrating contributors as well as editors, the volume has been underway for close to a decade. Some chapters were written at an early stage while others have been added more recently. This entails that some chapters could appear a little out-of-date – at least when it comes to the academic game of referencing (contributing problematically to the mounting tyranny of metrics in academia). We don't mind. The aim is not to provide a 'state of the art' of critical geography, but to collect geographically situated (and often personal) histories of the making of critical geographies.

Conclusion

This collection is one intervention into telling the stories of critical geographies – how these geographies came into being, how they existed in particular places and times and what they might look like in the future. Alternative approaches to storytelling could focus on particular groups, networks or assemblages, in order to highlight aspects of the making and circulation of critical geographies. Our purpose here was, in some ways, to challenge the spatial hierarchy of knowledge production. In order to do this, we focussed on particular places rather than on specific epistemological or ontological approaches to the development of critical geographies. In doing so, other socio-spatial hierarchies emerge, so this must be understood as a partial telling of the stories of critical geographies, open to situated interpretation and challenge.

Despite its spatial and social limitations, the book also serves a broader political purpose as an act of international solidarity. While critical geography may be prominent in the United Kingdom and the United States, with practitioners of critical geography awarded professional recognition and advancement, this is not the case in many other states and regions. Instead, critical geographers may be isolated or marginalized. Their experiences range from struggles to gain recognition and acceptance for their research and teaching to, in more extreme instances, threats to their livelihoods and physical and mental wellbeing. This collection makes the histories of critical geographies visible, and affords them a place in broader international disciplinary histories. In this way, it uses the hegemony of critical geographies in the Anglosphere to make space for critical geographies elsewhere.

We are heartened by the growing attention to the spatial hierarchies of knowledge production within the discipline of geography, and by the sustained efforts to reflect on the broader socio-spatial implications of how geographic knowledge is produced. This attention is clearly influenced by critical geographic approaches. Our collection provides an insight into how this has emerged, by foregrounding the histories of critical geographies and the grounded struggles of critical geographers who sought to make the discipline of geography, and their lived geographies, more inclusive and just. In this way, we show how a material geographic framing – in this instance, states and regions – offers a route for considering the implications of spatial difference for the socio-spatiality of knowledge production. We hope this is a starting point for others, in different places, to engage with, apply and further develop the emancipatory potential of critical geography.

References

Barnes, T.J. and E. Sheppard (eds.) 2019. *Spatial Histories of Radical Geography: North America and Beyond*, Oxford. Oxford: John Wiley & Sons.

Berg, L.D. 2004. Scaling knowledge: Towards a *critical geography* of critical geographies. *Geoforum* 35(5), 553–558.

Berg, L.D. and R.A. Kearns 1998. America unlimited. *Environment and Planning D: Society & Space* 16(2), 128–132.

Fregonese, S. 2017. English: Lingua franca or disenfranchising? *Fennia* 195(2), 194–196.

Gregory, D. and J. Urry (eds.) 1985. *Social Relations and Spatial Structures*. London: Palgrave (Macmillan).

Gregson, N., K. Simonsen and D. Vaiou. 2003. Writing (across) Europe: On writing spaces and writing practices. *European Urban and Regional Studies* 10(1), 5–22.

Lefebvre, H. 1986. *La production de l'espace*. Paris: Anthropos (English version *The production of space*, 1991, Oxford: Blackwell).

Livingstone, D.N. 1992. *The Geographical Tradition*. Malden, MA and Oxford: Blackwell Publishing.

Massey, D.B. and J. Allen (eds.) 1984. *Geography matters! A reader*. Cambridge: Cambridge University Press.

Müller, M. 2021. Worlding geography: From linguistic privilege to decolonial anywhere. *Progress in Human Geography*, DOI: https://doi.org/10.1177/0309132520979356.

Paasi, A. 2005. Globalisation, academic capitalism, and the uneven geographies of international journal publishing spaces. *Environment and Planning A: Economy & Space* 37(5), 769–789.

Simonsen, K. 1996. What kind of space in what kind of social theory? *Progress in Human Geography* 20(4), 494–512.

Timar, J. 2004. More than 'Anglo-American', it is 'Western': Hegemony in geography from a Hungarian perspective. *Geoforum* 35(5), 533–538.

2 The evolution of Palestinian critical geography in Palestine and beyond[1]

Ghazi-Walid Falah and Nadia Abu-Zahra

Geography – as a discipline, study, interest, practice and profession – has been defined as what geographers choose to do or not to do (Falah, 1994, 9; Kirby, 1992, 236). This definition helps to interpret how a small group of Palestinian scholars have created space within the international community of geographers to plant the seeds of a non- or anti-hegemonic discourse. This chapter begins with anecdotal snapshots illustrating this creation of space and nurturing of ideas amidst what was not always a uniformly hospitable environment.

Snapshot 1: in the years 1987–1990, professional geographers, high school teachers and University students from various parts of Palestine met in annual conferences in the Galilee, Deir al-Asad, Haifa and Nazareth, under the auspices of the Nazareth-based Galilee Center for Social Research. Out of these initiatives, in part, emerged a fourth conference inaugurating the Palestinian Geographical Society. Organized by two leading Palestinian geographers at the time, this last conference was held in Nicosia, Cyprus, in order to facilitate the attendance of Palestinian geographers who could not enter Palestine, such as those based in Arab countries or whose passports did not allow them to visit Palestine under Israeli control.

One of the first steps of the Palestinian Geographical Society was to seek full membership of the International Geographical Union (IGU). The application, at the 27th Annual Meeting in Washington, DC in 1992, was rejected because the main applicant at the time resided in the United States. The applicant, Ghazi Falah, was informed that the application had to come from Palestinian geographers or institutions, residing in Palestinian territory defined as the West Bank and Gaza Strip, where such territory "does not overlap with Israel's sovereign territory" – that is, the "domain" of the Israeli Geographical Society. The (forced) transnationalism and dispossession of the Palestinian people presented an insurmountable (and unmentionable) problem for IGU regulations.

Snapshot 2: in the same time period, a special issue of *GeoJournal* (vol. 21, 1990) dedicated to the geographies of Palestine met with

DOI: 10.4324/9781315600635-2

opposition. Contributing authors included Palestinians in the Galilee, Jordan, Gaza Strip and the United States, as well as two non-Palestinians residing in the United Kingdom. The special issue was openly opposed by an Israeli member of the journal's editorial board, who requested that the issue not be published. When the journal's editor-in-chief declined the request, the board member asked for his name to be deleted from that issue only. Thus, a thick black line now covers his name on the inside cover page of that particular issue.

Both events serve to demonstrate the anti-colonial struggles involved in establishing a space for any geographies of Palestine or Palestinians, be they merely for formal recognition or actual critical geographies. An underlying theme throughout these efforts is that the overpowering effects of space and place in Palestine and on Palestinians render much writing by non-geographers necessarily geographical. This is so much the case that non-geographers – when writing on Palestine – publish in journals such as Geopolitics and Political Geography, as their writing by necessity concentrates on movement restrictions, Israeli state efforts to narrow human living space for indigenous Palestinians and how daily life must continue under these conditions (Tawil-Souri, 2011a, 2012a). The best-known Palestinian writer in history, Edward W. Said, late Professor of Literature at Columbia University, popularized the term, "geographical imaginations", inspiring some of the most prominent publications to date in human geography, and especially critical geography (Gregory, 1994).

The notion that non-geographers contribute to and serve to inspire critical geography in Palestine is addressed most in the first section of this chapter. We argue that by the time Palestinian critical geographers came into their own in the 1980s, they had already been preceded in their thinking and approaches by a range of Palestinian critical thinkers, who themselves wrote on spatial and geographic issues. These early thinkers had concentrated their efforts on raising awareness of the detriments and de-development implied or imposed by colonial practices and discourse surrounding Palestine and Palestinians. Their works, described in brief here, formed a strong basis for subsequent research and critical analysis of inequalities, power struggles, human rights and citizenship and, of course, territory.

The second and third sections explore Palestinian critical geography. The focus of the second section is on how critical geographies have been used to document and challenge narratives and practices of colonization; while the focus of the third section is on how critical geographies have been used to inform and empower the struggles of marginalized groups. In the second section, we spotlight some of the flagship publications that document Palestinian life prior to British and Zionist military occupation, as well as the processes and effects of denationalization and ongoing dispossession. What makes these works "critical" – in both senses of the word – is that they stand against the strong current of a stream of pronouncements and later academic writing that at best blames Palestinians for their suffering or at worst negates their existence (Said, 1988, 1).

While a strength of Palestinian critical geography has been its capacity to counter such narratives within and beyond academia, a potential weakness in Palestinian critical geography is its ability or inability to engage publicly with Palestinians or others through joint endeavours. "Action" research or involvement that answers Said's (1994) call for "public intellectuals" admittedly has its own concrete obstacles in Palestine – restrictions on mobility and intensive Israeli surveillance being chief among them – yet the unique examples of such research could perhaps be more in number, even under these severe conditions.

The third section of this chapter looks briefly at how contemporary Palestinian critical geographies – across a wide range of sub-disciplines – offer new insights into the realities of everyday life and proffer radical alternatives for the present and future. The chapter concludes with a discussion of the potential futures of Palestinian critical geography and shows how Palestinian critical geography has developed and changed over time within the broader international context of critical geography.

The foundations of Palestinian critical geography

> Just as none of us is outside or beyond geography, none of us is completely free from the struggle over geography. That struggle is complex and interesting because it is not only about soldiers and cannons but also about ideas, about forms, about images and imaginings.
>
> (Said, 1993, 6)

Beyond our earlier-cited definition of geography, as simply "what geographers do", is the wider reality that many non-geographers have contributed to the study of geography in Palestine – and this study as critical exercise in particular. In the years after the Ottoman period, Palestinian nationalism had crystallized as both inclusive and cosmopolitan (Khalidi, 1997), accompanied by growing civil societies in Palestine's urban centres. The strong anti-colonial positioning of Palestinian writers of this period still informs much of today's critical geographers' writing and concerns. The near universally felt sense or "awakening" of regional unity across multiple and overlapping identities gave "encouragement and coherence to an otherwise disruptive modern history" (Said et al., 1988, 237).

Much of the work of this period was dedicated to landscape descriptions and traditional regional geo-histories of specific areas of Palestine, for example, works by the Nazareth-born Khalil Baydas (1874–1949), and journalists, politicians and historians like Muhammad Izzat Darwazeh, Najib Nassar, Khalil al-Sakakini, Fadwa and Ibrahim Tuqan and Abdelrahim Mahmoud. Those Palestinians who witnessed the unfolding of military regulations supporting open colonization of their land were able to widely publish their analyses and even to reach audiences ordinarily excluded from such discussions, such as the Palestinian landowners and peasantry. These publications, together with diaries, speeches and other written and oral forms of communication among those less politically active and visible at the time constitute the road map for a burgeoning critical geography *in situ*.

Documenting and challenging colonization

The dispossession and denationalization of 1.4 million Palestinians in 1948 (also called the "Nakba" in Arabic, meaning "catastrophe") and the continuing practices of systematic oppression, exclusion and discrimination form pivotal elements in Palestinian critical geography and anticolonial writings in general. Critical legal studies, critical legal geography and critical political sociology have all incorporated study of the regulatory framework that propped – and continues to prop – up colonialism in Palestine (Abu-Zahra and Kay, 2012, 162).

The most important spatial thinker of the immediate post-Nakba period was the historian Walid Khalidi. He chronicled the history of Palestinians in *Before Their Diaspora*, describing the diversity and "continuous intermingling" dating back to Roman and Byzantium times (Khalidi, 1984, 7). Khalidi further asserted in his research that the Zionist leadership had plans for expelling Palestinians from their homeland prior to the 1948 War. His article "Plan Dalet: Master Plan for the Conquest of Palestine" (Khalidi, 1961) became the cornerstone for subsequent work revealing that Zionist forces at the time were engaged in ethnic cleansing of Palestinians from Palestine, a process best summarized in 1988, when a group of intellectuals – including Edward W. Said, Ibrahim Abu-Lughod, Janet L. Abu-Lughod, Muhammad Hallaj and Elia Zureik – co-authored a chapter that brought home the demographic and ethnic cleansing aspects of the war (Said et al., 1988). A second important book of the time that also backed up Khalidi's original research was Benny Morris's (1987) *The Birth of the Palestinian Refugee Problem*. While Morris attributed expulsion to low-ranking officers' decisions – rather than on prior intent among the Zionist leadership – later evidence pointed to the latter, further confirming Khalidi's original work (Pappé, 2006). Palestinian geographers took note of this assault – and ones to follow after 1948 – on the Palestinian domestic and cultural landscape and its later use as the material basis for the denial of rights (Falah, 1996a).

Today, the consequences of dispossession and denationalization are still keenly felt, as shown in critical geographical literature among and about Palestinians.

In Palestine now, identity documentation not only signifies what a person is, but more importantly what a person is *not*, that is, a member of a privileged minority with state-protected human rights.

(Abu-Zahra and Kay, 2012, 163)

Geographers and legal scholars documented ongoing expulsions and developed a critical legal geography that was informed by their direct analysis of land expropriation (Abu Hussein and McKay, 2003; Falah, 1991, 2003; Jiryis, 1973). Historical geography blended into legal geography as researchers documented some 34 separate laws, ordinances and regulations designed to dispossess Palestinians within the Armistice lines (Abu Kishk, 1984, 31).

As Palestinian geographers worked to detail the long-time Palestinian presence and experience, they faced continuous efforts of erasure. The persistence of denial by the Israeli political class – of the very existence of Palestinian people in Palestine – has led to much analysis (Masalha, 2003), including by geographers. As Said et al. (1988, 241) stated, "By denying the existence of the Palestinian people, and by dehumanizing them, Zionists meant to hide from the world the intended victims of their colonization". Beyond simply refraining from criticism, some Israeli geographers have a crucial function in building an exclusionary discourse and reality:

As in many other cases of nation-building, Israeli geographers have played an important role in *manipulating of landscape and places* to form a modern Jewish Israeli national identity [emphasis added].

(Golan, 2002, 554).

Two of Israeli geography's "founding fathers" (Waterman, 1985, 195), Yehoshua Ben-Arieh and Yehuda Karmon, refrain from referring to the indigenous population as Palestinian, choosing instead to use labels derived from religion (see Ben-Arieh, 1975, 1976). This terminological elision is described in more detail in the following paragraphs.

Ben-Arieh's historical work relies on itinerant foreigners' accounts, and depicts the indigenous Palestinian population as prone to "exaggeration and fancy" (Ben-Arieh, 1972a, 99), with a penchant for "backsheesh" or bribes (Ben-Arieh, 1973, 18). Palestinians, who remain unnamed as such, are described as "bandits", filled with "hatred" and "anger" at "infidels" (Ben-Arieh, 1972b, 82–83), who render such fear as to keep "the people of Jerusalem locked inside the walls of the Old City" (Ben-Arieh, 1975, 262). Who exactly these "people of Jerusalem" are is not a subject of research; while centuries and millennia of Palestinian history are ignored, Ben-Arieh focusses on the "increasing activity of the Jews and Christians in the city" as the cause, "undoubtedly", of Jerusalem's "rising importance" to Ottoman and "European Powers" (Ben-Arieh, 1976, 59).

Yehuda Karmon, meanwhile, describes Palestinians as "marauding" enemies (Karmon, 1960, 156–157, 249), ignorant of the "secret of organized irrigation", who unnecessarily flood their crops and thereby delay their harvest (Karmon, 1953, 19, 22). Palestinian wariness of the interventions of occupying British forces is put down to backwardness, and their agricultural practices – uniquely developed to preserve the wetlands and wildlife of northern Palestine, where water buffalo were once a key feature of communal life and landscape (Abu-Zahra, 2007a, 369; Hütteroth and Abdulfattah, 1977, 48) – are characterized as lacking "the knowledge and the means to subdue nature" (Karmon, 1953, 24). If representation is a "site for domination and resistance" (Blomley, 2006, 91), then the work of Ben-Arieh and Karmon constitutes a clear challenge to those wishing to engage in critical geography and the deconstruction of "dominance" by class and race, as described in Sparke (2004, 778) and Hall (1980). This challenge is well met in geographical writing on Palestinian Bedouin, which squarely addresses the inferior status assigned to Palestinians after colonization of their country (Falah, 1983, 1985). Yet, Palestinian geographers found themselves personally targeted the moment they dared weigh in on Israeli policy (see Editorial Note, 1985, 421, in *Geoforum*).

For some Israeli geographers, their contact with Palestinian critical geography is truly a dialogue in search of a common truth. Newman and Portugali (1987) undertake the exigent task of examining Israeli geography and conclude:

> Israeli writers ...employing notions of 'exploration', 'colonialism', and 'industrial reserve army' are nevertheless careful to ensure that their criticism of the activities of Israel authorities will not imply the delegitimisation of Israeli society or the existence of the Israeli state.
>
> (Newman and Portugali, 1987, 325)

Criticism outside the guild, government and state is ill met, in a context that rewards "organic intellectuals" (Gramsci, 1971). This is not unique, and examples can be found across the world of how "dominant forms of social science have sought to legitimate and naturalise [prevailing cultures of prejudice], and related forms of oppression and injustice" (Blomley, 2006, 91; see also Blackwell, Smith and Sorenson, 2003). Indeed, this dynamic is what often gives rise and reason to critical geography as a modest academic contribution to global counterhegemonic efforts.

The practice of critical geography: countering erasure and advocating for alternatives

Although Palestinian geographers have not led the way in becoming "public intellectuals" as Said (1994) urged, they continue to engage with public efforts to counter erasure and advocate for alternatives. One of the most impressive fields in which Palestinian geographers have succeeded in engaging the wider public is cartography. Beginning with a mapping of the Palestinian people, culture and economy of the 1500s (Hütteroth and Abdulfattah, 1977), and culminating in the maps used by the Negotiations Support Unit of the Palestinian Liberation Organization Negotiations Affairs Department, Palestinian cartography has accomplished what narrative could not: it laid bare the dispossession and denationalization of the Palestinian people.

Chronologizing Palestinian life and land in cartographic form was achieved through conscious effort, for among Palestinians the sense of unity preceded the image of the map.

> The multifaceted vision is essential to any representation of us. Stateless, dispossessed, de-centered, we are frequently unable either to speak the 'truth' of our experience or to make it heard. *We do not usually control the images that represent us*; we have been confined to spaces designed to reduce or stunt us; and we have often been distorted by pressures and powers that have been too much for us [emphasis added].
>
> (Said, 1986, 6)

Carefully outlining Palestinian reality in Cartesian terms required constant fieldwork in archives, across the terrain and among the people of Palestine. Thus, mapping in Palestine was born of an awareness of the importance of an image to communicate with the international community; it operated in reverse of the pattern described by Benedict Anderson, in which the map-as-logo precedes a sense of collectivity (Anderson, 2006, 179). While a number of atlases stand out for their meticulousness and inclusive vision (Abu Sitta, 2010; PASSIA, 2002), the landmark study continues to be *All That Remains*, a collective project involving key Palestinian geographers and led by historian Walid Khalidi (1992). Beyond this seminal work is a ten-volume historical atlas (Dabbagh, 1991 [1965–1976]), as well as a catalogue of more than three decades of Palestinian struggle against colonialism, written by political geographer Basheer Nijim – one of the founders of the Palestinian Geographical Society – and Bishara Muammar (Nijim and Muammar, 1984).

As Palestinian geographers became progressively more confined in their movements by Israeli restrictions, and as colonization intensified in Jerusalem and the rest of the West Bank and Gaza Strip, a segment of Palestinian cartography came to concentrate on these areas. Some theorists posited that shifting control over resources embodied colonialism:

> Israeli colonization of the West Bank for over forty years has gradually evolved to differentiate it from other familiar forms of classical colonialism… colonization of the West Bank is less a matter of managing the population through a Foucauldian framework of biopolitics, and more a matter of controlling the resources (land, water, and airspace).
>
> (Zureik, 2011, 4–5)

Cartographically exposing colonial control over resources and the indigenous population

The imperative, therefore, was to document this shifting control over resources – land and water confiscation – in order to better understand economic domination, proletarianization, population expulsion and induced transfer, political suppression and denial of Palestinian rights.

This theoretical approach was paralleled by its own contradiction – that population control was key to resource control. Yet, for decades, and continuing into the present, translating resource control and dispossession into cartographic terms was a principal aim for Palestinian geographers. Khalil Tufakji engaged in a daunting project to document historical land tenure in Jerusalem and worked ceaselessly as the key cartographer to live and work in the city (Fischbach, 2003, 337). Michael Younan directed his efforts to the most recent changes in Jerusalem and across Palestine, even predicting what was yet to come (Halper and Younan, 2005). Walid Mustafa looked at Jerusalem from the 1800s to the present (Mustafa, 2000). The Applied Research Institute of Jerusalem and Land Research Center launched and sustained a project to monitor colonization, mapping expropriation as it happened with satellite imagery, geographic information systems, on-the-ground fieldwork and intimate knowledge of what was taking place on the ground (ARIJ and LRC, 2012).

Individual geographers used cartography to uncover the colonialism and siege that lay behind the façade of a "peace process" and the "two-state solution" (Falah, 2005). Yet, all the while, Israel had military power, communication technology, access to international media and "institutional mediation" (Newman and Falah, 1997, 114). From the 1970s and before, in works such as the *Atlas of Israel* (Amiran, 1970), and *The Land That Became Israel* (Kark, 1990), Zionist geographers continued to produce their own brand of "knowledge" about which land belonged to whom, while maintaining a conspicuous silence on indigenous rights.

The contest through cartography over resources is striking (Tawil-Souri, 2011a) but not unique to Israeli and Palestinian cartography (Murphy, 1990, 534). For Palestinians, however, the power differential and challenges for criticizing the hegemonic discourse are nearly prohibitive. Nevertheless, critical historical geographers and critical cartographers inside and outside Palestine have painstakingly chronicled the omissions in the Israeli discourse, for Palestine as a whole, and Jerusalem in particular (Mustafa and Abdul Jawad, 1987). They did so in the face of formidable obstacles, with restricted access to archives and archaically draconian regulations over mobility. In common with Palestinian researchers in law and human rights, geographers at times suffered imprisonment and interrogation – with torture methods such as prolonged positions and sleep deprivation – as research was criminalized (Falah, 2007). Despite and even because of this, Palestinian geographers' interventions countered Israeli cartographic hegemony on Palestine, bringing together stark parallels with colonial and anti-colonial experiences and cartography elsewhere in the world.

Advancing alternatives for justice, equity and inclusion

In addition to critical cartography, the applied research of urban planners, such as those at An-Najah University's Urban and Regional Planning Unit (in the Nablus region of the West Bank), is also significant. Urban and rural geographers' applied research breaks new ground in multiple directions: researching and advocating equal access to public services; exploring avenues for participatory planning using geographic information systems; developing strategies for affordable housing; integrating refugees and overcoming their physical separation in rural and urban areas; and considering ways to enhance daily life through rural and urban services and aesthetics (Abdelhamid, 2009, 2010).

These activities are all the more remarkable given that they take place in a context of ongoing military occupation and attack – a context which elsewhere would elicit a top-down, emergency management approach to planning, recovery and reconstruction. These few examples from the broad array of critical and participatory research taking place in An-Najah University demonstrate a unique emphasis on capacities (rather than solely inequalities), and on intra-Palestinian debates and strategies. The criticisms levelled at the Palestinian Authority are indirect and highly constructive – suggesting ways forward rather than merely taking aim with the academic's pen. In terms of Palestinian critical geography, the Unit's work is but one example, and is mirrored elsewhere – such as al-Khalil (Hebron), Nazareth, Gaza, Jerusalem or Jenin – but without additional fieldwork in these areas and in the absence of accessible publications, such wider work has yet to gain international prominence.

What have, by contrast, generated substantial publications beyond Palestinian institutions are projects that draw attention to ongoing colonialism in the West Bank and Gaza Strip; and that emphasize water rights, water exploitation and unequal distribution (Abu-Ayyash, 1976; Abu-Zahra, 1999; Alatout, 2008; Asadi, 1990; Dahlan, 1990; Elmusa, 1998; Falah, 1984, 1990; Hassan et al., 2010a, 2010b; Mustafa, 1986; Nijim, 1984, 1990; Saleh, 1990). In exposing inequalities and exploring the colonial and other roots of these inequalities, geographers have used the terrain of academia to encompass and augment the shrinking terrain of public, personal and private life among Palestinians living in Palestine.

Feminist, antiracist and transformative insights have also unfolded within Palestinian critical geography, and serve to "contest the hegemony of dominant representations" (Blomley, 2006, 91–92). A noteworthy element of Palestinian transformative geography is the gradual shift from identifying exclusionary thought and imaginaries within colonialism, toward promoting inclusionary thought and imaginaries for the present and future. This shift paralleled a shift in focus from land to population and was led by sociologists, poets and politicians – followed by critical geographers. Philosophy professor and former Palestinian politician in the Israeli government, Azmi Bishara, launched a campaign for inclusion irrespective of faith, ethnicity, colour or creed, under the banner, "a state of all its citizens". Alongside poet-politicians like Tawfik Zayyad, Emile Habiby and Samih al-Qassem, Bishara elucidated collective Palestinian sentiments of inclusion and social justice while remaining active in quotidian politics.

In sociology, Zureik's classification of the situation as "internal colonialism" matched his detailed study of peasant marginalization, intense land confiscation, political manipulation, residential and occupational segregation and a duality of economic and social relations (Zureik, 1979). Other sociologists also drew light toward inequalities and discrimination against the indigenous Palestinians (Ghanem, 2008). Geographers then followed with studies of how the practice and profession of planning in particular played a role in these dynamics (Khamaisi, 1997, 2011). Studies on urban destruction or "urbicide" in the West Bank and Gaza Strip bridged the subjects of land and population, bringing an emphasis on human suffering and daily life in refugee camps and neighbourhoods attacked and besieged by the Israeli military (Abujidi, 2013; Dahlan, 1990).

These and other studies eased out of the previous concentration of attention on land and into the realm of population and its control. Studies of resource expropriation and extraction became studies of colonialism, apartheid and ethnic cleansing, for example, with a focus on segregation and housing discrimination, as well as population control and land confiscation – internal colonialism against the indigenous Palestinians – in the Galilee (see, for example, Falah, 1993, 1996b). These works constitute some of the earliest and most influential pieces of writing on Israeli efforts to spatially isolate Palestinians remaining within the Armistice lines after 1948, pointing to parallels with South African racial policies of the same period (1948-1991, following on three centuries of colonization).

The physical obstacles to movement imposed by the Israeli military across Palestine ostensibly demanded scholarship (see Tawil-Souri, 2011c, 2012b); from 2002 onward, the Wall in the West Bank became an urgent focus for attention (Abu-Zahra, 2006, 2007b; Falah, 2003). Yet, in keeping with the trend toward examining population control rather than solely land control, new works emerged on identity documents and population registration. Geographers, like sociologists (Abu-Laban and Bakan, 2019), highlighted the administrative tools used to uphold and entrench inequality, while consistently demonstrating how equality and inclusion could have widely felt benefits (Abu-Zahra, 2008; Tawil-Souri, 2011b, 2012a).

Perhaps the key shift in the recent literature, however, is toward seeking insights into how these administrative tools and their consequent inequality render daily life asunder. Such insights have been labelled as a "double deconstruction", simultaneously addressing geopolitical and micropolitical forms of oppression and injustice (Sparke, 2004, 784). The work of Lina Jamoul (2004) garnered widespread praise for its ability to traverse the interior of the home and the expanse of the militarized landscape in a single breath. This approach is typical of Palestinian writing, and particularly of diarized transcriptions of the intimate effects of weaponized assaults on denationalized Palestinian citizenry (Hamzeh, 2000, 2001). Palestinians chronicle more than this, however, and detail the absurdities that arise in response to a constant Israeli – and at times international – denial of their identity, rights, and of course, geography (Abdel-Fattah, 2009; Peled-Elhanan 2008).

The remarkable work on coping mechanisms and altered geographies of daily life is perhaps one of the most valuable contributions that Palestinian researchers have made to current and future analysis of oppressive contexts; it constitutes a rare examination of how geopolitical and discursive hegemony – and the resistance to that hegemony – translates within everyday experiences (Taraki, 2008). Describing, for instance, her effort to move from place to place within the straight-jacket-like constraints across her native land, Birzeit University professor Lisa Taraki reflects on the hierarchies created through selective dispensation of Israeli military "permits" or apartheid-style passes (Taraki, 2003).

Yet another example of this "double deconstruction" is found in Abu-Nahleh's (2006) graphic fieldwork with a family who, 17 days after the death of the husband and father, is shelled by tank missiles in their home. After feeling the second missile strike their home as they huddled in one of the dark rooms, the mother "started feeling the mouths of my children to see if they were breathing" (Abu Nahleh, 2006, 113). The interplay between geographies is crucial in Palestinian writing, which touches on children's geographies, geographies of affect, urban geography (Taraki and Giacaman, 2006) and sociocultural geography, while displaying a constant awareness of geopolitical circumstances and interventions. The personal, the (geo)political and the everyday are merged into a lived reality that jumps from the page and grips the reader in an urgent call to intercede, to cease complacency and to speak, act or stand to challenge dominance and abuse.

The collective Palestinian experience of vulnerability provides a commonality in attitudinal response – a shared culture of resistance (Said, 1980). While regional differences can be found, household surveys confirm a prevalence of similarities in approach and outlook (Johnson, 2006, 59). This has given rise to a gentle sense of humour in certain Palestinian writing, as a means of transcending and transforming a conflictual and oppressively narrowing space. The popular works of Amiry (2005) and Abdel-Fattah (2009), as well as major strides forward in the media of film – an important format for participatory research (Pain, 2003, 654) – and websites, exemplify this humour, transcendence and transformative element of social/cultural geography.

Conclusion

While Palestinian geographers have indeed "chosen" what to do and what not to do, as suggested in our introduction to this chapter, much of this "choice" has been – and continues to be – a response to their circumstances, in particular, of dispossession, denationalization and denial of their history, presence and rights. Drawing upon the early works of non-geographers, which celebrated pre-expulsion diversity while taking a firm anti-colonial stance, subsequent geographers recreated the life and times of the over 500 villages and localities destroyed and depopulated in 1947–1949 (Al-Aref, 1956–1960; Khalidi, 1992). Unconsciously echoing the writings of critical cartographers of places like Turtle Island and Abya Yala (the Americas), Aotearoa (New Zealand) and other spaces of colonialism and decolonialism, Palestinian geographers countered the ongoing erasure of their place-names, and more crucially of their places themselves: erasure through mass bulldozing and demolition of homes, shops, markets, streets, orchards, villages and refugee camps. The scale from and for which this counter-narrative was projected stretched from the home to the supra-state region, with visions for regional unity and a non-ethnically-defined inclusivity.

Considering the critical geographies discussed in this chapter, we raise questions that are of relevance to critical geography anywhere. How has the discipline changed over time, and what implications does this carry? Does critique and criticism extend to Palestinian governance, society or scholarship? And to what extent have critical geographers succeeded in becoming public intellectuals?

To answer the first question, we began with a broad overview of how Palestinian critical writing documented historical events and village life, from perhaps a very public point of view. This public style of writing has since moved into the more private and personal, sometimes as a result of conscious feminist theoretical approaches (Fuller and Kitchin, 2004, 3. See also Abu-Zahra 2006, 2008; Jamoul 2004), sometimes as a result of circumstances particular to Palestine (Falah, 2007). Edward Said, for instance, encouraged "people's narratives [...] interviews, autobiographical reflections, oral testimonies, all of them by actors, some important some modest, in the various dramas of Palestine" (1988, 17). He explained how important these narratives were, not only to challenge stereotypes against Palestinians, but also to dispute "the prevailing research norms that require Western witnesses as the only dependable or credible evidence" (Said, 1988, 17).

Yet, in following these particularities of the Palestinian context, as well as critical research approaches that advocate "giving voice", the past, present and future of Palestine has become a personal story, with research entering the private spaces of people's lives. Such exposure is in part worrisome, given the degree to which ordinary Palestinians are already under surveillance in their homes, communications and personal spaces. Might the highly personal interview, once published, be used against individuals, who describe, for instance, their dependence on medical services, mobility or income – all of which are in the power of the Israeli military to cut off (Abu-Zahra, 2007b)? Palestinian critical geography has shifted from reserving its eye for issues of land and public symbols, to magnifying its lens onto individuals' personal and private lives. While this has its strengths, academically speaking, it may yet bear difficult consequences.

A second shift has been to examine more critically Palestinian governance and society. This has been part and parcel of the trend to inclusive thinking within Palestinian writing and activism. When Palestinian geographers publish inward-looking criticism, they tend to concentrate on issues of inclusivity, stressing the need for pluralism and healthy tolerance of opposition (e.g. Mustafa, 1998, 1999). Couching criticism in technical language – related, for example, to urban or rural planning – has been a key strategy in conveying critical messages to policy- and decision-makers (see, for example, Abdelhamid, 2009). Palestinian geographers, who urge that public governance emphasize participation and equality (e.g. Abdelhamid, 2010), are important contributors to Palestinian society, despite not often having or taking the opportunity to publish about their experiences in the English-language and international academic press.

The final irony of Palestinian critical geography is that it has likely faced wider success in making human rights issues an internationally public affair than in reaching out to the Palestinian public, even if broadly defined to encompass the 13 million Palestinians denationalized until today. International geographers' attitudes toward human rights in Palestine are a contemporary litmus test of right/left-wing standing, just as attitudes toward the Vietnam War or South African apartheid once were in past generations. The academic boycott of Israeli institutions – alongside but more so than the cultural, military or commercial boycotts – has stood as a lightning rod for debate in academic journals, such as *Borderlands* (Gordon, 2003; Reinhart, 2003), *Political Geography* (O'Loughlin, 2004; Slater, 2004) and *Settler Colonial Studies* (Salamanca et al., 2012, reproduced on the *Antipode* website). Despite the barrage of "anger" against them, Palestinians who had called for the boycott saw that they had denaturalized systematic inequality and oppression; they had shown that, "Israel is boycottable, like South Africa was boycottable" (Omar Barghouthi, quoted in Khouri, 2005). That particular quote was distributed to the Critical Geographers email list in 2005, amid close to 100 emails, between February and August 2005, debating the boycott.

In sum, Palestinian critical geography has grown in large part from multidisciplinary, anti-colonial roots, to branch into international critical discussions of academic activism at a time when those discussions are intensifying. While any such venture requires constant self-examination and critique,

Palestinian writers have become accustomed to being questioned and conducting some soul-searching to find answers:

> [N]othing – literally nothing – about Palestine can go without proof, contention, dispute and controversy...
>
> (Said, 1988, 11)

Perhaps the turn toward the personal and highly concrete or tangible, within Palestinian academic and popular writing, is one form of response to this constant questioning. Inside and beyond Palestine, it could be a negative symptom, of desperation, of individuals forced into more and more vulnerable positions, compelled to be more revealing, having to prove endlessly – and with more divulging of personal information – that Palestinians indeed are indigenous survivors of ongoing colonialism. Yet, alternatively, pulling together all the threads in this chapter, the increasing notion of international attitudes toward Palestine as a barometer of integrity, and the concomitant turn in Palestinian writing toward the personal, could also be interpreted as signs of more voices joining, and more confidence in speaking out on the intimate effects and dynamics of political violence, systematic oppression and discrimination. In the sense that new work in personal geographies (geographies of affect) can be more inclusive – emphasizing a common humanity rather than a geopolitical identity – these recent trends bode well not only for Palestinians, but also for all in the region, all who relate to these dynamics, and all who engage with this work.

Note

1 A version of this chapter has previously been published as G.W. Falah and N. Abu-Zahra (2014) The Evolution of Palestinian Critical Geography in Palestine and Beyond, *The Arab World Geographer* 17(4): 303–338. The chapter is published with permission from *The Arab World Geographer*.

Bibliography

Abdel-Fattah, R. 2009. *Where the Streets Had a Name*. Sydney: Pan Macmillan Australia.

Abdelhamid, A. 2009. Physical planning and spatial management in the Palestinian territories. *Siyasat* 9, 120–45 (In Arabic).

Abdelhamid, A. 2010. Towards enhancing an integrated strategy for the development of affordable housing policies in the Palestinian territories. *Siyasat* 13–14, 165–87 (In Arabic).

Abu-Ayyash, A. 1976. Israeli regional planning in the occupied territories. *Journal of Palestine Studies* 5(3–4), 83–108.

Abu Hussein, H. and F. McKay. 2003. *Access Denied: Palestinian Access to Land in Israel*. London: Zed.

Abujidi, N. 2013. *Urbicide in Palestine: Spaces of Oppression and Resilience*. London: Routledge.

Abu Kishk, B. 1984. Arab land and Israeli policy. *Al-Mawakib Cultural Magazine* (Nazareth) ½,3041. (In Arabic).

Abu-Laban, Y. and A.B. Bakan. 2019. *Israel, Palestine and the Politics of Race.* London: I.B. Tauris.

Abu Nahleh, L. 2006. Six families: Survival and mobility in times of crisis. In L. Taraki (ed.), *Living Palestine: Family Survival, Resistance, and Mobility Under Occupation.* Syracuse, NY: Syracuse University Press, pp. 103–184.

Abu Sitta, S.H. 2010. *Atlas of Palestine, 1917–1966.* London: Palestine Land Society.

Abu-Zahra, N. 1999. *Palestinian Water Management: Current Problems, Current Strategies, Future Considerations.* Toronto: University of Toronto (unpublished dissertation), 113 pp.

Abu-Zahra, N. 2006. Field note: Palestine. *Women's Studies Quarterly* 34(1/2), 242–49.

Abu-Zahra, N. 2007a. Environment: Change and natural resource extraction: Palestine. In, J. Suad, A. Najmabadi, J. Peteet, S. Shami, J. Siapno & J.I. Smith (eds.), *Encyclopedia of Women and Islamic Cultures, Volume 4: Economics, Education, Mobility and Space.* Leiden: Brill Publishers, pp.369–72.

Abu-Zahra, N. 2007b. Resisting the wall: An interview with Nazeeh Shalabi in the village of Mas'ha, West Bank, *Palestine: Arab World Geographer* 10(1), 38–56.

Abu-Zahra, N. 2008. Identity cards and coercion. In, R. Pain & S.J. Smith (eds.), *Fear: Critical Geopolitics and Everyday Life.* Aldershot: Ashgate Publishing Ltd., pp.175–92.

Abu-Zahra, N. and A. Kay. 2012. *Unfree in Palestine: Registration, Documentation and Movement Restriction.* London: Pluto.

Al-Aref, A. 1956–60. *The Catastrophe of Palestine and Paradise Lost (Nakbat Falasteen wal Firdous al-Mafqud).* Beirut: Dar al-Hoda, Volumes 1-10. (In Arabic).

Alatout, S. 2008. 'States' of scarcity: Water, space, and identity politics in Israel, 1948–1959. *Environment and Planning D: Society and Space* 26(6), 959–82.

Amiran, D.H.K. 1970. *Atlas of Israel.* Tel Aviv and Jerusalem: Survey of Israel and Ministry of Labour.

Amiry, S. 2005. *Sharon and My Mother-in-Law.* London: Granta.

Anderson, B. 2006 [1983]. *Imagined Communities.* London: Verso.

ARIJ and LRC (Applied Research Institute of Jerusalem and Land Research Center). 2012. *Monitoring Israeli colonizing activities in the Palestinian West Bank and Gaza.* http://www.poica.org/proj-objectives/proj_obj.php (5 November 2012)

Asadi, F. 1990. How viable will be the agricultural economy be in the new state of Palestine? *GeoJournal* 21(4), 375–83.

Ben-Arieh, Y. 1972a. Pioneer scientific exploration in the Holy Land at the beginning of the nineteenth century. *Terrae Incognitae* IV, 95–110.

Ben-Arieh, Y. 1972b. The geographical exploration of the Holy Land. *Palestine Exploration Quarterly* July-December, 81–92.

Ben-Arieh, Y. 1973. William F. Lynch's expedition to the Dead Sea, 1847–48. *The Journal of the National Archives.* Spring, 14–21.

Ben-Arieh, Y. 1975. The growth of Jerusalem in the nineteenth century. *Annals of the Association of American Geographers* 65(2), 252–69.

Ben-Arieh, Y. 1976. Legislative and cultural factors in the development of Jerusalem: 1800-1914. *Geography in Israel.* A collection of Papers Offered to the 23rd International Geographical Congress USSR, July-August, Jerusalem, 54–105.

Blackwell, J.C., M.E.G. Smith and J.S. Sorenson. 2003. *Culture of prejudice: arguments in critical social science.* Broadview Press.

Blomley, N. 2006. Uncritical critical geography? *Progress in Human Geography* 30(1), 87–94.

Dabbagh, M.M. 1991 [1965-1976]. *Biladuna Filastine [Our Homeland, Palestine]*. Beirut: Dar al-Tali'ah, 10 vols. (In Arabic).

Dahlan, A.S. 1990. Housing demolition and refugee resettlement schemes in the Gaza Strip. *GeoJournal* 21(4), 385–95.

Editorial note. 1985. Planned Bedouin settlement in Israel: Alternative views. *Geoforum* 16(4), 423.

Elmusa, S. 1998. *Water Conflict: Economics, Politics, Law and Palestinian-Israeli Water Resources*. Washington D.C.: Institute for Palestine Studies.

Falah, G. 1983. The development of the "planned Bedouin settlement" in Israel 1964-1982: Evaluation and characteristics. *Geoforum* 14(3), 311–23.

Falah, G. 1984. Recent Jewish colonization in Hebron. In, D. Newman (ed.), *The Impact of Gush Emunim: Politics and Settlement in the West Bank*. Kent, UK: Croom Helm, pp.231–246.

Falah, G. 1985. How Israel controls the Bedouin in Israel. *Journal of Palestine Studies* 14(2), 35–51.

Falah, G. 1990. Arabs versus Jews in Galilee: Competition for regional resources. *GeoJournal* 21(4), 325–336.

Falah, G. 1991. The facts and fictions of Judaization policy and its impact on the majority Arab population in Galilee. *Political Geography Quarterly* 10(3), 297–316.

Falah, G. 1993. *Galilee and the Judaization Plans*. Beirut: Institute for Palestine Studies. (In Arabic).

Falah, G. 1994. The frontier of political criticism in Israeli geographic practice. *Area* 26(1), 1–12.

Falah, G. 1996a. The 1948 Israeli-Palestinian War and its aftermath: The transformation and de-signification of Palestine's cultural landscape. *Annals of the Association of American Geographers* 86(2), 256–85.

Falah, G. 1996b. Living together apart: Residential segregation in mixed Arab-Jewish cities. *Urban Studies* 33(6), 823–57.

Falah, G-W. 2003. Dynamics and patterns of the shrinking of Arab lands in Palestine. *Political Geography* 22, 179–209.

Falah, G-W. 2005. Geopolitics of 'enclavisation' and the demise of a two-state solution to the Israeli-Palestinian conflict. *Third World Quarterly* 26(8), 1341–372.

Falah, G-W. 2007. The politics of doing geography: 23 days in the hell of Israeli detention. *Environment and Planning D: Society and Space* 25(4), 587–93.

Fischbach, M. 2003. *Records of Dispossession: Palestinian Refugee Property and the Arab-Israeli Conflict*. New York: Columbia University Press.

Fuller, D. and R. Kitchin. 2004. Radical theory/critical praxis: Academic geography beyond the academy? In, D. Fuller & R. Kitchin (eds.), *Radical Theory/Critical Praxis: Making a Difference Beyond the Academy?* Kelowna, BC: Praxis (e)Press, pp. 1–20.

Ghanem, A. 2008. *Palestinians in Israel: Indigenous Group Politics in the Jewish State*. Ramallah: Madar. (In Arabic).

Golan, A. 2002. Israeli historical geography and the Holocaust: Reconsidering the research agenda. *Journal of Historical Geography* 28(4), 554–65.

Gordon, N. 2003. Against the academic boycott. *Borderlands e-Journal* 2(3).

Gramsci, A. 1971. *Selections from the Prison Notebooks*. Translated by Maore, Q. and Smith, G. N. New York: International Publishers.

Gregory, D. 1994. *Geographical Imaginations*. Oxford: Blackwell.

Hall, S. 1980. Race, articulation and societies structured in dominance. In, UNESCO (ed.), *Sociological Theories: Race and Colonialism*. Paris: UNESCO, pp. 305–45.

Halper, J. and M. Younan. 2005. *Obstacles to Peace: A Reframing of the Palestinian-Israeli Conflict* (3rd *Edition*). Bethlehem: Palestine Mapping Center.

Hamzeh, M. 2000. *Ordinary days in Dheisheh: Is the world watching?* Paris: Editions 00h00.

Hamzeh, M. 2001. *Refugees in our own land: Chronicles from a Palestinian refugee camp in Bethlehem*. London: Pluto.

Hassan, M.A., K. Shahin, B. Klinkenberg, G. McIntyre, M. Diabat, A. Tamimi and R. Nativ. 2010a. Palestinian water II: climate change and land use. *Geography Compass* 4, 139–57.

Hassan, M.A., G. McIntyre, B. Klinkenberg, A. Tamimi, R.K. Paisely, M. Diabat, M.K. Shahin. 2010b. Palestinian water I: Resources, allocation and perception. *Geography Compass* 4, 118–38.

Hütteroth, W.-D. and K. Abdulfattah. 1977. *Historical Geography of Palestine, Transjordan and Southern Syria in the late 16ᵗʰ Century*. Erlanger Geographische Arbeiten, Sonderband Erlangen, Germany: Vorstand der Fränkischen Geographischen Gesellschaft.

Jamoul, L. 2004. Palestine: In search of dignity. *Antipode* 36(4), 581–95.

Jiryis, S. 1973. The legal structure of the expropriation and absorption of Arab lands in Israel. *Journal of Palestine Studies* 2, 82–103.

Johnson, P. 2006. Living together in a nation in fragments: Dynamics of kin, place, and nation. In L. Taraki (ed.), *Living Palestine: Family Survival, Resistance, and Mobility Under Occupation*. Syracuse, NY: Syracuse University Press, pp. 51–102.

Kark, R. 1990. *The Land that Became Israel: Studies in Historical Geography*. New Haven: Yale University Press.

Karmon, Y. 1953. The settlement of the Northern Huleh Valley since 1838. *Israel Exploration Journal* 10, 4–25.

Karmon, Y. 1960. An analysis of Jacotin's Map of Palestine. *Israel Exploration Journal*, 10, 155–73, 245–53.

Khalidi, R. 1997. *Palestinian Identity: The Construction of Modern National Consciousness*. New York: Colombia University Press.

Khalidi, W. 1961. Plan Dalet—The Zionist Master Plan for the conquest of Palestine. *Middle East Forum* 37(9), 22–8.

Khalidi, W. 1984. *Before their Diaspora: A Photographic History of the Palestinians, 1876-1948*. Washington DC: Institute for Palestine Studies.

Khalidi, W. (ed.). 1992. *All that Remains: The Palestinian Villages Occupied and Depopulated by Israel in 1948*. Washington DC: Institute for Palestine Studies.

Khamaisi, R. 1997. Israeli use of the British Mandate planning legacy as tool for the control of Palestinians in the West Bank. *Planning Perspectives* 23(3), 321–40.

Khamaisi, R. 2011. Territorial dispossession and population control of the Palestinians. In, E. Zureik, D. Lyon & Y. Abu-Laban (eds.), *Surveillance and Control in Israel/ Palestine*. Abingdon and New York: Routledge, pp. 335–52.

Khouri, R.G. 2005. Seeking alternatives to a third Palestinian intifada. *Daily Star*, May 30. http://www.dailystar.com.lb/article.asp?edition_id=10&categ_id=5&article_id=15481 (18 December 2012).

Kirby, A. 1992. Editorial comment: Publishing deca(ye)de. *Political Geography* 11(3), 235–7.

Masalha, N. 2003. *The politics of Denial: Israel and the Palestinian Refugee Problem.* London and Sterling VA: Pluto Press.

Morris, B. 1987. *The Birth of the Palestinian Refugee Problem, 1947-1949.* Cambridge: Cambridge University Press.

Murphy, A.B. 1990. Historical justifications for territorial claims. *Annals of Association of American Geographers* 80, 531–48.

Mustafa, W. 1986. The West Bank and Gaza Strip under occupation. *Palestinian Affairs* 162–63, 15–37. (In Arabic).

Mustafa, W. 1998. The necessity of pluralism and opposition in the Palestinian society. *Bethlehem University Journal* 17. (In Arabic).

Mustafa, W. 1999. Death penalty: Lessons of execution. *Sawt Al Watan* 50. (In Arabic).

Mustafa, W. 2000. *Jerusalem: Population and Urbanization 1850–2000.* Jerusalem: Jerusalem Media and Communication Center.

Mustafa, W. and S. Abdul Jawad. 1987. *The Collective Destruction of Palestinian Villages and Zionist Colonization 1882-1982.* London: Center for Development Studies.

Newman, D. and G. Falah. 1997. Bridging the gap: Palestinian and Israeli discourses on autonomy and statehood. *Transactions of the Institute of British Geographers NS* 22, 111–129.

Newman, D. and J. Portugali. 1987. Israeli-Palestinian relations as reflected in the scientific literature. *Progress in Human Geography* 11, 315–32.

Nijim, B.K. 1984. Israeli Jewish settlements in the West Bank, 1967-1980. *Asian Profile* 12, 257–69.

Nijim, B.K. 1990. Water resources in the history of the Palestine-Israel conflict. *GeoJournal* 21(4), 317–23.

Nijim, B.K. and B. Muammar. 1984. *Toward the De-Arabization of Palestine/Israel 1945-1977.* Dubuque, Iowa: Kendall/Hunt Publishing Company.

O'Loughlin, J. 2004. Academic openness, boycotts and journal policy. *Political Geography* 23(6), 641–3.

Pain, R. 2003. Social geography: On action-orientated research. *Progress in Human Geography* 27(5), 649–57.

Pappé, I. 2006. *The Ethnic Cleansing of Palestine.* Oxford: Oneworld Publications.

PASSIA. 2002. *The Palestine Question in Maps 1878–2002.* Jerusalem: PASSIA: The Palestinian Academic Society for the Study of International Affairs.

Peled-Elhanan, N. 2008. The denial of Palestinian national and territorial identity in Israeli schoolbooks of history and geography, 1996–2003. In, R. Dolón & J. Todolí (eds.), *Analysing Identities in Discourse.* Amsterdam: John Benjamins, pp. 77–110.

Reinhart, T. 2003. Academic boycott: In support of Paris IV. *Borderlands e-Journal* 2(3).

Said, E.W. 1980. *The Question of Palestine.* London: Routledge & Kegan Paul.

Said, E.W. 1986. *After the Last Sky* (with photographs by Jean Mohr). New York: Pantheon Books.

Said, E.W. 1988. Introduction. In, E.W. Said & C. Hitchens (eds.), *Blaming the Victims: Spurious Scholarship and the Palestinian Question.* London: Verso, pp.1–20

Said, E.W. 1993. *Culture and Imperialism.* New York: Alfred A. Knopf.

Said, E.W. 1994. *Representations of the Intellectual.* New York: Vintage.

Said, E.W., I. Abu-Lughod, J.L. Abu-Lughod, M. Hallaj and E. Zureik. 1988. A profile of the Palestinian People. In, E.W. Said and C. Hitchens (eds.), *Blaming the Victims: Spurious Scholarship and the Palestinian Question.* London: Verso, pp. 235–96.

Salamanca, O.J., Qato, M., Rabie, K., and Samour, S. 2012. Past is present: Settler colonialism in Palestine. *Settler Colonial Studies* 2(1), 1–8.

Saleh, H.A.K. 1990. Jewish settlement and its economic impact on the West Bank, 1967-1987. *GeoJournal* 21(4), 337–348.

Slater, D. 2004. Editorial comment: Academic politics and Israel/Palestine. *Political Geography* 23(6), 645–6.

Sparke, M. 2004. Political geography: political geographies of globalisation (1) – dominance. *Progress in Human Geography* 28(6): 777–94.

Taraki, L. 2003. Hot days in Ramallah.*Counterpunch*, July 14. http://www.counter-punch.org/taraki07142003.html (3 July 2020).

Taraki, L. 2008. Urban modernity on the periphery: A new middle class reinvents the Palestinian city. *Social Text* 26(2(95)): 61–81.

Taraki, L and R. Giacaman. 2006. Modernity aborted and reborn: Ways of being urban in Palestine. In, L. Taraki (ed.), *Living Palestine: Family Survival, Resistance, and Mobility Under Occupation*. Syracuse, NY: Syracuse University Press, pp. 1–51.

Tawil-Souri, H. 2011a. Review essay: Mapping Israel-Palestine. *Political Geography* 30(8), 57–60.

Tawil-Souri, H. 2011b. Colored identity: The politics and materiality of ID cards in Palestine/Israel. *Social Text* 107, 67–97.

Tawil-Souri, H. 2011c. Qalandia checkpoint as space and non-place. *Space and Culture* 14(1), 4–26.

Tawil-Souri, H. 2012a. Uneven borders, colored (im)mobilities: ID cards in Palestine/Israel. *Geopolitics* 17(2), 1–24.

Tawil-Souri, H. 2012b. Digital occupation: The high-tech enclosure of Gaza. *Journal of Palestine Studies* 42(2), 27–43.

Waterman, S. 1985. Not just milk and honey—now a way of life: Israeli human geog-raphy since six-day war. *Progress in Human Geography* 9(2), 194–234.

Zureik, E.T. 1979. *Palestinians in Israel: A Study of Internal Colonialism*. London: Routledge & Kegan Paul.

Zureik, E.T. 2011. Colonialism, surveillance, and population control: Israel/Palestine. In, E. Zureik, D. Lyon & Y. Abu-Laban (eds.), *Surveillance and Control in Israel/Palestine*. Abingdon and New York: Routledge, pp. 3–46.

3 Social change and the (re)radicalization of geography in South Africa

Brij Maharaj and Maano Ramutsindela

Introduction

For almost six decades radical geographers were responsible for initiating a great deal of intellectual ferment in the discipline with regard to philosophies, theories, concepts and societal relevance; demands for more theoretically informed research which took cognisance of social, economic and political realities; and contributed to increasing respect and recognition for the discipline. Critical human geographers in particular did not only bring human conditions – such as inequalities, social and environmental justice, gender/race oppression and exploitation – to the forefront of geographic inquiry, but also helped create a degree of consciousness and sensitivity that has seen some scholars and their journal outlets taking explicit positions on matters related to the discipline. Examples of these positions that have been widely publicized are the Reed Elsevier's involvement in arms and the implications of this on the morality of Elsevier's geography journal outlets (Murakami Wood, 2009; Chatterton and Featherstone, 2006). Equally, the critique on the international status of geography journals has seen a number of changes – some cosmetic – to practices in geography (Paasi, 2005). More recently, the neo-liberal globalization era in northern and southern contexts has developed into one of the main reference points for a critically engaged geographic scholarship.

The broad templates for critical scholarship outlined above have been calibrated to suit local geographic inquiries as the various chapters of this volume illustrate. In South Africa the radicalization of geography can be ascribed to three main sources, namely, the social and political conditions created by apartheid; international exchanges; and the post-apartheid neo-liberal environment. Those broader conditions were referred to by Nelson Mandela, who commented that:

> When I go to the place and area of my birth ... the changed geography of the place strikes me with a force that I cannot escape. And that geography is not one of mere landscapes and topography, it is the geography of the people ...one was saddened at the poverty of the people – poverty

DOI: 10.4324/9781315600635-3

lived out in the geography of the place. It is the geography of women and young people, walking miles and miles to find the paltriest pieces of wood for fire to cook ... and to keep a shelter warm.[1]

In this chapter we argue that these conditions combined to engender a radical geography that, while drawing from concepts and thinking in geography elsewhere, developed a local character that has the potential to contribute to contemporary critical geographic thought. We claim that social change has consistently been a crucial factor in both the rise and decline of radical geography in the country. To be sure, while the need to transform society from apartheid gave impetus to the emergence of critical geography in South Africa, the emergence of democracy since 1994 has been accompanied by a marked silence; a product of opportunities offered by a transforming society. In the concluding part of this chapter, we paint a picture of the futures of critical geographic scholarship in the country and how this might be useful to think about mapping the future of the discipline in general. We divide the chapter into three main sections that capture the trajectory of radical geography in South Africa. The first part provides background on the colonization of the discipline in order to illustrate how both internal and external factors contributed to a docile discipline in the country. In this first part, we highlight that the Eurocentric dominance was very evident. Emphasis on quantification and spurious objectivity meant that the injustices of apartheid were seldom challenged and in fact were often legitimated by geographers. The dire conditions of deprivation that Mandela referred to above had no immediate meaning to a highly segregated white geography community. In the second part, we pay attention to the rise of critical scholarship in South Africa in the 1980s and the call for the decolonization of geography and the need for the indigenization of the content and direction of the discipline. This is followed by the third part in which we discuss conditions for its decline. We argue that the lure of financial opportunities offered by the growing consultancy industry together with the benefits of political relevance weakened the critical scholarship that had emerged in the 1980s. We also ascribe this weakening to the inability of geographers to set new research agendas after the demise of apartheid. With the benefit of hindsight, the critical geography that emerged in the country had its own weaknesses that offer lessons for a broader agenda for geographers.

Conservative roots in South African geography

Geographers like all members of humanity are shaped by their own geography, their location in this world of inequalities, and their production of geography is influenced accordingly.

(Taylor, 1989, 103)

Research in South African geography emulated the intellectual trends in the western world, albeit with a considerable time lag. Like most colonial societies, the origin of geography as an academic discipline was shaped by imperial agendas that unfolded over time. The discipline developed within the environmental determinism paradigm that legitimated the superiority of the European colonizers and the associated racist discourse on a supposedly scientific basis. Wesso (1994, 326) captured this paradigm aptly when he maintained that:

> Environmental or geographic determinism was, therefore, a significant ideological buttress for imperialism and racism, and it found fertile breeding ground in South Africa after the establishment of Union in 1910, particularly in view of the concerted effort by the government to eradicate the British-Afrikaner divide in the South African society, and to replace it with a black-white divide. It is perhaps not too far fetched to argue that it was the environmentalist paradigm which rescued geography from its peripheral position within the South African education system, and which gave it some respectability among academics.

Although, under apartheid, the Afrikaners generally rejected colonial education as a result of their opposition to imperial Britain, they invoked a Christian Calvinist curriculum and accepted the environmental determinist paradigm which legitimated notions of racial superiority (Wesso and Parnell, 1992). The rejection of colonial education was by no means a progressive agenda as in fact South African education, and geography curriculum in particular, was based on internal colonialism and its superiority/inferiority complexes. Much of what the discipline offered was nonetheless still based on external concepts and paradigms. For example, it paraded and approached geography as a spatial science in the late 1960s and early 1970s when this paradigm was being challenged by Anglo-American academics who were initially responsible for this trend. The spatial science approach promoted a "utilitarian geography" which "buttressed a society based on inequalities of race, class and gender" (Wesso, 1994, 333). Reflecting on this period, Rogerson and Beavon (1988, 84) noted that:

> Geographers were mesmerised by the delights of delineating the nature and characteristics of work and business places, fascinated by central place theory and the search for continua or hierarchies. In addition, they were inexorably committed to diffusionist notions of development which asserted that by operationalising growth poles, and promoting decentralisation, one would achieve a magical hierarchical spread of development from the affluent core areas of 'white' South Africa into the poverty-stricken, 'backward' peripheral areas ... The major preoccupation of the community of human geographers was of retaining an

academic respectability, defined by the standards and current band-wagons rolling in North America and Western Europe.

The consequences of adopting these approaches and their theoretical and ideological underpinnings were that most of the research legitimized the status quo (see Davies, 1971); the discipline deployed its geographical skills to service the apartheid state (Crush, 1984, 1991); misinterpreted poverty and inequality through Anglophone diffusionist models and through racist lenses (Rogerson and Beavon, 1988; Wellings and McCarthy, 1983); ignored black communities and their problems (Beavon and Rogerson, 1981). By so doing, the discipline "never permitted [the black community – Africans, Coloureds and Indians] a concerted geographic voice. The[se communities] were colonised into the belief that geography mattered not to them, that geography was made by whites for whites" (Wesso, 1994, 331). In fact, between 1917 and 1980, only two articles published in the journal of the South African Geographical Society focussed on black townships (Beavon, 1982). The research focus has been almost exclusively "white social space", and it was tacitly assumed that there was a harmonious social order and no conflict (Rogerson and Browett, 1986, 224). These studies were theoretically quiescent, and there was a tendency to ignore the historical, political and ideological forces which structured apartheid space. This was amazing, given that apartheid constituted an unparalleled example of state-directed socio-spatial structuring, with a "special fascination for the geographer ... Indeed, it would not be too much of an exaggeration to describe apartheid as the most ambitious contemporary exercise in applied geography" (Smith, 1982, 1). These observations laid the ground for the development of a radical geography in South Africa. McCarthy's (1982, 53) warning that "unless professional geography in South Africa wishes to be judged by history as having been totally incapable of addressing the fundamental issues of its times it should act immediately to redress the current impasse" was followed by discussions on the need for a decolonized South African geography to which we turn in the next section.

De-colonization and the emergence of progressive approaches

The scholarly project towards decolonization was twofold. It entailed a shift away from Eurocentric paradigms towards the development of more indigenous approaches to understand socio-spatial processes in line with critiques of imperialism in geography. In this regard, Professor Ron Davies' inaugural lecture at the University of Cape Town in 1976 represented the first major shift in approach where he suggested that South African cities have more in common with the colonial city than with the patterns of the advanced capitalist, western city (Davies, 1976). New emphasis was thus placed on the specificity of South African cities and their peculiar character (Beavon, 1982; Simon, 1984). Decolonizing geography also entails bridging the gap between ivory towers and communities, and to empower the poor and the disadvantaged (Crush, 1992). This attempt resonated with calls for academic activism by radical geographers more broadly. The peculiar apartheid project made decolonizing geography more appealing to critical geographers in the country and the local experience fed into the international dialogue on critical scholarship (Mather, 2007).

After the Soweto riots in 1976, researchers, of necessity, became more aware of the need to investigate problems experienced by the underprivileged and exploited masses. Almost simultaneously, geography also came under the influence of a "new school" of neo-Marxist historiography developed by historians, sociologists, and political scientists. In terms of the "new school", South African society could not be explained purely in terms of race, but rather required a more sensitive analysis, "informed by historical materialism, on the themes of class, capitalism, racial domination and the institutions of labour repression" (Rogerson and Beavon, 1988, 85). Such an analysis revealed that "South Africa's landscapes of social control and domination are imbued with a history of interaction and struggle between black and white, labour and capital, class and class" (Crush, 1986, 3).

As a result of the influence of the political turbulence and the "new school" the housing and employment problems of Blacks began to receive increasing attention from geographers, who also adopted a more critical stance towards explanation (Beavon and Rogerson, 1979; Lincoln, 1979). This momentum was accelerated in the 1980s when a number of articles appeared, both locally and internationally, critically evaluating the development of geography and geographers in South and southern Africa (e.g. Beavon and Rogerson, 1981; Crush and Rogerson, 1983; Wellings, 1986; Wellings and McCarthy, 1983). According to Scott (1986, 45), "the fluid and tempestuous currents of contemporary social thought and escalating conflict in South Africa society has bred a self-consciousness amongst geographers". This period has been characterized by a methodological and epistemological shift in urban research towards the political economy paradigm (Beavon and Rogerson, 1981). Human geography in South Africa finally began "to break out of the limited confines of apartheid apologism and 'liberal' analysis which have long held sway" (Simon, 1983, 309). Associated with this trend, the difficulties and inequities of living under apartheid became the primary concern of geographic research (Smith, 1982; Wellings, 1986). By the mid-1980s, Rogerson (1986, 128) could report:

> Happily, the new geography breaks from the sort of academic 'impartiality' formerly prevalent ... which included among its lesser virtues the fact that it hardly ever said anything upsetting or politically controversial. With the growth of political awareness, commitment and sensitivity, no longer are South African geographers eschewing confrontation with the controversial and politically sensitive issues raised by the workings of apartheid. Rather, the research outpourings of many

geographers are contributing to debunking the mythologies of the State's propaganda machinery and exposing the injustices, harshness and pain inflicted by the 'inhuman' geography of apartheid.

In a review of "apartheid human geography in the 1980s", largely in response to an impending academic boycott, Rogerson and Parnell (1989, 13) argued that "a major development in human geography has been the strengthening and consolidation of a critical research prospectus". However, an important concern about the more critical geography emerging was that it did not address the question of empowerment, emancipation and transformation (Wellings, 1986).

Two observations should be made about "the breed of critical geographers" in the 1980s. First, these were based in English-speaking universities which were generally seen as liberal. Their exposure to academia overseas and the networks they had developed across the Atlantic allowed them to import concepts and paradigms, some of which were paraded as "local intellectual innovations". It must be noted, though, that socio-political conditions in South Africa provided the platform on which a critical outlook could be nourished with a great deal of local relevance. Second, as in much of liberal thinking in South Africa, these critical geographers spoke on behalf of the black people whose scholarship they had silenced. In 1989 Mabin (1989, 124) contended that the fact that the number of publishing black geographers referred to in the Rogerson and Parnell (1989) review "could be counted on the fingers of one hand" reflected poorly on the geography community in South Africa, and illustrated the "relative weakness of the discipline in addressing pressing scholarly, pedagogic and political concerns in contemporary South Africa". This was an indictment on the development of geography as a "collective enterprise" in South Africa. Wellings (1986, 124) called for a "restructuring of the academic labour process" which, at the minimum, will lead to a waning of white middle class control in the discipline. This call was emphasized by Soni (1991, 1) who argued that the "conspicuous absence of black geographers in major research agendas and publications has raised questions of academic hegemony". Thus, the apparent critical geography had a serious racial bias.

Institutionally, academic geography in South Africa was organized and administered by two organizations based on language: the liberal English South African Geographical Society (SAGS) and the conservative Afrikaans Society for Geography. Black geographers were relegated to the periphery of the discipline. This was emphasized by Professor Lindisizwe Magi in his inaugural lecture at the University of Zululand in 1990 in which he emphasized that, "no African has ever held an elected position in ... national and regional geography councils [and] No African has served on the editorial boards of their journals" (Magi, 1990, 7).

There were few black geographers who were based mainly in ethnic universities; institutions that were burdened with large numbers of ill-equipped students and inequitable distribution of resources. The results were privileged groups of geographers in white universities while black universities were more like what Mamdani (1999, 131) calls "detention centres for black intellectuals". According to a survey of the state of the geography discipline in South Africa, blacks "experienced exclusion either by careless omission or intent. This fact stands to the shame of the discipline as a whole" (Fairhurst et al., 2003, 12).

The transformation of the discipline was initiated at the Conference of the Society for Geography held at the University of Pretoria in 1989. An invitation was sent out to black geographers attending the conference to meet informally:

> To initiate a discussion of whether there is a role for the South African Geographical Society to play in boosting the image of geography in black communities, and - perhaps more importantly - in helping black geographers overcome the legacy of discrimination in our society. A further issue may be the particular difficulties of teaching geography and conducting research on the predominantly black campuses.[2]

Black geographers considered the invitation as a move to prevent the looming international boycott of South African geographers; as part of the state's conspiracy to 'win hearts and minds'; and as paternalistic and demeaning. They noted that it required the Soweto 1976 uprising to enable some English-speaking geographers to discover South Africa's 'common people'. Black geographers wished to set the record straight that while some white English-speaking geographers have spent valuable energy and resources researching black poverty and other inequalities stemming from the policies of the apartheid state, they (black geographers) have quietly, systematically and in an unsung manner advanced the development of geography and geographical education at the grassroots level (Black Geographers Discussion Group, 1989, 3–5). For example, geographers at the University of the Western Cape had established a study group which aimed to "further the study of geography and to promote the societal recognition of its value by debating issues regarding a geography for a post-apartheid society" (Wesso, 1989, 17).

Black geographers argued that the problems they experienced emanated directly from the apartheid system, and in a memorandum challenged the SAGS and Society for Geography to publicly petition the state to abandon its apartheid policy, and to provide free, compulsory and equal education for all; to publicly call on tertiary institutions to reject the "quota system" and accept all students; and to appeal to all education institutions, and especially geography departments, "to adopt equal opportunity, affirmative action programmes in the appointment of staff" (Black Geographers Discussion Group, 1989, 4).[3]

At the Pretoria conference, Wesso (1989, 10) called for the development of a people's geography which was "committed not only to alleviate the plight of the oppressed masses in this country, but, above all, to instil in all people the values and the ideal of a truly just, non-racial and democratic society". At the Quadrennial Conference of the South African Geography Society in Potchefstroom in 1991, Soni (1991) argued that both geography and geographers need to substantially redefine their roles if they wanted to be relevant and recognized as progressive forces in the transformation towards a democratic, non-racial and non-sexist society.

The issues raised by black geographers at the Pretoria Conference in 1989 and at Potchefstroom in 1991 led to the SAGS convening a Geography Forum at the University of Natal in Pietermaritzburg in September 1991. The Forum facilitated frank and critical discussions relating to the problems experienced by black geographers, the need for one national geography society and the high level of prejudice (race and gender) within the South African geographical community. As a result of the mandate from the Forum, the SAGS and the Society for Geography disbanded and the new Society of South African Geographers was formed in 1994 and held its first conference at the University of Durban-Westville in 1995 under the theme "Geography in a Changing Society: Critical Choices for Change in Southern Africa".

In spite of many problems, by the end of the 1980s and early 1990s, the voices of black geographers were beginning to be heard in the publishing world, nationally and internationally (e.g. Khosa, 1989; Maharaj, 1989a, 1989b; Magi, 1989; Seethal, 1991; 1992; Soni and Maharaj, 1991). This was related to the fact that the first cohort of black geographers had completed their higher degrees, mostly under the influence of mentors outside South Africa. However, an indictment against black and white geographers was their failure or reluctance to engage with geography and geographers in Africa. The end of formal apartheid ushered in a new environment which impacted on critical scholarship in the country.

Post-apartheid capitulation to neo-liberalism

McCarthy (1991, 23) opined that South Africa's post-apartheid reconstruction will depend in part on a sensitive understanding of the "geographical legacy of apartheid and the scars it has left behind, and also to the complex local, regional and environmental diversity that characterises the South African whole". Indeed, the dominant research themes in the early 1990s were in line with the political agenda for a transforming South African society, and these were reflected in research on socio-spatial restructuring related to the redrawing of municipal and regional boundaries, land reform, urban and rural development, environmental hazards, access to basic needs and gendered perspectives. Subsequently, most progressive geographers became policymakers and consultants to the new government. Duncan and Vale captured this move aptly when they maintained that many "among the country's best and most published intellectuals migrated, or mutated into a new post-apartheid industry, consulting" (Duncan and Vale, 2004, 6). In this regard Said's (1994, 63) warning about the implications of this trend for muting critique and pressurizing intellectuals to conform is worth noting:

> In a more consistent and sustained way ... intellectuals who are close to policy formulation and can control patronage of the kind that gives or withholds jobs, stipends, promotions tend to watch out for individuals who do not toe the line professionally and in the eyes of their superiors come to exude an air of controversy and noncooperation.

Bekker (1996, 20) has similarly argued that consultants and "policy makers are not free agents. Their activities are confined both by their political masters and by budgetary allocations that oblige then to prioritise often against their better judgement". Under such circumstances, Said (1994, 64) has argued, "the temptation to turn off one's moral sense ... or to curtail skepticism in favour of conformity, are far too great to be trusted". Hence, Seekings (2000, 833, 835) laments that lucrative consultancy contracts have inevitably led to a decline in critical scholarship in South Africa, noting that:

> In their engagement with policy design, researchers have tended to remain isolated from theoretical work in the various relevant disciplines and failed to locate their work in the comparative literature of cities elsewhere in the world ... The combination of new decision-makers in the state and well connected academics outside of it has resulted in a booming but largely uncritical policy studies industry.

By marketing "themselves to capital or the state to facilitate their interventions in space" (Blomley, 1994, 383), geographers risked locking "the discipline into the same trends that obtained in the past" (Ramutsindela, 2002, 8). A further indictment against policymakers and consultants is the view that very few of their policies have "challenged the apartheid geography or have come to terms with the changing social patterns of post-apartheid South African cities" (Bremner, 2000, 87), and have in fact reinforced race-class inequalities.

As the post-apartheid rainbow mirage became more apparent, there were calls for more radical critiques of reconstruction strategies (see Bond, 1998, 2000a, 2000b, 2002; Bond and Tait, 1997; Maharaj and Ramballi, 1998; Maharaj and Narsiah, 2002; Narsiah, 2002; Seethal, 2002). Although many radical scholars of the 1980s have been seduced by neo-liberalism in the 1990s, leftist intellectuals should not despair: urban and rural "South Africa offers fertile case study material in the application of political-economic theory to contemporary policy and practice" (Bond, 1995, 149). This will be illustrated briefly with reference to the slide to neo-liberalism, and especially the impacts on the poor.

The first wave of post-apartheid planning and development strategies was driven by the Reconstruction and Development Programme (RDP) which had a strong basic needs and social justice orientation. However, there was concern that the RDP was attempting to reduce poverty without transforming the "economic policies and practices that reproduce poverty and inequality" (Marais, 1998, 192). This weakness in policy formulation became apparent in the government's macro-economic policy, especially the adoption of the neo-liberal Growth, Employment and Redistribution strategy (GEAR) in June 1996. This apparent World Bank infested "home-grown structural adjustment programme" (Lehulere, 1997, 73) encountered an aberration with the basic-needs-oriented RDP (Bond, 2000a, 2000b). GEAR represents the view that the "market is more efficient than government at providing basic services" (Bakker and Hemson, 2000, 6), hence its orientation towards the privatization of basic services in keeping with the dictates of the neo-liberal agenda (Narsiah, 2010) and "reflect ... inadequate subsidies targeted at the poor" (Parnell and Pieterse, 1999, 75).

The privatization of services has far-reaching geographical implications. Under apartheid the access to services had a distinct spatiality. Townships were inadequately serviced, if at all, while the racially privileged enjoyed access to services comparable to those in the first world (Turok, 1994). The privatization of basic services militates against the aim to build an inclusive society. The provision of a minimum level of service to disadvantaged areas re-enforces apartheid boundaries in the geography of service distribution (Bakker and Hemson, 2000).

As John Saul argues, the poor are being "sacrificed on the altar of the neoliberal logic of global capitalism" in what he calls the "post-apartheid denouement" (Saul, 2001, 1). There has been an uneasy response from the South African academic community to this situation. According to Ramphele (1999, 205), the "silence of academics on questions of redistribution of wealth and opportunity ... is tantamount to connivance". The state's response to critique of its policies is to label protagonists as racists or reactionaries or clever Blacks, and Jansen's (2003, 11) reflections are perceptive:

> The vocation of the intellectual in South Africa has fallen on hard times. Persons are under attack, reputations are muddied and lives are even threatened. Courageous voices have been severely attacked by politicians, academics and the general public for daring to pose uncomfortable questions about health, education, warfare and the presidency itself. In this fragile democracy, it is more important than ever to be vigilant to the conditions under which public intellectuals speak and are compelled to speak.

The close association between progressive intellectuals and state has had a damaging effect on critical scholarship in the country: it has led to co-option, compliance and conformity and the silence of critical scholars (Desai and Bohmke, 1997; Jansen, 2003), with black intellectuals "appearing too hungry for status to be angry, too eager for acceptance to be bold, too self invested in advancement to be defiant" (West, 1993, 38; see also Said, 1994). Harvey (2001, vii) has emphasized that progressive intellectuals must "combat the presumptions, prejudices and political predilections that at any time constrain thinking in ways which may at best be understood as repressive tolerance and at worst merely repressive".

Prospects for renewal and relevance

Any critical theory should take as its starting point the emancipatory role and "a commitment to socially transformative research" (Fien and Hillcoat 1996, 28; Painter, 2000). In the context of South Africa, there are three pathways for a critically engaged geographic scholarship. First, the groundswell of protests caused by disappointments with the policies of the post-apartheid state provides avenues for an emancipatory geography. A sustained intellectual project that dissects the implications and impact of state policies on the marginalized in society is most likely to advance critical geography in South Africa. Critical geographers in the country argue that social and spatial inequalities in terms of access to basic services and amenities have become more pronounced in recent years as the impact of neo-liberal strategies become evident in South African cities (Bond, 2002; McDonald and Page, 2002).

Second, the acknowledgement in the past that geographical activities cannot be confined to ivory towers augurs well for geography in the 21st century. There is a need for progressive geographers to put their talents at the service of disadvantaged groups outside academia (Chouinard, 1994; Kitchin and Hubbard, 1999; Oldfield, 2008; Routledge, 1996). This is more crucial given that the "postapartheid state is advancing the interests of national or international capital at least as strongly as its apartheid predecessor" (Seekings, 2000, 835). As inequalities escalate critical intellectuals are likely to be challenged to consider alternative theoretical and practical interventions (Gibson, 2001).

Third, the Society of South African Geographers has its own flagship journal, the *South African Geographical Journal*, which could be used to promote a critical scholarship that could be easily silenced by other outlets. The journal could be used as an avenue for local and international scholars working on South African issues and to integrate local voices into a broader community of concerned geographers.

Concluding remarks

In recent years there have been many reviews of radical geography (e.g. Blomley, 2007; Wills, 2006). However, there is no reference to any radical work outside the Anglo-American network (see Minca, 2003). This reflects the "parochialism of much contemporary western scholarship" (Robinson, 2003, 275). This is a serious omission, especially in the internet era when radical literature from the South can be easily accessed. Of course, radical

Anglo-American geographers writing about issues in the developing world were acknowledged. Southern sites are used for empirical data to test Northern theories. It was imperative that the geography discipline "resign its role in the service of 'Western' imperialism" (Shaw, Herman and Dobbs, 2006, 272). Robinson (2003, 273) warned ominously that "a geography whose intellectual vision is limited to the concerns and perspectives of the richest countries in the world has little hope of effectively participating in the debates that will matter in the twenty-first century". A major issue is the extent to which geographers outside the Anglo-American world "were again being asked to be map-takers rather than mapmakers" (Chaturvedi, 2003, 149). Notwithstanding serious resource constraints, the challenge for scholars in the South "is to ensure that we write and reflect on our work so as to contribute to theoretical debates directly" (Oldfield et al., 2004, 295).

The ultimate challenge for geography in South Africa is to maintain and sustain a critical intellectual agenda; to survive as a discipline in an era of commodification, restructuring and institutional mergers, when even adherents are beginning to lose faith; to engage in socially relevant research which is sensitive to the stresses and strains of transformation; and to ensure the Africanization of the discipline, both in terms of institutional and organizational structures, as well as the curriculum. This will entail a shift from Eurocentricism to Afrocentricism, and will require greater engagement with colleagues and literature emerging from north of the Limpopo River than has otherwise been the case. The responsibility of the university is to provide the space for critical intellectuals to develop and ideas to flourish in what some have described as the discipline for the 21st century. In conclusion, South African geography has come a long way from the dark days of environmental determinism, colonialism and apartheid. The need to sustain a critical, independent intellectual prospectus in geography in particular, and academia in general, cannot be overemphasized, given South Africa's repressive apartheid legacy and Africa's postcolonial record.

Notes

1 Extract from Nelson Mandela's address to the Regional Conference of the International Geographical Union, Durban, 4 August 2002, on receiving the Planet and Humanity Award.
2 Letter from Dr Alan Mabin, University of Witwatersrand, to Mr C.E.P. Seethal, University of Bophuthatswana, 29 May 1989.
3 In his inaugural address at the University of Fort Hare, Professor Cecil Seethal revealed that he was the author of this document (Seethal, 2006).

References

Bakker, K. and D. Hemson. 2000. Privatising water: BoTT and hydropolitics in the new South Africa. *South African Geographical Journal* 82, 3–12.
Beavon, K.S.O. 1982. Black townships in South Africa: Terra incognita for urban geographers. *South African Geographical Journal* 64, 3–20.

Beavon, K.S.O. and C.M. Rogerson. 1979. Hawking and the 'urban poor': how not to make out in the contemporary South African city. Paper presented at the South African Geographical Society Conference, University of Cape Town.

Beavon, K.S.O. and C.M. Rogerson. 1981. Trekking on: Recent trends in the human geography of Southern Africa. *Progress in Human Geography* 5, 159–89.

Bekker, S. 1996. The policy making predicament. *Indicator SA* 13, 17–20.

Black Geographers Discussion Group, 1989. A Response to A. Mabin, T. Hart and P. Hattingh. *Zululand Geographer* 8, 3–5.

Blomley, N.K. 1994. Editorial: Activism and the academy. *Environment and Planning D: Society and Space* 12, 383–5.

Blomley, N. 2007. Critical geography: Anger and hope. *Progress in Human Geography* 31, 53–65.

Bond, P. 1995. Urban social movements, the housing question and development discourse in South Africa. In, D.B. Moore and G.J. Schmitz (eds.), *Debating Development Discourse - Institutional and Popular Perspectives*. London: Macmillan, pp.149–77.

Bond, P. 1998: Privatisation, participation and protest in the restructuring of municipal services. *Urban Forum* 8, 19–41.

Bond, P. 2000a. *Elite Transition: From Apartheid to Neoliberalism in South Africa*. London: Pluto Press.

Bond, P. 2000b. *Cities of Gold, Townships of Coal: Essays on South Africa's New Urban Crisis*. Trenton: Africa World Press.

Bond, P. 2002. *Unsustainable South Africa: Environment, Development and Social Protest*. Pietermaritzburg: University of Natal Press.

Bond, P. and A. Tait. 1997: The failure of housing policy in post-apartheid South Africa. *Urban Forum* 8, 115–42.

Bremner, L. 2000. Post-apartheid urban geography: a case study of Greater Johannesburg's rapid land development programme. *Development Southern Africa* 17, 87–104.

Chatterton, P. and D. Featherstone. 2006. Intervention: Elsevier, critical geography and the arms trade. *Political Geography* 26, 3–7

Chaturvedi, S. 2003. Towards a critical geography of partition(s): some reflections *on* and *from* South Asia. *Environment and Planning D: Society and Space* 21, 148–54.

Chouinard, V. 1994. Reinventing radical geography: Is all that's left right? *Environment and Planning D: Society and Space* 12, 2–4.

Crush, J. 1984. Any space for apartheid? *Canadian Journal of African Studies* 18, 623–6.

Crush, J. 1986. Towards a people's historical geography for South Africa. *Journal of Historical Geography* 12, 2–3.

Crush, J. 1991. The discourse of progressive human geography. *Progress in Human Geography* 15, 395–414.

Crush, J. 1992. Beyond the frontier: The new South African historical geography. In, C. Rogerson and J. McCarthy (eds.), *Geography in a Changing South Africa: Progress and Prospects*. Cape Town: O.U.P., pp. 10–73.

Crush, J. and C. Rogerson. 1983. New wave African historiography and African historical geography. *Progress in Human Geography* 7, 203–31.

Davies, R.J. 1976. *Of Cities and Societies: A Geographer's Viewpoint*, Inaugural Lecture, University of Cape Town.

Davies, W.J. 1971. Patterns of non-white population distribution in Port Elizabeth with special reference to the application of the group areas act. *Special Publication No.1*, Institute for Planning Research, University of Port Elizabeth.

Desai, A. and H. Bohmke. 1997. The death of the intellectual, the birth of a salesman. *Debate* 3, 10–34.

Duncan, J. and P. Vale. 2004. The best way to connect new dots. *Mail and Guardian*, 30 January – 5 February 2004, pp. 6–7.

Fairhurst, U.J., R. J. Davies, R.C. Fox, P. Goldschagg, M. Ramutsindela, U. Bob and M. M. Khosa. 2003. *Geography: The State of the Discipline in South Africa*. Pretoria: Society of South African Geographers.

Fien, J. and J. Hillcoat. 1996. The critical tradition in research in geographical and environmental education research. In, M. Williams (ed.), *Understanding Geographical and Environmental Education*. London: Cassell, pp. 26–40.

Gibson, N. 2001. Transition from apartheid. *Journal of Asian and African Studies* 36, 65–85.

Harvey, D. 2001. *Spaces of Capital – Towards a Critical Geography*. Edinburgh: Edinburgh University Press.

Jansen, J. 2003. Hard times: The (self-imposed) crisis of the black intellectual. *Indicator SA* 20, 11–6.

Kitchin, R.M. and P.J. Hubbard. 1999. Research, action and 'critical' geographies. *Area* 31, 195–8.

Khosa, M. 1989. 'Dipalangwang': Black commuting in the apartheid city. *African Urban Quarterly* 4, 251–9.

Lehulere, O. 1997. The political significance of GEAR. *Debate* 3, 73–86.

Lincoln, D. 1979. Ideology and South African development geography. *South African Geographical Journal* 61, 99–110.

Mabin, A. 1989: Does geography matter? A review of the past decade's books by geographers of contemporary South Africa. *South African Geographical Journal* 71, 121–6.

Magi, L.M. 1989. Cognition of recreation resources through photographic images. *South African Geographical Journal* 71, 67–73.

Magi, L.M. 1990. Geography in society: The site of struggle. *Inaugural Lecture*, University of Zululand.

Maharaj, B. 1989a. Residential renovation in public sector housing. *Development Southern Africa* 6, 43–57.

Maharaj, B. 1989b. Residential immobility – the case of tenants in 'outhouses.' *South African Journal of Sociology* 20, 241–7.

Maharaj, B. and K. Ramballi. 1998. Local economic development strategies in an emerging democracy: The case of Durban in South Africa. *Urban Studies* 35, 131–48.

Maharaj, B. and I. Narsiah. 2002. From apartheid apologism to post-apartheid neo-liberalism: Paradigm shifts in South African urban geography. *South African Geography Journal* 84, 88–97.

Mamdani, M. 1999. There can be no African renaissance without an Africa-focused intelligentsia. In, M.W. Makgoba (ed.), *African Renaissance: The New Struggle*. Sandton: Mafube, pp. 125–36.

Marais, H. 1998. *South Africa: Limits to Change - the Political Economy of Transformation*. London: Zed Books.

Mather, C. 2007. Between the global and local: South African geography after apartheid. *Journal of Geography in Higher Education* 31, 143–60.

McCarthy, J.J. 1982. Radical geography, mainstream geography and southern Africa. *Social Dynamics* 8, 53–70.

McCarthy, J.J. 1991. Theory, Practice and Development. Inaugural Lecture, University of Natal, Pietermaritzburg.

McDonald, D.A. and J. Page. 2002. *Cost Recovery and the Crisis of Service Delivery in South Africa.* Pretoria: HSRC Publishers.

Minca, C. 2003. Critical peripheries. *Environment and Planning D: Society and Space* 21, 30–8.

Murakami Wood, D. 2009. Spies in the information economy: academic publishers and the trade in personal information. *ACME: An International E-Journal for Critical Geographies* 8(3), 484–93.

Narsiah, S. 2002. Neoliberalism and privatisation in South Africa. *GeoJournal* 57, 3–13.

Narsiah, S. 2010. The neoliberalisation of the local state in Durban, South Africa. *Antipode* 42, 374–403.

Oldfield, S., S. Parnell and A. Mabin. 2004. Engagement and reconstruction in critical research: negotiating urban practice, policy and theory in South Africa. *Social and Cultural Geography* 5, 285–299.

Oldfield, S. 2008. Who's serving whom? partners, processes and products in service learning projects in South Africa. *Journal of Geography in Higher Education* 32, 269–85.

Paasi. A. 2005. Globalisation, academic capitalism, and the uneven geographies of international journal publishing spaces. *Environment and Planning A* 37, 769–89.

Painter, J. 2000. Critical human geography. In, R.J. Johnston, D. Gregory, G. Pratt and M Watts (eds.), *The Dictionary of Human Geography* (4th edition). Oxford: Blackwell, pp. 126–8.

Parnell, S. and E. Pieterse. 1999. Developmental local government and post-apartheid poverty alleviation. *Africanus* 29, 61–84.

Ramphele, M. 1999. The responsibility side of the academic freedom coin. *Pretexts: Literary and Cultural Studies* 8, 201–6.

Ramutsindela, M. 2002. The philosophy of progress and South African geography. *South African Geographical Journal* 84, 4–11.

Robinson, J. 2003. Postcolonialising geography: Tactics and pitfalls. *Singapore Journal of Tropical Geography* 24, 273–89.

Rogerson, C.M. 1986. South Africa: Geography in a state of emergency. *GeoJournal* 12, 127–8.

Rogerson, C.M. and J.G. Browett. 1986. Social geography under apartheid. In, J. Eyles (ed.), *Social Geography in International Perspective.* London: Croom Helm, pp. 221–50.

Rogerson, C.M. and K.S.O. Beavon. 1988. Towards a geography of the common people in South Africa. In, J. Eyles (ed.), *Research in Human Geography - Introductions and Investigations.* Oxford: Basil Blackwell, pp. 83–99.

Rogerson, C.M. and S.M. Parnell. 1989. Fostered by the laager: apartheid human geography in the 1980s. *Area* 21, 13–26.

Routledge, P. 1996. The third space as critical engagement. *Antipode* 28, 399–419.

Said, E. 1994. *Representations of the Intellectual.* London: Vintage.

Saul, J. 2001. Cry for the beloved country: The post-apartheid denouement. *Monthly Review*, (January), 1–51.

Scott, D. 1986. Time, structuration and the potential for South African historical geography. *South African Geographical Journal* 68, 45–66.

Seekings, J. 2000. Introduction: Urban studies in South Africa after apartheid. *International Journal of Urban and Regional Studies* 24, 832–40.

Seethal, C. 1991. Restructuring the local state in South Africa: Regional services councils and crisis resolution. *Political Geography Quarterly* 10, 8–25.

Seethal, C. 1992. The transformation of the local state in South Africa (1979-1991): Group areas, property "super-taxation" and civic organisations. *Urban Geography* 13, 534–56.

Seethal, C. 2002. Regenerating rural economies: The case of Limehill, South Africa. *GeoJournal* 57, 61–73.

Seethal, C. 2006. Traversing landscapes, transforming places: The politics of space in South Africa. *Inaugural Lecture, University of Fort Hare*, 23 February.

Shaw, W.S., R.D.K. Herman and G.R. Dobbs. 2006. Encountering indigeneity: Re-imagining and decolonising geography. *Geografiska Annaler* 88(3), 267–76.

Simon, D. 1983. Review of D.M. Smith (ed.) Living under apartheid: Aspects of urbanisation and social change in South Africa. *Tijdschrift voor economische en social geografie* 74(4), 309.

Simon, D. 1984. Comment: The apartheid city. *Area* 16, 60–62.

Smith, D.M. (ed.) 1982. *Living under Apartheid*. London: George Allen and Unwin.

Soni, D. 1991. Knocking at the door: The future of geography and geographers in South Africa. Paper presented at the Conference of the South African Geographical Society, University of Potchefstroom.

Soni, D. and B. Maharaj. 1991. Emerging rural urban forms in South Africa. *Antipode* 23, 47–67.

Taylor, P.J. 1989. Editorial comment: Geographical dialogue. *Political Geography Quarterly* 8, 103–105.

Turok, I. 1994: Urban planning in the transition from apartheid part 1: The legacy of social control. *Town Planning Review* 65, 243–59.

Wellings, P. 1986. Editor's Introduction - Geography and development studies in Southern Africa: A progressive prospectus. *Geoforum* 17(2), 119–31.

Wellings, P.A. and J.J. McCarthy. 1983. Whither southern African human geography. *Area* 15, 337–45.

Wesso, H. 1989. People's Education: Towards a people's geography in South Africa. Paper presented at the Society for Geography Conference, 3–6 July.

Wesso, H. 1994. The colonisation of geographic thought: The South African Experience. In, A. Godlewska and N. Smith (eds.), *Geography and Empire*. Oxford: Blackwell, pp. 316–32.

Wesso, H. and S. Parnell. 1992. Geography education in South Africa: Colonial roots and prospects for change. In, C. Rogerson and J. McCarthy (eds.), *Geography in a Changing South Africa*. Cape Town: Oxford University Press, pp. 186–200.

West, C. 1993. *Race Matters*. Boston: Beacon Press.

Wills, J. 2006. The left, its crisis and rehabilitation. *Antipode* 38, 907–15.

4 The emergence of critical geographies in USA and Anglo-Canada*

Linda Peake and Eric Sheppard

Purporting to provide a historical account of the evolution of Anglophone radical/critical geography in North America is a hazardous proposition (for Francophone Canada, see Klein, 2020). First, the sheer quantity of radical/critical geography within this relatively confined area of the globe is enormous: spanning more than 40 years (and arguably much longer), and innumerable individuals, organizations, activities, and academic and non-academic writings. This cannot possibly be captured adequately in a single chapter. Second, many such accounts are possible, each marked by the situated knowledge of the narrators. Our perspective is that of two relative latecomers to what at that time was known as radical geography; neither of us participated in the early years.[1] In writing this chapter, we have attempted to plumb the recollections of early participants we could identify, but inevitably what we write is not what they would have written. Our particular predilections about what is significant inevitably shape this account. This is not, then, the definitive story, but a provocation: one particular account that can only be enriched as others react to, correct, and differently narrate these events. Third, as critical scholars we must be alert to the occlusions made possible by already existing narratives of the emergence of radical geography in Anglophone North America—accounts that become canonical simply by dint of the lack of alternatives. In particular, we interrogate the conventional wisdom, today, that radical geography emerged out of Clark University with the publication of *Antipode* in 1969, and was primarily Marxist. This is the case, but there also was much more. Fourth, as geographers we must be alert to the geography of knowledge production. In interrogating conventional wisdom, therefore, we begin to disinter both the theoretical/ideological variegation, characterizing the field from its beginnings but unevenly through time, as well as outlining the complex spatialities connecting the United States with Anglo-Canada and beyond.

Our account is structured chronologically. First, we examine the spectral presence of radical/critical geography in North America prior to the mid-60s. Second, we narrate the emergence of both radical and critical geography between 1964/1969 until the mid-1980s, when key decisions were taken that moved radical/critical geography into the mainstream of the discipline.

DOI: 10.4324/9781315600635-4

Third, we examine events since the mid-1980s, as radical geography merged into critical geography becoming in the process something of a canon in mainstream Anglophone human geography.

Hauntings: radical geographers avant la lettre

It might be expected that we begin in the tumultuous times of the mid and late 1960s, dating the birth of North American radical geography to 1969 (Castree, 2000). The story has been widely told of a staid discipline, caught between a Hartshornian regionalist past and its aspirations to be recognized as a value-free spatial science, whose disengagement with the world was disrupted by the impact of the May 1968 student revolts in Europe and the anti-Vietnam War and civil rights movements in the United States (Peet, 1977, 2000). But this would be to ignore the precursors of that time and those individual scholars already mapping out potential routes toward radical/critical geography. Four men are widely acknowledged as radical geographers whose political and academic contributions predated 1969, and have since been assessed and analysed: the Russian anarcho-communist Pyotr Kropotkin (1842–1921), French anarchist Élysée Reclus (1830–1905), the German and American Sinologists Karl Wittfogel (1896–1988), and Owen Lattimore (1900–1989) (Clark and Martin, 2004; Dunbar, 1978; Galois, 1976; Harvey, 1983; Peet, 1985).[2] Yet, others' contributions have been overlooked. For example, Mary Arizona (Zonia) Baber (1862–1956), a founder of the Chicago Geographical Society, was committed to peace, antiracism and conservation, and worked closely with Puerto Rican suffragist movements (Monk, 2004). No doubt, Baber is just one of a number of radical geographers, otherwise gendered, and racialized, whose contributions have been sidelined and then elided by the "1969 story", and who await recovery in the name of tracing multiple histories, situated knowledges, counternarratives, silences, and lacunae.

By the mid-1960s, a handful of radical geographers were percolating the field, also shaping what was to happen in 1969. Bill Bunge, a communist when he wrote his paean to quantitative geography (Akatiff, personal communication; Bunge, 1966), co-founded the Detroit Geographical Expedition and Institute (DGEI) with the African American community leader Gwendolyn Warren in 1968. The Expedition and Institute was committed to practicing the kind of radical pedagogy and activist research widely espoused today by post-structural and feminist geographers; it operated at the University of Michigan, and then Michigan State, until the latter closed it at the end of 1970 (Heyman, 2007; Horvath, 1971). Clark Akatiff had entered UCLA's graduate program in 1960 as a Marxist, was hired as an Assistant Professor at Michigan State University in 1966 and participated in the Detroit Expedition (Akatiff, 2007). Jim Blaut, life-long leftist, joined the faculty at Clark in 1967, having returned from five years in the Caribbean, including a stint in Puerto Rico where he joined its Movimiento Pro Independencia

(Mathewson, 2005; Santana, 2005). In October 1967, Blaut and co-conspirator David Stea flew Jim's plane to Washington DC, to observe the countercultural attempt to "raise the Pentagon", scaring the living daylights out of fellow passenger Dick Peet (2000).[3] When Ben Wisner arrived at Clark in 1968, radicalized by the anti-war movement and experiences in Tanzania, a small group of radical geographers was already active.[4] While not self-identified as radical, several sympathetic geographers helped create space in a hostile disciplinary and political environment, across both Canada and the United States, including Jim Lemon, Robert McNee, Richard Morrill, Phil Wagner, Julian Wolpert, and Wilbur Zelinsky.

Of course, what is striking about this list is its exclusivity. Neither women nor people of colour feature in accounts of this period of North American radical academic geography, yet the seeds of their participation were also sown in this period and they were to figure increasingly in developing both radical and critical geography. Their absence in reflections on this period speaks strongly to the ways in which the production of knowledge reflected the social demographics and political preoccupations of the overwhelmingly white, male, and middle-class North American academic geographers of that time.

Radical geography: 1969–1986

It was in 1969 that radical geography gained visibility as a "center of calculation" (Latour, 1987) within US geography, through two events. First, at the AAG meetings in Ann Arbor, Michigan, radical geographers for the first time defined themselves as a group (the meeting was relocated from Chicago, in response to opposition to meeting in the city where Richard Daley had repressed protests outside the Democratic National Convention—Ann Arbor being chosen because Michigan geographers were particularly vocal in this opposition (Eichenbaum, personal communication). Clark Akatiff (2007, 7) recalls: "three busloads from the [Detroit] Expedition brought the presence of the black streets to the walls of Academe. There were acts of showy militancy. Free Huey was scribed on the wall. Militant interventions were forced on staid academic panels about the 'Problem'". For the 1970 meetings, symbolically in San Francisco, the Detroit Expedition was accorded a plenary session led by Bunge, divided into an academic session and the peoples' geography, of mixed success: "it was like the original Acid Tests in the sense that we knew something was happening, but we didn't know what it was..." (Akatiff, 2007, 10–11). By the 1971 Boston meeting, radical geographers had pushed the AAG business meeting to pass a "strongly worded resolution opposing the Indochina War..., along with others concerned with the status of women, graduate students, and Spanish-speaking minorities, and a resolution calling for the release of Angela Davis...was only narrowly defeated" (Smith, 1971, 155).

Second, *Antipode: A Journal of Radical Geography* was initiated at Clark in 1969, at the end of David Stea's graduate seminar (Mathewson and Stea, 2003). It was a student-led initiative: "a reaction against the Vietnam War, racism and pollution...The key to *Antipode's* origin is the term 'radical.' We were groping for root causes of the problems, contradictions, inconsistencies, and hypocrisies with which we had grown up.... The "specter that stalks Europe" that Marx made famous didn't come first to mind because of who we (mostly white and male and middle class) young Americans were" (Wisner, 2010).

Antipode's emergence was the relational effect of multiple conditions of possibility, but it created visibility, and a place, for radical geography by dint of being a concrete and recognized academic object (a journal), drawing others into the orbit of Clark University where it was physically located. The early issues were eclectic, reflecting those who were aware of it and bound together by a shared no—rejection of the US societal status quo—and diverse yeses. Articles were included on imperialism, poverty, ghettoes and African Americans, geography's whiteness, women, American Indian geography, the environment and nature, remote sensing, migration, and a map projection. The progenitors of radical thinking at Clark, until they left, were Blaut and Stea. They catalyzed the radical politics of Ben Wisner and the other students who started *Antipode*, but also of Peet, who had arrived as a new faculty member with a freshly minted Berkeley Ph.D. in 1967. Peet took over the editorship of *Antipode* in 1970 (with volume 2), and co-produced it until 1985 with generations of Clark students—a number of whom went on to influential academic careers. David Harvey coincidentally arrived in the United States in 1969 from Bristol, where his experiences in Baltimore triggered his philosophical shift from logical positivism to social justice and then to Marxism. Harvey (personal communication) recalls this period as a "collision between the more book-wise UK trained geographers (like me) and the street-wise down with the people orientation of some of the US animators.... I certainly learned the importance of being in the street from Bunge.... [T]here was a joint exploration of anarchism and Marxism".

Harvey visited Clark in 1970, and in 1972 published the first explicitly Marxist paper in *Antipode* (indeed, in Anglophone geography. Harvey, 1972). As Wisner (2010) recalls: "the earliest days of *Antipode* were not informed by rigorous political economy. Only later, under Dick Peet's editorship and the frequent contributions of Jim Blaut and David Harvey, did we benefit from a systematic exploration of capitalism, its logic, and imperialism – its highest stage". As Peet has described it: "From 1972 onwards the emphasis...changed from an attempt to engage the discipline in socially significant research to an attempt to construct a radical philosophical and theoretical base...increasingly found in Marxian theory, which...many US geographers began reading in the early 1970s" (Peet, 1977, 17).[5]

By the mid-1970s, reading Marx had become *de rigeur* for radical North American geographers, and *Antipode* was taking an increasingly Marxist tone—publishing a number of classic theoretical treatises in Marxist geography, although other kinds of approaches continued to be represented.[6] It was also a progressively masculine discourse, dominated by confident (Harvey), assertive (Peet), imposing (Blaut, Soja, Bunge), and difficult (Bunge) personalities. Yet, radical geography also was active beyond Clark, Worcester, MA. Although the DGEI had been terminated, with Bunge fired by Wayne State, new expeditions sprang up. Geographical Expeditions were set up in Toronto (by Bunge), Vancouver and Sydney, Australia (by Ron Horvath, who Akatiff had persuaded to move to Michigan State from UC Santa Barbara in 1967), and these ideas were carried to London, England (by Bob Colenutt) and to Worcester (Peet, 2010). In 1971 Larry Wolf and Wilbur Zelinsky founded a second critical geography group, the Socially and Ecologically Responsible Geographers (SERGE), publishing the mimeographed journal *Transition*, until 1986. A significant further development, with a strong Canadian footprint, was the founding of the Union of Socialist Geographers (USG).

The idea for a Union of Socialist Geographers emerged from a group of graduate students in Vancouver, catalyzed by the geographical expeditions and the critical support of Michael Eliot-Hurst, who had become chair of Geography at Simon Fraser University.[7] There were some 40 students working on the expedition; Ron Horvath had been hired at Simon Fraser after termination of the DGEI. Simon Fraser's geography graduate students included active exiles radicalized through political struggles in Ireland and South Africa, Britain and the United States. Eliot-Hurst's presence was vital, having himself been radicalized after experiencing state surveillance of teaching at California State University—Northridge in the 1960s: he "almost single-handedly oversaw the creation of a virtual graduate school in radical/Marxist geography at SFU in the early 1970s" (Breathnach, personal communication).[8] The first meeting of the USG was held in Toronto on May 26–28 1974, in parallel with the Canadian Association of Geographers meeting and "under the roof" of the Toronto Geographical Expedition (see Figures 4.1 and 4.2).[9] Hurst provided vans enabling a group of Simon Fraser students to travel across the country to attend.

The mandate of the USG was as follows:

> The purpose of our Union is to work for the radical restructuring of our societies in accord with the principles of social justice. As geographers and people, we will contribute to this process in two complementary ways:
>
> 1 Organizing and working for radical change in our communities, and
> 2 Developing geographic theory to contribute to revolutionary struggle.
>
> (Akatiff, 1974b, 1)

Figure 4.1 Members of the first meeting of the Union of Socialist Geographers (USG) in Toronto, May 1974. Photo credit: Clark Akatiff and Ron Horvath (used with permission).

Tensions between theory and practice are evident in these USG goals, and initial debates occurred about whether to use Socialist, Marxist, or Radical to delineate the Union. There were plans for a "regional hierarchy of communicants" to distribute mailings, coordinated by Clark Akatiff, to publish in the *Association of American Geographers Newsletter, Transition,* and *Antipode,* and plans to meet independently of, but parallel to, national CAG and AAG meetings, as well as regional meetings. Thirty-three people are listed as attending the first meeting (four being women; the racial and gender bias is visible in Figure 4.1). In addition, two sessions were organized at the CAG meeting. In 1975 regional meetings were held, with a subsequent national meeting hosted by Blaut in Chicago, alongside the AAG national meeting in Milwaukee. Thereafter, the USG held meetings alongside the CAG and the AAG in alternate years until 1981. Regional meetings of the USG were also held in the US Midwest and Québéc in the late 1970s.

At the invitation of Peet, the Simon Fraser USG local edited one issue of *Antipode* in 1976 (8#3), including articles on anarchism, environmentalism, Ireland, and Latin America. This contrasted with the much more

Figure 4.2 Partial list of participants in the USG founding photograph (Figure 4.1), from left to right: GUNNAR OLS – Gunnar Olsson; Wilbur Zelinski; JIM LION – Jim Lyons; NATAN EDELS – Nathan Edelson; Big Jim Blaut – James Blaut; DAVID HARVEY; VdV – Edward Vander Velde Jr; WILD BILL – William Bunge; ACK – Clark Akatiff; Susan Bunge; HORVATH – Ron Horvath; Judy Stamp Humphrey; Richard Peet; Tom Edison, Sue Cozzens; Colum Regan? – Colm Regan; Charl IPCA – Charles Ipcar.

Source: Clark Akatiff and Ron Horvath

Marxist orientation of the other issues of that year. In the meantime, the USG began to publish its own regular journal, *The USG Newsletter*, again initiated from Vancouver. Volume 1 appeared as five issues, the last in summer 1976. It set out to provide a venue for reporting on USG meetings, seminars, bibliographies, course outlines, book commentaries, and event announcements. The second issue included an academic article, on Marx's theory of circulation, and several others followed. The two issues of volume 2 (1976–1977) were edited by USG locals at Johns Hopkins (Baltimore) and McGill University (Montreal), successively. Québéc became an active second Canadian local, its participants (including Damaris Rose and Sue Ruddick) catalyzed by street protests and the left-leaning Parti Québécois. John Bradbury (1942–1988) was a vital figure in Montreal, heavily involved in local and international aspects of the USG. His radicalism was formed under the influence of Keith Buchanan and Terry McGee in New Zealand, and then at Simon Fraser before joining the McGill faculty.

By the 1977 meeting in Regina (attended by Eric), the trappings of an academic organization were in place: an executive committee, membership structure, financial statements, connections with parallel organizations in other countries, instructions to authors, and stillborn plans for a textbook and monograph series. Volume 3 was edited successively in Vancouver, Montreal, and Toronto and then Minnesota (on anarchism). At the January 1978 IBG meetings, a USG local was formed in the United Kingdom (members of the Vancouver collective also presented papers at the 1978 IBG meeting in Manchester). By the 1978 meeting alongside the AAG in New Orleans, USG locals also existed in Ireland, Denmark, and Australia—the latter where Horvath had returned to teach. The editing of subsequent issues of the newsletter alternated between Simon Fraser, McGill, and Sydney (Australia), before settling at Minnesota where the local collective edited, printed and distributed it (1979–1981, including issues submitted from Queens University (Canada), London (UK), and Montreal branches). Over time, academic articles became increasingly prominent, with a persistent diversity of theoretical and substantive approaches (including anarchist, gay, and feminist geography). By 1981, the USG had 180 North American members, and Midwest, Ontario, east coast, west coast, and Quebec local collectives. *The USG Newsletter* had become a third radical geography journal, alongside *Antipode* and *Transition*.

During the 1970s, forces of conservatism within the discipline and the academy posed continual barriers to the presence of a revolutionary radical geography in North America. Akatiff and Bunge were denied tenure, and in other cases the USG helped catalyze campaigns when others faced a similar threat (e.g., Dick Walker at Berkeley). Eliot-Hurst was replaced at Simon Fraser University, and the new chair set about dismantling radical geography. Horvath left, and when Eric interviewed at SFU in March 1976 the new regime plainly did not know how to react to a quantitative geographer enthusiastically supported by students because of his radical leanings. In 1977, on the Peace Bridge at Fort Erie, Canadian customs seized copies of *Antipode* from the possession of Dick Peet and Phil O'Keefe, on the grounds that they were not "really geography" (USG Newsletter, 3#2, 1978–1979, 5). By the late 1970s, however, radical geography was less preoccupied with breaking away from than breaking into the institutional structures of the discipline. This catalyzed extensive debate within the USG about whether to retain its independence—reinforced by declining subscriptions from an ever-expanding membership. There were also discussions about whether to formalize the relationship between the USG and *Antipode*.

Eric recalls a particularly intense debate, at a USG annual meeting in 1980, about whether to disband the USG, at a time when some USG members (including Eric) had taken the initiative to create a Socialist Geography Specialty Group (SGSG) within the AAG (which only required 100 AAG members' signatures). Eric recalls Neil Smith strategically in favour of shifting energy to the SGSG, with Jim Blaut energetically opposed. The crux of the argument was whether incorporation within the AAG would blunt radical geography's radicalism.[10] The SGSG's officers came from the USG; its stated purpose was: "To examine geographical phenomena critically, questioning the implications of geographical research for the well-being of social classes. To investigate the issue of radical change toward a more collective society" (Socialist Geography Specialty Group, 1980, 1). The SGSG grew from 105 to 177 members during the 1980s: the USG was dissolved in 1981. In 1986, SERGE was also disbanded (its long-time leader and editor of *Transition*, Larry Wolf, retiring from the University of Cincinnati). Finally, in 1985, ongoing discussions about what the future of *Antipode* should look like, and whether it should go commercial, came to a head. Dick Peet took Eric aside and, much to the latter's surprise, asked whether he would be willing to become co-editor of *Antipode*, with Joe Doherty (St. Andrews University, Scotland), and negotiate a transition to commercial publication with Basil Blackwell. After difficult negotiations with Blackwell, "who had us over a barrel, we published the first commercial issue in 1986, with the stated intention that *Antipode* will continue to be a forum for the publication of significant contributions to a radical (Marxist/socialist/feminist/anarchist) geography...we will seek to maintain traditional areas of strength in environmental questions, urban political economy and development issues. We intend to improve the journal's coverage of feminist approaches" (Sheppard, 2010). A "Debates and Reports" section, introduced to facilitate polemic and discussion alongside substantive articles, thrived.

Critical geography: 1964–1986

The year 1986 marked not only the commercialization of *Antipode* but also the demise of *Transition*; it was the turning point when radical geography in Anglophone North America became reframed as "critical". As radical geography entered into the mainstream, it merged with a nascent but, we would argue, *already existing* critical geography. The late-1960s to mid-1970s saw a flourishing of different voices in *Antipode, Transition,* and the USG newsletters; socialist, feminist, anti-racist, anarchist, and environmentalist approaches to studying social problems and advocating social change were all evident. This reflected the multivalent, intersecting protest and social movements unleashed by a 1960s politics of radicalism, anti-racism, sexual liberation and emancipation, in which various protagonists were involved in multiple ways, and the complex linkages between these and academic trajectories. That these voices were progressively less heard with a hardening of orthodoxy, as a Marxist critique of capitalism came to dominate *Antipode*, is hardly surprising. That they existed and, in some cases, continued to exist, in times and places within or alongside radical geography and in others apart from it, is indicative not only of the transversal and unpredictable intellectual and spatial paths of the evolution of Anglophone North American critical geography, but also of the impossibility of attempts to explore its evolution through a core (Clark, SFU) versus periphery (everywhere else) model of knowledge dissemination. This period was especially important for the establishment of the emerging fields of African American/ anti-racist geographies, feminist (Marxist, liberal, and other variants) geographies, and (although slightly less so) for geographies of sexualities.

In the 1960s the United States was in crisis; anti-war fervour and the civil rights demands of African Americans dominated the decade. The discipline of geography was also in crisis, and not only because of its focus on intellectual framings that had their roots in the mathematical abstractions of mechanical social engineering rather than the material realities of social transformation (Peet, 2000), but also there was another crisis that encapsulated North American geography, one that was not recognized as such. It was a crisis of an "absent presence" of whiteness, a normalization that went unquestioned; geography was a segregated and institutionally racist discipline. A national survey of geography departments in 1970 found that a total of only 12 African American faculty were employed in the United States (with only one at the level of full professor) and only two of these were employed at predominantly white universities. A full five years before the advent of radical (Marxist) geography, in 1964, the first attempts to address this segregation were made by Saul Cohen, a professor at Boston University (who in 1965 was to become Director of the Graduate Program at Clark). Invited to become the Executive Officer of the AAG in 1964, he toured a number of southern (traditionally African American) colleges in an exercise aimed at identifying talented students who could join geography training programs, arriving coincidentally at Albany State in Georgia on the day of the riots (Darden et al., 2006). A proposal was developed as a result of his findings to establish within the AAG a Commission on Geography and Afro-America (COMGA) to support the recruitment of African American geographers and research into issues facing African Americans (Deskins and Siebert, 1975).

Five universities that offered graduate programmes in geography—Clark, Chicago, Michigan, Syracuse, and Wisconsin—were selected to take part in the COMGA project of recruiting undergraduates to enroll in summer institutes. In 1968 Don Deskins at the University of Michigan became COMGA's first director (Darden et al., 2006). Small but significant gains were made. Two surveys were conducted in 1968 on the low levels of participation by African Americans in geography. Two more surveys were conducted in 1970 and 1974 and the *Southeastern Geographer* and *Economic Geography* each dedicated a special issue to geographic research on African Americans, in 1971 and 1972, respectively (Dwyer, 1997). A small number of African American geographers began to document experiences of the social problems and issues of political participation they faced, as well as the (under) development of black residential areas. They also investigated and reported upon their marginalized position in the discipline (Deskins, 1969; Deskins and Speil, 1971; Donaldson, 1969; Horvath et al., 1969; Rose, 1970; Wilson and Jenkins, 1972). However, this level of activity was not sustainable and by the late 1970s COMGA was defunct, with the small gains of the early 1970s being eroded in the 1980s.[11]

During the early 1970s, radical geographers' interest in poverty was also inflected with concerns of race as much as those of class. Bill Bunge's DGEI had been formed (in 1968) in conjunction with African American community leaders to highlight the racism and poverty under which daily urban life was lived by African Americans. He and other radical geographers were publishing in *Antipode* on the conditions of life in urban ghettoes in the United States and in the developing world (Blaut, 1974; Bunge, 1971, 1976; Elgie, 1974; Harvey, 1972; Smith, 1974), as were a few African American scholars (Darden, 1975; Donaldson, 1971).[12] By the mid-1970s, however, Marxist concerns largely turned away from race and racism, an unfortunate turn of events that led somewhat to studies in the global urban north reducing understandings of race to an effect of class, and in the global south to their incorporation into underdevelopment and imperialism. Race and racism—theoretical objects of study so central to the inception of radical geography—disappeared from the agenda. Interestingly, they were just starting to appear in a third trajectory of studies of race carved out by North American humanistic geographers interested in the everyday lives of racialized communities (Ley, 1974), an approach that was eventually to lead to the new cultural geography.

Just as studies of race and racism exhibited distinct and somewhat separate paths, so did feminist approaches (Kobayashi, 2003). Engaging in social and political movements for civil rights and against war was not only the prerogative of radical geographers. For women entering the discipline, it was not only these experiences but also their engagement in the Peace Corps and second-wave feminism that marked a changing context (Monk, 2004). This was the period when the institutional framing was put in place that would allow feminist approaches to prosper, notwithstanding that in the late 1960s and early 1970s women made up an incredibly small percentage of faculty members. In 1973, for example, women accounted for only 3.4% of the faculty in the United States and Canadian graduate Geography departments, with only one female full professor (Monk, 2006). Notwithstanding the very small number of women in the discipline, the Committee on the Status of Women in Geography (CSWG) of the AAG was formed as early as 1971, although more by accident than design (Monk, 2004). In 1979, the Geographic Perspectives on Women (GPOW), the AAG specialty group on research on women and gender, was launched, and the AAG adopted a bylaw on affirmative action. A similar picture was played out in Canada a few years later, not least because a number of Canadian feminist scholars who were to play important roles in radical/critical geography were outside Canada (mostly in the United Kingdom and the United States) pursuing Ph.Ds in the 1970s, not returning until the early 1980s, when the Canadian Women and Geography Study Group (CWAG) of the CAG was formed (in 1982).[13]

The eclecticism that has come to characterize North American feminist geography was already evident in the early 1970s, with studies simply describing and mapping the differential geographies of men and women, registering their disparate access to services, employment, and facilities (Caris, 1978; Mazey and Lee, 1983), studies of the status and position of women within the discipline (Rubin, 1979; Zelinsky 1973), and studies that initially explored gender roles but quickly moved on to feminist approaches exploring the politics and economics of gender relations (Burnett, 1973; Bruegel, 1973; Mackenzie and Rose, 1983). The latter field was especially influenced by Marxist and socialist feminist analyses (although not dominated by them, as with the case in the United Kingdom) and was indicative of the strong transnational linkages that existed between a number of feminist geographers in Canada, the United States, and the United Kingdom. As Damaris Rose (personal communication) states, "the thinking of some of us was shaped interactively rather than sequentially by influences on both sides of the Atlantic". Indeed the establishment of the Women and Geography Study Group of the Institute of British Geographers in 1980 was predominantly due to the presence in the United Kingdom of the Canadian socialist feminist Suzanne Mackenzie. Suzanne was a charismatic presence whose warmth, wit, and expansive nature influenced a whole generation of women geographers to embrace feminism.

A third significant dimension of critical geography was also to emerge in this period, although without any institutional recognition. It was in the USG newsletters and *Antipode,* but also other journals (Winters, 1979) that studies of gay geographies first emerged in the late 1970s, along with informal meetings of gay and lesbian geographers at the annual conferences of the AAG (Ketteringham, 1979). These tentative forays mark the origins of the sexuality and space studies that from the mid-1990s onwards have been an integral element of critical geography.

Critical geography: Mid-1980s onwards, ever onwards or back to the future?

While the progress of radical geography through the late 1960s and 1970s into the mid-1980s was an assertive one, that of critical geography was less assured, more hesitant. And while radical and critical voices were growing, they had yet to gain widespread acceptance from the mainstream, which was still a decade or so away. Nonetheless, it was evident that this intensely political period in geography was providing a new intellectual leadership. Many of its earliest practitioners were progressing through the academic ranks, to become not only full professors but also in the process internationally renowned scholars, developing new fields of study, occupying prestigious chairs, and becoming presidents of geographical associations, editors of journals, and medal winners. It was in this period, in 1983, for example, that the critical geography journal *Environment and Planning D: Society and*

Space was launched, to become incredibly popular with a wide range of disciplinary scholars. Unquestionably, this groundwork set the stage from the 1990s onwards for a remarkable increase in the number of younger scholars who were to identify as being on the "Left".

The early 1980s had been a period of internal critique for radical geographers—to maintain a commitment to revolutionary ideals or to join in the mainstream and accept a more fluid conception of praxis. For feminist geographers and geographers of sexuality studies, it was a period of consolidation, and for anti-racist geographers one of retrenchment. In short, it was a mercurial time for radical/critical geography, only to become increasingly turbulent as postmodernism and poststructuralism started to make their impact felt on the discipline in the mid-to-late-1980s. The widespread adoption of these philosophical approaches across the social sciences and humanities (and beyond) has been identified in geography as the so-called critical/cultural turn. It was to lead to a convergence of interests in the recognition that race, gender, and sexuality, as class, were social and cultural constructions, charged through with power lines, social meanings, and identities, meritorious of their own theoretical framings and united through a shared relational epistemology (Peake, 2009). Since the critical/cultural turn, with its prioritization of the cultural (often at the expense of the political-economic) and the rise of identity politics (often to the occlusion of class politics), North American radical/critical geography has diversified into a prolific twisting and geographically expanding skein of ways of knowing, sharing a progressive politics and activist bent (in theory, if not necessarily in practice). As a result, by the late 1990s radical/critical geography was to become the new canon, the new mainstream, accompanied by a proliferation of practitioners and publications.

Notwithstanding these achievements, the 1980s and 1990s cannot be characterized as a glowing affirmation of the relentless rise of radical/critical geography. In the late 1980s, in many respects the institutional picture remained virtually as dismal as in the preceding two decades. It was, for example, proving very difficult to diversify the academic profession in North America. Although there had been an increase in the number of women gaining faculty positions, they were still significantly underrepresented. In 1988–1989 in Canada women faculty comprised only 8.1% (up from 3.4% in 1973), and there was still only one female full professor (Mackenzie, 1989).[14] A 1987 survey of geography departments in North America also found just over 5% (n = 73) of academic geographers were people of colour (African Americans, Hispanics, Native Americans, and Asians) (Shrestha and Davis, 1989).[15]

The 1990s witnessed a number of significant developments in radical/critical geography that helped it gain ground in the mainstream. *Antipode* not only remained viable but increased its number of issues. There were significant debates taking place about the future direction of Marxist geography between those like Harvey, whose focus remained primarily on class, and detractors such as Gibson-Graham (1996) whose Marxist analyses were also informed by feminist and queer theory. The turning point for feminist geography to enter the mainstream came in 1994, when Susan Hanson gave her presidential address at the AAG on feminist geography and *Gender, Place and Culture: A Journal of Feminist Geography* was established. In 1996 a Sexuality and Space Specialty Group was finally founded within the AAG, and in 1998 a workshop was held at University of Kentucky on race and racism that led not only to two special journal issues (*Social and Cultural Geography* 2000, 1(2) and *Professional Geographer* 2002, 54(1)), but also the reinvigoration of a new generation of scholars and studies on the social construction of race and critical race theory, helping the formulation of a Diversity Task Force within the AAG, in 2003. The increasing influence of critical geography was further signalled by various name changes. In 1997 the first international group to officially adopt the title of critical as opposed to radical was the International Critical Geography Group (ICGG). Shortly after, in 2002, the AAG's Socialist Geography Specialty Group changed its name to the Socialist and Critical Geography Specialty Group.

Critical geography was unquestionably flourishing, but for many it was also losing sight of its alternative nature and political purpose, and thereby its viability. By the late 1990s, in response to a neoliberal problematic that had haunted radical/critical geography throughout the decade, namely the relationship between professionalization and activism, new radical/critical spaces began to emerge beyond the academic mainstream. In 1997, the ICGG formed out of the Critical Geography Forum (an international list-serv) with its mission to develop:

> The theory and practice necessary for combating social exploitation and oppression. We have formed this international association to provide an alternative to the increasingly institutionalised and corporate culture of universities. We believe that a "critical" practice of our discipline can be a political tool for the remaking of local and global geographies into a more equal world.
>
> (https://internationalcriticalgeography.org, accessed August 23 2021)

The ICGG has held eight international conferences thus far, the first being in Vancouver, followed by Daegu (2000), Bekescsaba (2002), Mexico City (2005), Mumbai (2007), Frankfurt (2011), Ramallah (2015) and Athens (2019). In 1999 there was an attempt, led by Don Mitchell of Syracuse University, to engage in another democratic project of knowledge production. Named the People's Geography Project, its major goal was "to popularize and make even more relevant and useful to ordinary people the important, critical ways of understanding the complex geographies of everyday life that geographers have and continue to develop" (http://www.peoplesgeographypro-ject.org/, accessed August 23 2021). The short-lived nature of this exercise may well speak to its lack of a popular base (unlike Bunge's, also short lived but grass-roots, in your face DGEI) and the replacing of engagement and activism with professionalization, that has resulted in the North American academy (and beyond) in what Castree (2000) refers to as academicization, namely a reluctance to engage in activism at the risk of not progressing up the academic ladder.

Most recently, in the 2000s there has been the launch of two new journals. *ACME: An International E-Journal for Critical Geographies* was launched in 2002. The journal's purpose, like that of *Antipode*, is to provide a:

> Forum for the publication of critical and radical work about space in the social sciences .... Analyses that are critical and radical are understood to be part of the praxis of social and political change aimed at challenging, dismantling, and transforming prevalent relations, systems, and structures of capitalist exploitation, oppression, imperialism, neo-liberalism, national aggression, and environmental destruction.
> (https://acme-journal.org/index.php/acme,
> accessedAugust 23 2021)

Yet, it differs from *Antipode* and other radical journals in two significant ways. First, it is disseminated for free (although supported by Canadian SSHRC funding, which made possible its hosting at UBC Okanagan) and has a circulation far beyond that of commercially published journals. Secondly, it publishes in languages other than English, thereby challenging the Anglo-American hegemony of critical/radical geography.[16] The newest introduction, in 2008, was *Human Geography: A New Journal of Radical Geography*. Its underlying rationale is the "need to retain control of the value produced by academic labor" (www.hugeog.com, accessed August 23 2021). In many ways, including its founding editor (Peet) and geographical location (Clark), *Human Geography* harks back to the early days of *Antipode*. Presumably in reaction to the trajectories of a commercialized *Antipode*, *Human Geography* was opposed to the removal of publishing from the hands of academics into those of publishers owned by a few multinational media conglomerates, seeking to address "the wide range of urgent social and political issues...hardly mentioned in the existing journals" (www.hugeog.com, accessed August 23 2021). *Human Geography* is run on a shoe-string budget as a non-profit organization, publishes a hard copy journal, is explicitly Marxist in tone, and publishes in Spanish as well as English. These examples of ongoing experiments in critical and radical academic geographic publishing speak to the ongoing struggles to carve out space for radical/critical geography in the contemporary world.

Conclusion

Little is known of the precursors to the last four decades of radical/critical geography in the United States and Anglo-Canada—who, and where, were the players, the networks, the catalysts. What we know is patchy at best, and documented overwhelmingly in favour of white males. What we also know is that the stakes of not engaging with the multifarious historical geographies of radical geography are too high; these unearthed genealogies of radical/critical geography demand inter-rogation. That so many accounts start not in 1964 but in 1969 speaks to the as yet uninterrogated whiteness that pervades the field, a field that has as yet not addressed the trampling underfoot of anti-racist efforts in establishing the grounds on which radical/critical geographies have arisen. Although there has been interest in excavating the advent of the DGEI and *Antipode*, huge gaps remain in our understanding of these now historical moments, and most especially those of the USG.[17] Founded respectively at the University of Michigan, Clark University, Massachusetts, and Simon Fraser University, Vancouver, these three operations formed discernable nodes of radical geography whose spa-tialities spread out tentacle-like, transcending not only the US-Canada border but also going beyond to Europe and Australasia. The last four decades have seen the coarse thickening of these now transnational net-works and of many others, as (some) academic appointments are increas-ingly opened to (some) non-nationals, as attendees at (some) national conferences are increasingly multinational, and as technological devel-opments aid the development of multinational teams of researchers. The spatialities of the United States and Canadian radical/critical geography increasingly crisscross the globe, albeit unevenly and still heavily biased towards Anglo-America.

The seemingly inexorable march forward of journals and organizations in the 21st century would appear to indicate that radical/critical geogra-phy in the United States and Canada is alive and well; it has succeeded in its aim of advancing critical geographic theory. As we have stated, it is now canonical in mainstream Anglophone human geography. But has radical/critical geography succeeded in its aims of increasing access to the means of knowledge production, through both pedagogy and research, to become a peoples' geography that is grounded in a desire for working towards change through a critical geography practice?[18] On this question we think the jury is out. We know that many individuals are politically engaged and some discipline-wide interventions have been extremely successful, such as that of people's mapping, which through GIS and other technologies have been of use to local community groups and in major environmental disasters. We both believe that much has been achieved, but also that surficial agreements to differ have taken the place of vigorous debate; that methodological progress overall has been stultifying; that engagement with epistemology has been often at the expense of praxis; and that diversification and pluralism are leading to a lack of a common purpose (see Barnes and Sheppard, 2010 on economic geography). While we do not think the terms "critical" and "radical" have become so widespread and scatological as to be meaningless, chaotic concepts, we also do not think that their uneven progress can be left unquestioned.

There is also too much at stake in not continuing to question and to break down Anglo-American hegemony in critical/radical geography. Our geographical imaginaries, and the labour they do of democratizing the relations of knowledge production, risk being diminished if we continue to reproduce ourselves in our own image. If critical/radical geography is about where we can see geography from; how far we can see; and where we can learn geography from (Harvey, 2000, 254), then North American geographers may have the furthest distance to travel.

Acknowledgements

We wish to thank the following for their comments on an earlier draft: Clark Akatiff, Lawrence Berg, Ulrich Best, Nathan Edelson, Jack Eichenbaum, Alison Hayford, Audrey Kobayashi, Jan Monk, Jim Overton, Dick Peet, Damaris Rose, Sue Ruddick, Neil Smith, Alan Wallace, and Ben Wisner. Beyond this, we want to thank all those folks who participated in and facilitated the development of critical/radical geographies in North America. Our delving back into its past was an immensely enjoyable one, albeit one that has made us more concerned than ever that its documentation is sparse and the artefacts and memories of that time are in danger of disappearing. We urge those so minded to take up its recording as a matter of urgency.

Key Sources (as of August 2021)

Newsletters

- https://antipodeonline.org/supplementary-material/the-union-of-socialist-geographers-newsletter-1975-1983/

Journals

- *Antipode:* http://www.antipode-online.net
- *ACME:* https://acme-journal.org/index.php/acme/index
- *Environment and Planning D: Society & Space:* https://journals.sagepub.com/home/epd
- *Gender, Place and Culture: A Journal of Feminist Geography:* https://www.tandfonline.com/toc/cgpc20/current
- *Human Geography: A New Radical Journal*: http://www.hugeog.com

AAG/CAG Groups

- Socialist and Critical Geography Specialty Group of the AAG (SCGSG): https://community.aag.org/communities/community-home?CommunityKey=3603da9b-2fe0-4cd1-a07c-ffcac465e5b8

- Sexuality and Space Specialty Group of the AAG. Now known as the Queer and Trans Geographies Specialty Group: https://qtgaag.wordpress.com/about/
- Disabilities Specialty Group of the AAG: https://community.aag.org/geogable/home
- Geographic Perspectives on Women Specialty Group (GPOW) of the AAG: Now known as Feminist Geographies Specialty Group: https://feministgeographies.org
- Women and Geography Study Group of the CAG (CWAG). Now known as the Feminist Intersectional Solidarity Group (FIGS): https://www.cag-acg.ca/figs

Listservs

- Leftgeog listserv: http://www.lsoft.com/scripts/wl.exe?SL1=LEFTGE-OG&H=LSV.UKY.EDU
- Critical Geography Forum: https://jiscmail.ac.uk/cgi-bin/wa-jisc.exe?A0=CRIT-GEOG-FORUM

Websites

- Peoples Geography Project: http://www.peoplesgeographyproject.org/

International Organizations

- International Conference of Critical Geography: https://international-criticalgeography.org/conferences/
- IGU Commission on Gender and Geography: https://igugender.wixsite.com/igugender

Notes

* This chapter has previously been published as L. Peake and E. Sheppard (2014) The Emergence of Radical/Critical Geography within North America, *ACME: An International Journal for Critical Geographies* 13(2): 305–327, published under the Creative Commons "Attribution/Non-Commercial/No Derivative Works" Canada licence. The text is published with permission from the authors.

1 Eric's first exposure to radical geography was as an undergraduate at Bristol in 1971 when the newly hired lecturer Keith Bassett, freshly returned from Penn State, brought a stack of *Antipodes* to one of his lectures. Linda's radical awakening also came in the United Kingdom in the late 1970s, courtesy of her lecturers at Reading University. Sophie Bowlby took her to her first Women and Geography Group meeting, introducing her to Suzanne Mackenzie, among others, while John Short and Andrew Kirby introduced her to other radical work in the discipline.

2 In 1952, former communist Wittfogel denounced Lattimore during the McCarthy hearings (Lattimore, 1950).

3 On October 21, 1967, the Youth International Party (Yippies) attracted over 100,000 people to an event billed as levitating the Pentagon (Akatiff, 1974a).

4 One of their activities was to produce "a 'peasant farm' by slashing and burning and hand-tilling some land at Martyn Boyden's house. Blaut insisted we pay a huge rent in kind to the landlord (Boyden) to make it realistic". (Wisner, personal communication)

5 Interestingly, many of the early generation of radical geography faculty, including Bill Bunge, Ron Horvath, Dick Peet, David Harvey, Ed Soja, Gunnar Olsson, Doreen Massey, and Michael Eliot Hurst, were refugees from a spatial science whose empiricism and methodological individualism had not been able to match up to their progressive politics (Sheppard, 1995).

6 Wisner and O'Keefe (personal communication) lament inadequate attention to the environment notwithstanding the early influence of environmental activism over radical geography: the emergence of political ecology in the 1980s, catalyzed by Michael Watts, Piers Blaikie, and Harold Brookfield, enrolled such issues into a Marxian framing.

7 We are grateful to be able to draw, here, on the recollections of Clark Akatiff, Nathan Edelson, Alison Hayward, Alan Wallace, Audrey Kobayashi, Jim Overton, Damaris Rose, Sue Ruddick, Jack Eichenbaum, David Stea, Proinnsias Breathnach, Ben Wisner, Neil Smith, and David Harvey.

8 Eliot-Hurst was one of the first North American geographers to declare his homosexuality.

9 See *Antipode* (1975 7(1): 86) on the first annual meeting of the USG.

10 Smith (personal recollection) recalls: "Jim's prognosis was surely correct, and an independent socialist group was obviously ideal, the USG had worked well but wasn't especially building, wasn't expanding quickly, and the political mood was dissipating. We wanted organization for the USG but were resistant in our post-60s way to heavy organization, and this made the USG a bit haphazard". At the time, Eric sided with Neil; on reflection, he wonders what also was lost.

11 Although short-lived, COMGA paved the way for the establishment of the AAG's Standing Committee on Affirmative Action and Minority Status, now known as the Diversity and Inclusion Committee (Patricia Solis, personal communication).

12 There was evidence of some collaboration between members of these two groups, particularly by Bunge and Horvath.

13 The Canadian Women and Geography Study Group is one of the more influential study groups in the CAG, its founders including Fran Klodawsky, Audrey Kobayashi, Suzanne Mackenzie, Damaris Rose, Sue Ruddick, and Pamela White. Its members have played leading roles in the CAG; past presidents of the CAG include Alison Gill and Audrey Kobayashi. It also sponsors one of the CAG's two annual lectures: The Suzanne Mackenzie Memorial Lecture. Unfortunately, CWAG, like GPOW, has no written history of the organization.

14 This figure is for universities with doctoral programmes. When those universities with master's programmes are added the figure rises to 10.6% (Mackenzie, 1989).

15 Given the relatively low response rate to this survey of just over 40% it is likely that this figure is inflated. Non-responses were geographically concentrated in predominantly white areas of the United States where most probably no people of colour were employed.

16 Debates about this linguistic hegemony have been a defining aspect of radical/critical geography since the early 1990s, influencing practices in a number of critical journals of geography. When Linda was managing editor of *Gender, Place and Culture* the journal began publishing abstracts in Spanish, and now also in Cantonese, making at least this journal more visible to the non-Anglo-speaking academic worlds.

17 It was in collecting background information for this chapter that we realized there was no publicly available collection of a complete set of the USG newsletters. We therefore decided to collect copies from previous USG members to scan and provide an e-version of them, which the journal *Antipode* has kindly agreed to host on its website.

18 David Harvey defines a people's geography as: "The geography we make must be a peoples' geography, not based on pious universalisms, ideals and good intents, but a more mundane enterprise that reflects earthly interests, and claims, that confronts ideologies and prejudice as they really are, that faithfully mirrors the complex weave of competition, struggle, and cooperation within the shifting social and physical landscapes of the twentieth [and twenty-first] century. The world must be depicted, analyzed, and understood [as] the material manifestation of human hopes and fears mediated by powerful and conflicting processes of social reproduction. Such a peoples' geography must have a popular base, be threaded into the fabric of daily life with deep taproots into the well-springs of popular consciousness. But it must also open channels of communication, undermine parochialist worldviews, and confront or subvert the power of the dominant classes or the state. It must penetrate the barriers to common understandings by identifying the material base to common interests" (Harvey, 1984, 7).

References

Akatiff, C. 1974a. The march on the Pentagon. *Annals of the Association of American Geographers* 64, 26–33.

Akatiff, C. 1974b. Union of Socialist Geographers: Preamble and minutes. Toronto. (manuscript available from author).

Akatiff, C. 2007. The roots of radical geography: AAG convention, San Francisco 1970. A personal account. *Association of American Geographers*. San Francisco. (Manuscript available from author).

Barnes, T.J. and E. Sheppard. 2010. 'Nothing includes everything': Towards engaged pluralism in Anglophone economic geography. *Progress in Human Geography* 34(2), 193–214.

Blaut, J.M. 1974. The ghetto as an internal neo-colony. *Antipode* 6(1): 37–41.

Bunge, W. 1966. *Theoretical Geography*. Lund: C.W.K. Gleerup.

Bunge, W. 1971. *Fitzgerald: Geography of a Revolution*. Cambridge, Mass.: Schenkman Publishing Company.

Bunge, W. 1976. Racism in geography. In, R. Ernst & L. Hugg (eds.), *Black America, Geographic Perspectives*. New York: Anchor Books, pp. 4–8.

Bruegel, I. 1973. Cities, women and social class: a comment. *Antipode* 5(3), 62–63.

Burnett, P. 1973. Social change, the status of women and models of city form and development. *Antipode* 5(3), 57–62.

Caris, S. 1978. Geographic perspectives on women: A review. *Transition: Quarterly Journal of the Socially and Ecologically Responsible Geographers*8(1), 10–14.

Castree, N. 2000. Professionalisation, activism, and the university: Whither 'critical geography'? *Environment and Planning A* 32, 955–70.

Clark, J.P. and C. Martin (eds.). 2004. *Anarchy, Geography, Modernity: The Radical Social Thought of Elysée Reclus*. Lanham, MD: Lexington Books.

Darden, J. 1975. Population control or a redistribution of wealth: A dilemma of race and class. *Antipode* 7(2), 50–2.

Darden, J. et al. 2006. *Final Report: An Action Strategy for Geography Departments as Agents of Change. A Report of the AAG Diversity Task Force.* Washington D.C.: AAG.

Deskins, D.R. 1969. Geographical literature on the American Negro, 1949–1968: A bibliography. *Professional Geographer* 21, 145–49.

Deskins, D.R. and L. Siebert. 1975. Blacks in American geography: 1974. *Professional Geographer* 27(1), 65–72.

Deskins, D.R. and L. Speil. 1971. The status of blacks in American geography: 1970. *Professional Geographer* 23, 283–89.

Donaldson, O.F. 1969. Geography and the black American: The white papers and the invisible man. *Antipode* 1(1), 17–33.

Donaldson, O.F. 1971. Geography and the black American: The white papers and the invisible man. *Journal of Geography* 70(3), 138–49.

Dwyer, O.J. 1997. Geographical research about African Americans: A survey of journals 1911-1995. *Professional Geographer* 49(4), 441–51.

Dow, M.W. 1976. William Bunge interviewed by Donald G. Janelle. In *Online Geographers on Film Transcriptions (1998)*, 3 November. https://www.loc.gov/item/mbrs01844974/, accessed September 28, 2021

Dunbar, G. 1978. *Elisée Reclus, Historian of Nature.* Hamden, CT: Archon Books.

Galois, B. 1976. Ideology and the idea of nature: The case of Kropotkin. *Antipode* 8(3), 1–16.

Elgie, R. 1974. Geography, racial equality and affirmative action. *Antipode* 6(2), 34–41.

Gibson-Graham, J.K. 1996. *The End of Capitalism (As We Knew It): A Feminist Critique of Political Economy.* Oxford UK and Cambridge USA: Blackwell Publishers.

Harvey, D. 1972. Revolutionary and counter revolutionary theory in geography and the problem of ghetto formation. *Antipode* 6(2), 1–13.

Harvey, D. 1983. Owen Lattimore: A memoire. *Antipode* 15(3), 3–11.

Harvey, D. 1984. On the history and present condition of geography: A historical-materialist manifesto. *Professional Geographer* 36(1), 1–11.

Harvey, D. 2000. *Spaces of Hope.* Edinburgh: Edinburgh University Press.

Heyman, R. 2007. "Who's going to man the factories and be the sexual slaves if we all get PhDs?" Democratizing knowledge production, pedagogy, and the Detroit Geographical Expedition and Institute. *Antipode* 39(1), 99–120.

Horvath, R. 1971. The "Detroit Geographical Expedition and Institute" experience. *Antipode* 3(1), 73–85.

Horvath, R., D. Deskins and A. Larimore. 1969. Activity concerning black America in university departments granting M.A. and Ph.D. degrees in geography. *Professional Geographer* 21(3), 137–9.

Ketteringham, W. 1979. Gay public space and the urban landscape. Paper presented at the Annual Meeting of the Association of American Geographers.

Klein, J-L. 2020. Radical geography goes francophone. In T.J. Barnes and E. Sheppard (eds.) Spatial Histories in Radical Geography. Oxford: Wiley. pp. 273–300.

Kobayashi, A. 2003. The construction of geographical knowledge: racialization, spatialization. In K. Anderson, M. Domosh, S. Pile and N. Thrift (eds.), *Handbook of Cultural Geography.* London: Sage. pp. 544–56.

Latour, B. 1987. *Science in Action.* Cambridge, MA: Harvard University Press.

Lattimore, O. 1950. *Ordeal by Slander.* New York City: Little Brown & Company.

Ley, D. 1974. *The Black Inner City as Frontier Outpost; Images and Behavior of a Philadelphia Neighborhood.* Washington, DC: Association of American Geographers.

Mackenzie, S. 1989. The status of women in Canadian geography. *The Operational Geographer* 7(3), 2–8.

Mackenzie, S. and D. Rose. 1983. Industrial change, the domestic economy and home life. In, J. Anderson, S. Duncan and R. Hudson (eds.), *Redundant Spaces? Studies in Industrial Decline and Social Change*. London: Academic Press, pp. 155–200.

Mathewson, K. 2005. Jim Blaut: Radical cultural geographer. *Antipode* 37, 913–26.

Mathewson, K.and D. Stea. 2003. In memorium: James M. Blaut (1927-2000). *Annals of the Association of American Geographers* 93(1), 214–22.

Mazey, E. and D. Lee. 1983. *Her Space, Her Place: A Geography of Women*. Washington, D.C.: Association of American Geographers.

Monk, J. 2004. Women, gender, and the histories of American geography. *Annals of the Association of American Geographers* 94(1), 1–22.

Monk, J. 2006. Changing expectations and institutions: American women geographers in the 1970s, *Geographical Review*, 96(2), 259–77.

Peake, L. 2009. Gender, race and sexuality. In. S. Smith, R. Pain, S. Marston and J.P. Jones III (eds.), *The Handbook of Social Geography*. London: Sage, pp. 129–95.

Peet, R. 1977. The development of radical geography in the United States. In, R. Peet (ed.), *Radical Geography: Alternative Viewpoints on Contemporary Social Issues*. Chicago: Maaroufa Press, pp.6–30.

Peet, R. 1985. Introduction to the life and thought of Karl Wittfogel. *Antipode* 17(1), 3–21.

Peet, R. 2000. Celebrating thirty years of radical geography. *Environment and Planning A* 32, 951–3.

Peet, R. 2010. Reminiscing the early Antipode (1970-1985). http://www.antipode-on-line.net (2 August 2021).

Rose, H.M. 1970. The development of an urban subsystem: The case of the Negro ghetto. *Annals of the Association of American Geographers* 60, 1–17.

Rubin, B. 1979. Women in Geography revisited: present status, new options. *The Professional Geographer* 31(2), 125–34.

Santana, D.B. 2005. Jim Blaut: ¡Presente! Puerto Rico: Theory, solidarity, and political practice. *Antipode* 37, 1023–6.

Sheppard, E. 1995. Dissenting from spatial analysis. *Urban Geography* 16, 283–303.

Sheppard, E. 2010. Eric Sheppard (1986-1991). http://www.antipode-online.net (2 August 2021)

Shrestha, N. and D. Davis, Jr. 1989. Minorities in geography: Some disturbing facts and policy measures. *The Professional Geographer* 41(4), 410–21.

Smith, D.M. 1971. Radical geography: The next revolution? *Area* 3(3), 153–7.

Smith, D.M. 1974. Race-space inequality in South Africa: A study in welfare geography. *Antipode* 6(2), 42–69.

Socialist Geography Specialty Group. 1980. Annual Report.

Stephenson, D. 1974. The Toronto geographical expedition. *Antipode* 6(2), 98–101.

Wilson, B.M., and H. Jenkins. 1972. Symposium: Black perspectives on geography. *Antipode* 4, 42–3.

Winters, C. 1979. The social identity of evolving neighborhoods. *Landscape*, 23, 8–14.

Wisner, B. 2010. Notes from Underground: The Beginning of Antipode (1969-1970). http://www.antipode-online.net (2 August 2021)

Zelinsky, W. 1973. The strange case of the missing female geographer. *The Professional Geographer* 25(2), 101–105.

5 Latin American critical geographies

Blanca Ramírez, Gustavo Montañez, and Perla Zusman

Introduction

Critical geography has a distinct meaning in each and every context where it has developed. There are general elements that overlap, but there are also characteristics that are contrasting and differentiate each approach. Among these elements it is possible to identify the historical development of geography itself as a discipline in different countries, especially how epistemological reflection has evolved in the various nation-states and how this evolution has been expressed in geographical research and thought in different times and contexts. Likewise it is necessary to differentiate between the conditions of critical geography as a rebellious representation of the reality lived in a given country, and the development of what is a critical element within the scientific discipline of geography and its institutionalization in each Latin American nation.

It is for this reason that, rather than considering Latin American critical geography as a whole, this chapter is based on the premise that different critical geographies contain common and transversal characteristics, but also that there are other elements that are specific to each region. Within these parameters, this chapter searches for both dimensions that constitute the historical development of critical geography: firstly, the common, and secondly, the disparate elements of geographical thought and research in the context of the different nations of Latin America. For this purpose, this analysis will take various epistemological criteria into account that relate to critical geography in its link with social sciences, along with contextual criteria that can be observed in every historical moment on the different stages in Latin America. Finally, the analysis will consider the way that these criteria are linked with institutionalized geography more broadly.

More than simply reaching a singular consensus position on Latin American critical geography, it is necessary to consider the importance critical geography has constantly had in the areas where it has been developed. In the 1960s and 1970s, in Argentina, Brazil, and Uruguay, this position was directed by epistemological trends like Marxism, phenomenology, or existentialism. Between the 1960s and the 1980s, in Mexico, critical geography

DOI: 10.4324/9781315600635-5

was presented as an alternative to positivist geography either as a regional proposal directed towards the development of the country or by adopting classic Marxism. Currently, there are various positions within critical Latin American geography. On the one hand, there are those who share the tradition of the Frankfurt school and can, on occasions, move towards activism through distinct movements such as feminism, queer activism, or environmentalism (Best, 2009). On the other hand, there are those who are more closely linked with the Latin American reality based on national liberation movements or transformations into modernity. In more general terms, critical geography can serve as an umbrella term to integrate the wide range of visions and subjects that are not tackled in classic geography.

If arriving at a definition of Latin American critical geography is in itself a complex task, identifying its different territorial manifestations makes the task of analysing and explaining it even more difficult. Firstly, pinpointing the contexts that define critical geography's development, institutionalization and links with other positions requires a specific characterization of each nation-state or region. Secondly, different political developments have without doubt influenced the way in which geography has been linked with various visions of critical analysis. Lastly, the different ways that countries have opened themselves to external influence in the field has also defined the ways in which geography has been interlinked with other traditions and theories in each specific case.

For these reasons, this chapter is divided into three sections discussing the development of critical geography in a specific Latin American region. The first section presents the development of critical geography in the northern region of the continent through analysing Mexico, and how this nation has adopted a unique vision of critical geography that is not shared by the other Central America countries, where there are few traces of any critical thinking in geography and limited interchange of ideas with the rest of the continent. The second section tackles the development of critical geography in the region encompassing the Andean countries of northern South America, including Colombia, Venezuela, and Ecuador. Lastly, the third section analyses ways in which critical geography has been interpreted in the Southern Cone. This section includes a closer look at the geography of Brazil and the visions promoted by Argentina, Chile, and Uruguay, which have impacted the continent as a whole.

Critical geography in Mexico and Central America

As opposed to the situation in other regions of the continent, Mexico and Central America have demonstrated a distinct development in the generation of geographical knowledge. This is exemplified by the very late participation in (or, on occasion, absence from) Meetings of Latin American Geographers (Encuentros de Geógrafos de América Latina, EGALES) (Box 5.1). It was for many years difficult to find attendees from Central

---

Box 5.1

Meetings of Latin American Geographers (EGALES)

Since 1987, a biannual EGAL has been held. These events, held at public universities, are open to researchers, teachers, and professionals from the region, and are characterized by the large number of different perspectives and diversity in the addressed issues. The number of participants has increased over the years. For example, while 137 speakers participated in the fourth meeting held in Mérida (1993), more than 3,000 works were presented at the tenth meeting in Montevideo (2009). Discussions on theory and research in geography, education in geography, and professional geography practice are general features at these events. Since 1993, there has also emerged a distinct interest in evaluating definitions and changes in Latin American territories deriving from increased integration of the world and internal transformations. Empirical studies have dominated over theoretical studies, and particularly work relating to the application of geographical information systems (GIS) has increased. The majority of the presentations are guided by theoretical perspectives developed in Europe and North America, although many turn to the spatial theory of Milton Santos. Although the 1989 Montevideo meeting made it clear that there was an interest in developing a Latin American critical geography, the meetings today are not only representative of this trend. Rather, the meetings offer a panorama of current advances in Latin American geography, and they provide a space for the circulation and exchange of ideas.

America, and it was not until 2011 that a group of Central American geographers participated in the Costa Rica EGAL, one of the specific critical spaces shared by different geographers of Latin America. In contrast, Mexican geographers had participated in the EGALES since the first meeting. Central American countries have their own ways of generating geographical knowledge with a limited focus on critical thinking.

Despite the fact the region shared the practice of geography developed by Alexander von Humboldt in the nineteenth century, historically speaking, there are three areas in which the development of geography in the northern region has coincided and differed. First, the field of geography was in some countries from an early stage linked to engineering and the need to map territories and resources. Second, the institutionalization of geography has in different ways been linked to the training of teachers to impart courses at university level. This continues to be of vital importance for the majority of countries. This has been achieved by the institutionalization of the science under different circumstances specific to each country, at different moments in the development process. The third relates to the specialization and independence of geography in relation to other disciplines which also has happened in a particular way in each country. Panamanian geography gained independence from history around 1980 (McKay, 2006, 30), for example, whereas Cuban geography became autonomous with the victory of the revolution up to 1959. In the last case, geography dissociated itself from the field of history and built links with the geosciences and Soviet-style academia (Dory and Douzant-Rosenfeld, 1995, 63–64). Today, these fields of geography – mapping, nature and resources, and the teaching of geographical history – have varying origins. Opportunities to carry out postgraduate studies abroad differ from one case to the next, and these possibilities influence the preference for one tradition or the other.

Puerto Rico and Cuba have been important in developing critical geography in the northern region. In the former case, Puerto Rico's link to the United States as a free associated state has meant that the development of geography has been directly connected with the trends coming out of the United States (Cruz Báez, 2007, 2–4). Also, as a result of relations with the United States, critical geographers have been directly involved in the search for the political independence of the Caribbean island. With regards to Cuba, institutionalization of geography happened earlier, and the field can be more clearly defined than in other countries (Dory and Douzant-Rosenfeld, 1995, 63–64). At the beginning, geography was associated with the French and German traditions of the nineteenth century and linked with history. From the revolution onwards, geography was studied independently. It was linked with geosciences and developed into two branches: physical geography and economic geography (Dory and Douzant-Rosenfeld, 1995. 65), both strongly influenced by Soviet geography, economic planning, and methods of spatial analysis. With an independent school and faculty, Cuban geography was put at the disposal of the revolution and Cuba's economic planning, with PhD student research topics being orientated towards meeting the needs of the State (Dory and Douzant-Rosenfeld, 1995, 64–73). If Cuban geographers had a smooth relationship with the Soviet Union prior to 1980, with the majority of national geographers obtaining their PhDs from institutions in that country, the relationship with the rest of Latin America has not been so easy. This can be shown in two ways: first, by looking and working on specific projects where Cuban institutions have participated as the generators of knowledge, and second, considering express invitations from other Latin American countries that would cover the expenses of Cuban assistants.

Little is known about the development of critical thought within geography in the rest of Central America, at least by those contributing to this chapter. This remains an area that deserves to be studied in further depth. For that reason, the rest of the contribution will be developed exclusively through the experience of the Mexican situation of critical geography.

In relation to Mexico, it is possible to trace a tradition of critical geography back to the Mexican Revolution of 1910. At the very least, it could be expected that an event as significant as the revolution would have brought about a change in the geography perspective established during the 40-year dictatorship of Porfirio Díaz, from the end of the nineteenth to the start of the twentieth century, which was characterized by visions imposed by the State and the modernization of the ruling classes (Ramírez, 2007, 77). Nevertheless, geography was only present as a science linked to various branches of engineering during the first 30 years of the twentieth century. It was not until the end of the 1930s that the College of Geography was set up in the Faculty of Sciences of the National Autonomous University of Mexico (Universidad Nacional Autónoma de México, UNAM). At the beginning the College was linked with natural and technical sciences, but it was subsequently, in 1941, moved to the Faculty of Philosophy and Letters. These events meant that geography emerged as a divided science, since the research of the Geography Institute remained in the Faculty of Sciences (Moncada, 1994), whereas university teaching of geography was carried out in the Faculty of Philosophy and Letters.

Towards the end of the 1930s, a period of strong nationalism, the presidency of Cuauhtémoc Cárdenas (1934–1940) pushed through a series of reforms. In response, geographical research began to be shaped by State planning focussing on a redistribution of land to put an end to large estates and the expropriation of oil industries (then in the hands of British companies). The importance president Cárdenas placed on planning meant that regionalization became a vital instrument for developing strategies for modernization and for the industrial and agricultural changes the country needed. This regionalization was derived from research carried out in UNAM. During this period, it could be said that Mexican geography took a critical approach, given the State's need to provide structural ways to modernize the country stemming from the national policy being carried forward. The critical thinking here had two specific perspectives. On the one hand, due to the need for economic independence from other countries, regional thinking gave, at least potentially, the possibility of modernizing and developing. On the other hand, there was a need to separate geography from engineering by developing a more humanistic perspective. The geographical and political backgrounds of the two pillars of Mexican geography, Jorge A. Vivó (1906–1979) in teaching and Angel Bassols (1925–2012) in regionalization, helped direct research and teaching in applied geography towards change and modernization, particularly through regionalization.

After the Cardenas period, the pioneering academic works in regionalization were Angel Bassols' *La división económica regional de México* [*The economic regional division of Mexico*] (1967) and Claude Bataillon's *Las regiones geográficas en México* [*The geographical regions of Mexico*] (1969). Tackling important issues such as agrarian reform proved difficult for political reasons (the opposition of the landlords) and due to the absence of adequate maps for detailed work to underpin redistribution of resources; this point was important for researching rural areas as well (Bataillon, 2008, 45). Also, the separation between the College (teaching) and the Institute of geography (research) in UNAM impeded the integration of a joint vision of geography and a critical approach that could be shared.

The arrival of Spanish exiles from the civil war to Mexico during the Second World War critically impacted fields such as sociology, politics and, to a lesser extent, geography (Bataillon, 2008, 73). Many of the exiles in Mexican geography worked from the perspective of their academic backgrounds as biologists, naturalists, astronomers, or economists, which they applied to their geographic studies. Nevertheless, some geography or cartography teachers were among them, such as Felipe Guerra Peña, Marcelo and Miguel Santaló, Leonardo Martín Echeverría, Josefina Oliva Teixell, and Carlos Sainz de la Calzada made a name for himself in medical geography (Bassols, 2008, 57–62), and, as we shall see, he and Bassols pushed for the creation of a particular critical geography group.

Discussions that linked geography to the social sciences, influenced by Marxism, arrived late, and only partially, in Mexico. Important texts were translated into Spanish and could have had direct bearing on the formation of a critical geography in Mexico. This included works such as Lefebvre's *El pensamiento marxista y la ciudad* [*Marxist thought and the city*], *De lo rural a lo urbano* [*From rural to urban*], and *El derecho a la ciudad* [*The right to the city*] (all published in 1973), and *Espacio y Política* [*Space and politics*], published three years later (Lefebvre, 1973a, 1973b, 1973c, 1976). The works of Pierre George were also translated, such as: *Geografía rural* [*Rural geography*] in 1969, *Panorama del mundo actual* [*A panorama of the current world*], and *Geografía económica* [*Economic geography*] in 1970, *Sociología y Geografía* [*Sociology and Geography*] in 1974, and *Geografía Activa* [*Active Geography*] in 1975 (George, 1969, 1970a, 1970b, 1974, 1975). Nevertheless, such readings were not incorporated into the educational program offered at the College of Geography. They could only be consulted outside the institution, and they were far from being recognized as examples of critical geography. The contributions of Anglo-Saxon geography remained unknown in the College, and hence there was no possibility of incorporating them into the academic curriculum.

The important students' movement of 1968, which demanded democracy and the liberation of political prisoners, had little impact on the field of Mexican geography. It was not until 1977 that a group of professors joined the political movement of the trade union of the UNAM, participating with other workers in the university strike that year. In this context, teachers and students started to feel the need to take a more open and critical perspective, and to be socially committed. An independent group called *Geógrafos Progresistas de México* [Progressive Mexican Geographers] was formed in May 1978 through the initiative of Carlos Sainz de la Calzada, the Spanish exile, and a group of students and teachers (Angel Bassols among them). They organized to push for an orientation of geography towards research, and to put an end to the academic isolation of geography. This was to be achieved through closer links to other sciences, particularly the social sciences and Marxism. But the association also managed to improve international relations, not just with the International Geographical Union (IGU), where there were existing links, but also with the editors of *Hérodote* in France, and *Antipode* in the United States. In 1983, the group was consolidated and created the critical journal *Posición* [*Position*]. This semiannual journal publically acknowledged the existence of a crisis in Mexican geography, due to the program of studies that had worked without any modification for more than 30 years, and that was impeding its development and delaying its actualization and consolidation as a science.

Posición had a significant impact. It brought together students and teachers to tackle the marathon task of extricating Mexican geography from the crisis in which it found itself. The topics covered were diverse and gave rise to two lines of action within the emerging field of critical geography: epistemological discussions and the redefining of geography in Mexico; and discussions of geography and education. The first line was centered on an epistemological search to define geography, starting from its links with social sciences but also discussing relationships of geography with studies of ecology and natural resources. This highlighted the development of the discipline at the time and helped to define the natural and/or social character of geography The discussions also opened up the possibility of making political and critical analyses of issues such as the centenary of the death of Marx and the victorious Nicaraguan revolution in 1979 (Mondragón, 1988)

Evidently, other issues were included in the journal, such as the economic crisis sweeping across Mexico from 1970–1980, the nationalization of the banks in 1982, the deterioration or expansion of the metropolitan zone of Mexico City, the problems with peasants and native communities, and the crisis developing in the Mexican agricultural sector. In all issues of the journal, the region, as a concept identifying natural and economic characteristics for development, was considered as an instrument of change and, therefore, the use of regionalization, as the technical way to identify different potentialities within Mexican space, was considered an important geographical tool in order to eliminate backwardness of the country. This was critical in the sense that geographical instruments were used in order to achieve more development and change within the country and not only used in the positivist manner.

The second critical line of action in the progressive group around *Posición* centered on geography and education. On the one hand, this concerned the need to change the curriculum of the geography degree, which had not been updated since 1960. On the other hand, engagements with education also represented a continuation of a struggle started in 1977 for the creation of a Geography Faculty, which would be independent of other degrees and specializations within the humanities. The death of Vivó in 1979 brought to a halt the activities of a commission set up to modify the academic curriculum and brought the commission face to face with two different standpoints that prevailed for a long time. The first was the position of those who acknowledged the existence of problems in the vision of geography maintained at the time, but who also acknowledged that the students fighting for change were too unaware of recent developments in geography in Mexico and in the rest of the world to contribute to changing the curriculum (Berenberg, 1988, 37) justifying the need to stick with the existing curriculum. The second position was promoted by a group supported by the *Progresistas,* the progressive geographers, students, and teachers, congregated in *Posición*, which questioned the existing curriculum. They considered themselves to be familiar with changes in international geography, and to understand the problems of the discipline in Mexico. This group insisted on the need to bring the science up to date (Fernández Christlieb, 1988, 39). They wanted to introduce courses such as History of Geographical Thinking and Epistemology of Science, to engage more with Anglo-Saxon or European geographers that were unknown in Mexico, and to include content on Marxism and Geography, all of which were absent from the curriculum at that time.

The reluctance among important sections of the College of Geography to change the curriculum, the lack of clear direction on how to teach research methods, and the need for technological updating, impeded the movement to renew the vision of geography. At the same time, recent graduates were distancing themselves from the university to pursue jobs or professional careers outside of the institution, or to specialize abroad. These factors, together with the economic crisis present in the country since 1980, caused the movement to fragment and the journal to close in 1990 when the last issue appeared. Almost three decades later there have been some updates to the curriculum, but there remains a lack of discussion of advances on geographical thinking in the twenty-first century.

Despite these setbacks, globalization, technology, and the work of some critically-minded geographers, who joined the open university of UNAM, meant that communication in the early 1990s was established with international networks, such as the EGALES. Very late in the day, this exposed Mexican geographers to Milton Santos' thinking, for example (Box 5.2).

Towards the end of the twentieth century, research and teaching in Geography was mainly concentrated in the UNAM, up to the creation of one School in Toluca (State of México) and more recently others in some States of the country. Even though close relations have been developed with schools and researchers outside Mexico, positivist thinking is still pre-eminent, and a lack of critical thinking is still a characteristic.

Some Mexican critical geographers, who teach and research in planning at other institutions, such as the Autonomous Metropolitan University (Universidad Autónoma Metropolitana, UAM), have also helped develop connections with critical thought from English-speaking and other regions of the world. Likewise, at the Iztapalapa campus of UAM, a new degree program in Human Geography was launched in September 2002 that is

Box 5.2

Milton Santos (1926–2001)

Milton Santos was one of those responsible for the development of critical geography in Brazil and in Latin America. He was born in San Salvador de Bahía and studied for his PhD in Strasbourg (France) under the direction of Jean Tricart. After the 1964 coup d'état in Brazil, Santos lived in exile in France where he worked as a teacher at different universities. In France he wrote two books that show his thoughts on the third world: *Le métier de géographe en pays sous-développés* (*The Job of the Geographer in Underdeveloped Countries*) and *Les villes du tiers monde* (*The Cities of the Third World*), both from 1971. Difficulties with renewing his contract in France took him to North America, firstly to the Massachusetts Institute of Technology (1971–1972), then to Toronto (1972–1973), and finally to Columbia University (1976–1977). It was at this time that he collaborated with the journal *Antipode*, participating in the coordination of two issues dedicated to the subject of underdevelopment (1977, 9(1), 9(3)). Santos also lived in Venezuela, working in planning, and in Tanzania (1974–1976) where he worked at the University of Dar es Salaam at a time when the country was going through a revolution. Santos was hired as a professor at the Federal University of Rio de Janeiro in 1979, joining the group of intellectuals that participated in the renewal of Brazilian geography. Within this context, he wrote *Por uma Geografia Nova* (*For a New Geography*) (Santos 1978/1996). Santos described this book as a systematic critique of the trends that had made the field of geography uncritical of capitalism and imperialism. In contrast, Santos presented geography as a social science, useful for understanding and transforming society. In 1984, he became a professor at the University of São Paulo, where he developed his work on the technical-scientific-informational *milieu* and the role of the Brazilian urban network within the framework of globalization. In 1996, Santos published *A Natureza do espaço* (*The Nature of Space*), which summarized his epistemological reflections since *Por uma geografia nova*; and in 2000 he published *Por uma outra globalização* (*For Another Globalization*). His perspective on space as a totality in constant movement due to the interaction between systems of objects (socially produced through work) and systems of actions (results of intentional human behaviour) has influenced how many Latin America geographers understand and articulate space in their analysis. His understanding of urban economies from the interaction between the "upper circuit" (banks, modern urban businesses, and industries) and the "lower circuit" (small businesses, small-scale manufacturing, and handicraft production) has been incorporated in the analysis of world cities (Zusman, 2002). His characterization of globalization established a basis for considering another possible type of globalization that is more humane and democratic.

orientated towards training specialists in relevant theories from the French and Anglo-Saxon traditions (Ramírez, 2007, 79). Furthermore, some postgraduate programs in Human Geography have been opened at institutions dedicated to regional anthropological studies, like the College of Michoacán (Colegio de Michoacán, COLMICH). With a very open and all-embracing vision of geography, researchers of this College have, in 2007 and 2010, organized two events aimed at clarifying standpoints and visions revolving around the link between geography and social sciences, and geography in practice, from a Mexican perspective. These standpoints were also laid out in the text titled *Geografía humana y ciencias sociales: una relación reexaminada* [*Human geography and the social sciences: a relationship reexamined*] (Chávez et al., 2009), and in *El Espacio en las Ciencias Sociales, Geografía, Interdisciplinariedad y Compromiso* [*Space in Social Sciences: Geography, Interdisciplinarity and compromise*] (Chávez and Checa, 2013).

Criticism of neo-liberalism, its brutal introduction in Mexico and in Latin America more generally, is one of the most important subjects in contemporary Mexican critical geography. Another topic is critical epistemology, which intends to face classic theoretical-methodological problems, such as the separation between human geography and physical geography, conceptions in geography, movement and transformation within physical or social geography, scales and their integration in space, space and imaginaries, cultures, gender, and contemporary subjectivity. It should be noted, however, that such issues still are addressed by other fields of social study than geography. In fact, research on gender is only carried out in a limited fashion in Mexican geography, whereas these studies are very important within sociology. In the same way, new conceptions of space, territory, and landscape are discussed more among urban sociologists than among geographers (Ramírez and López Levi, 2015).

Despite the fact that globalization has evidently contributed to the sharing of Mexican geography with other countries and the opening of Mexican geography to external influences, the problems critical geography has experienced in Mexico still exist. Now it shows a clear dependence on theory from abroad (critical and non critical), leading to the introduction of categories and theories without any substantial analysis of their pertinence and adequacy to the local reality, while the positivist approach is still preeminent in geography more generally. The separation of research and teaching between the Institute and the College in UNAM, and the reluctance to look further into new forms of studying geography, or to see space, place, or territory in a more dynamic way, continue to weigh down on students and to reduce the possibilities of Mexican geographers taking an alternative path that would lead to a significant new channel for Mexican critical geography.

Critical geography in the Andean countries of Northern South America

The emergence of what became critical geography in the Northern South American countries had its roots in pre-modern geography (Montañez, 2000), specifically the ideas and practices of diverse New World inhabitants prior to the emergence of modern geography in the American continent. Pre-modern geographers developed and used spatial knowledge to inform their patterns of mobility and settlement. Later, they developed a "rebellious geography" in opposition to Spanish invasion and colonization. One form of resistance was evident in the frequent escapes of African slaves from colonial controlled land to the tropical rainforest where they established their home and collective life places as *palenques.* Another form of resistance involved a challenge to the racist colonial division of space, into white and indigenous territories. Later, the struggle for independence from Spain gave rise to new forms of "rebellious geographies". For some criollos leaders – for example, Francisco José de Caldas in Colombia and similar figures in Ecuador (Keeding, 2005) and Venezuela – formal and informal geographical knowledge was an important foundation for their later involvement in independence movements. Caldas, for example, understood the significance of geographical knowledge for securing the future prosperity of an independent nation.[1] Venezuelan Simon Bolivar and others (McFarlane, 2014) used territorial war strategies in their guerrilla war against Spanish power. When independence from Spain was secured, the ideas of the integrationist Venezuelan Francisco de Miranda were put into practice by Bolívar.[2] This involved the formation of a nation-state called *Gran Colombia* (Great Colombia), encompassing the current three countries of Colombia, Venezuela, and Ecuador. However, this project lasted only a few years. Its failure can be attributed to the growing disagreement and contrasting interests among the elites, settled and dispersed among different territories.

At the end of the colonial era, between 1799 and 1803, Humboldt visited Northern South America and travelled inside some of these territories of present-day Colombia, Venezuela, and Ecuador. During these travels, he recorded detailed observations about nature and people in the New World. These materials were organized, analysed, and published some decades later in French and German languages. Later on, these books were translated and published in Spanish (Humboldt, 1848–1858, 1982, 1991). The travels of Humboldt through Northern South America was an early announcement of the development of modern geography, advanced by Humboldt in *Cosmos* (1848–1859). This work helped secure the position of geography, with its positivist approach, within the modern sciences. However, there was little engagement with these ideas in the territories visited by Humboldt. This is evident in a letter sent by Humboldt from Cumaná, located in the Captaincy General of Venezuela, on 17 October 1800, stating "The only thing I might regret in this solitude is being deprived of the advantages that result from the exchange of ideas" (in Perez and Alberola, 1993, 73–75). In the same letter, Humboldt also commented on the contradictions he observed among *criollos* who advocated for liberty and independence:

> Very often, I have found some men who speak with their mouth full of beautiful philosophical principles, however they denied themselves those principles with their own behavior: they give bad treatment to their slaves with the Raynal in their hands, and they speak with enthusiasm about the cause of liberty but at the same time they sell the sons of their slaves just few months after their birthdays. What desert will not be preferable to the behavior of these philosophers?
>
> (in Perez and Alberola, 1993, 73–75)

Thus, the independence of Colombia, Venezuela, and Ecuador meant a political break with Spain, but this did not lead to a change in dominant ideological approaches. Rather, independence was initially characterized by a surprising continuity of previous dominant thought, and by the persistence of tremendous social inequities in these three Latin America countries.

After independence, during the first half of the nineteenth century, Colombia and Venezuela shared another European visitor who was also influential in shaping modern geography in the region. The Italian military engineer, Agostino Codazzi, was invited first by the Venezuelan government and ten years later by the Colombian government, to oversee mapping the territories of the two emerging republics. This deliberate project to develop basic cartography for the two States was part of a Ruling class project to build and govern the new countries. The purpose was to consolidate the foundations for territorial knowledge and to create a geography of the nation-state, as an instrument of constructing collective and nationalist identity, the one the elites wanted to establish.

By the nineteenth and into the early twentieth century, Colombia, Venezuela, and Ecuador – similar to many other countries in the continent – had three main streams of geographic thought. These were: (1) state military and administrative geography; (2) modern Humboldtian geography; and (3) Reclus-inspired geography. The first stream was originally linked to military activities, involving both the definitions of internal and external frontiers and the need for administrative information such as cadastral information. Initiated by Agostino Codazzi, army general staff actively sought to map state territorial features from the early nineteenth century on. Later, geographical military institutes were established in each of these countries. These circumstances drove later on to an opposition from civilian geographers in order to transform those military institutes into civilian ones. The military geographical institutes were transformed into civil entities in Colombia (1950) and in Venezuela (2000). In contrast, the Geography Institute in Ecuador still remains under military control.

The second stream was represented by modern geography and influenced by the ideas of Humboldt (1848–1859). Its challenge was to build a new representation of nature, and to explore and understand the relationships within nature viewed as a diverse unity, including human living beings as a part of it. This was a form of critical geographic thought at that time. The growing emphasis on a new positivistic epistemological approach turned the explanation from God to Nature, in order to pose questions and seek answers about geographical phenomena.

The third stream was also influenced by European geography, specifically by the ideas of Reclus. This approach is most similar to what is now called critical geography, though these ideas passed through Northern South America without being valued or adopted as critical thought at that time. Reclus[3] travelled as an explorer through the mountain range of Sierra Nevada of Santa Marta, in Colombia, and left an important legacy of geographic reflection, but this remained unknown until the second part of twentieth century when critical geographers from various nations rescued his name as an anarchist geographer. Today, he is considered one of the pioneers of critical geography developments in the world. Although Reclus exchanged ideas with the distinguished Colombian geographer Francisco Javier Vergara y Velasco for over ten years, between 1888 and 1897 (Ramírez, 2007), this fact was not generally known at the time. These three countries just began to open themselves to the contributions of Reclus in the last quarter of the twentieth century, through the influence of French critical geography, headed by Pierre George and Yves Lacoste. Thus Reclus became better known in Venezuela and Colombia during the 1970s, and at a later stage by a small number of Ecuadorian geographers.

It is during the emergence of modern geography, and especially during the academic and professional consolidation of geography in Northern South America, that some manifestations of contemporary critical geography became more evident. Modern geography approaches took off rather late in this region compared to other countries of Latin America, such as Argentina, Uruguay, Brazil, Chile, or Mexico. The institutionalization of geography as a subject of professional formation at university level occurred first in Venezuela. The first undergraduate Venezuelan geography program began in 1956 at the Central University of Venezuela and the second in 1964 in The Andes University. Colombia began graduate training in 1985 and undergraduate training in 1992 in public universities. In the 2000s, two Colombian private universities also opened undergraduate and graduate geography programs. Ecuador was even later, since it just started undergraduate geographic professional formation in 1989, even though CEPEIGE [The Panamerican Center of Geographical Studies] had been operating in Ecuador since 1973. This international institution is supported by the American State Organization and offers short graduate courses annually for students coming from most American countries. However, there is still no graduate program in geography in Ecuador.

Critical approaches in geography have been gaining in strength in the three countries since the 1970s. This last wave coincides with neoliberal globalization and post-neoliberal era, and the related widening of political, economic, and spatial inequalities and injustices. More traditional modern geography has had greater contact with various lines of critical geography discourse, and links to other social sciences and the growing participation in international geography networks, publications, or events have contributed to strengthening the prospects for critical geography in this sub-region.

The development of Marxist critical geographic thought in this part of the continent occurred in the form of two main waves. The first wave, in the 1960s and 1970s, occurred in conjunction with strong revolutionary movements in most of Latin America. It was influenced by the Cuban Revolution and the Vietnam War, and promoted ideas of political and social change. The second wave of Marxist critical geography, from the 1990s onwards, has been in response to neoliberal globalization policies, the persistence of growing social and territorial inequalities, and the spread of resource extraction.

Marxist Anglo-Saxon critical geography was present in the three countries at the end of 1960s. The work of David Harvey, William Bunge, David Slater, and others began to influence young self-taught geographers. Some of this work was read in English, while some – for example *Social Justice and the City* (Harvey, 1973), *Radical Geography* (Slater, 1978), and "The New Left" (Peet, 1969) – was translated into Spanish. The radical geography journal *Antipode* was also influential, as some university libraries in Colombia and Venezuela subscribed to it. However, the Brazilian geographer Milton Santos, who received a PhD from a French university and was influenced by Marx and Lefebvre, became the most influential figure in Marxist critical geography in Latin America from the 1970s onwards. This began with the publication of *Geografía y economía urbanas en los países subdesarrollados* [*Geography and urban economies in underdevelopment countries*] (Santos, 1973), which demonstrated the importance of relations between developed and underdeveloped countries in the process of urbanization in the region. In the same decade, a small group of geographers organized the first meetings of the Southern Cone geographers, held in Uruguay in 1973 and Argentina in 1974. From this point on, with the idea of building a "New Geography", the seed was sown for what would eventually bloom into a constellation of critical geographies in Latin America. The EGALES played a central role in making this happen under the leadership of Milton Santos.

At the same time, the Andean countries of Northern South America experienced significant social mobilization, not only due to the demands of different segments of society, but also because of the emergence of new political projects. This was particularly true in the case of Colombia, which was one of the few countries that did not go through a military dictatorship, but instead experienced a long-lasting and intense guerrilla war. Social mobilization at this time was based on deep and extensive political and epistemological debate in universities. From these reflections, the question of the social and political role of traditional geography also arose, which opened the door to further debate on this issue.

The presence of dictatorships in a large part of the Southern Cone had, paradoxically, a positive effect on the development of critical geography in the Northern South American Region. This positive effect resulted from the dispersion of intellectuals from the Southern Cone who were forced into exile. This meant that a small but significant group of geographers arrived in Venezuela, bringing with them critical perspectives on geography. The most notable members of this group were the Uruguayan Germán Wettstein (1934-) and the Chilean Pedro Cunill (1935-) who migrated to Venezuela. The Uruguayan Daniel Vidart (1920-2019) went first to Venezuela, then to Colombia, and finally returned to his country of birth.

As in most other countries, critical geography in Colombia, Venezuela, and Ecuador has been growing in strength and influence during the recent period of neoliberal globalization. The practice of globalization has provoked intellectual and political resistance, in this region and elsewhere. However, paradoxically, it also enhanced access to international geographical literature and to critical geography networks around the world. These new sources of geographical thought allowed a diversification in the practice of geography, particularly in relation to the further consolidation of critical geography. In particular, critical geographers in this region drew from the tradition of Marxist thought, but were also inspired to develop and explore new critical perspectives by local and regional issues of concern. These issues included human interrelations with nature; human rights, especially for minority groups; resource extraction, specifically mining; the challenges posed by climate change; and the impacts of the so-called post-neoliberal governments of Latin America. It is likely that these issues will lead to new approaches to critical geographies in this region in the near future. However, different approaches to critical geography have been prioritized in the three countries in this recent period. These are discussed in turn.

In Colombia, the new political constitution of 1991 expressly mentioned the term territory. Public universities responded by promoting and developing geography programs. There are now six undergraduate programs, three masters programs, and two doctoral programs in Geography in Colombia, and this growth has been crucial to the development of geography in general, and to the expansion of approaches to critical geography among new generations of Colombian geographers. For example, some Columbian critical geographers have focussed on taking up and reinterpreting theoretical debates on geography in general and critical geographies in particular (Delgado Mahecha, 2003; Montañez, 2009a, 2009b). Others have emphasized the relationships between capitalist accumulation processes and the social construction of regional territories, with an emphasis on globalization (Espinosa, 1997; Montañez, 2005). Certain studies have also considered the potential of critical historical geography to provide insights into social territorialities and the space control of colonial areas in specific regions of present-day Colombia (Herrera, 2002), and on the organization of space and relationship to nature in African-descent communities in the Pacific region of Colombia (Oslender, 2008). In addition, other works in geography give critical insights into diverse topics such as the promotion of social participation in the use and design of geographic information systems (Barrera Lobatón, 2009); new methodologies for the analysis of territorial conflicts (Peña Reyes, 2008); and the use of ecological perspectives to assess current forms of production and commercialization of food (Rodríguez, 2003). Critical geography perspectives have also used subnational geopolitical analyses, for example of infrastructure megaprojects in Latin America, to contribute to broader debates in international geopolitics (Sánchez Calderón, 2008). Critical geography has also been enhanced by the activities of GeoRaizAL (Critical Geography from Latin American Root), predominantly based in and administered by teachers and students of the Geography Program at the Colombian Externado University. GeoRaizAL has organized several national and international events on critical geography topics since 2011.

In Venezuela, critical geography is above all represented by the geo-historical approach, which has been pre-eminent both in geographical research and geography education (Ceballos García, 2001). In this respect, the works of Tovar (1986) stand out with the development of the philosophical and methodological basis for research on the historical dynamics of geographical areas, at different scales. Along the same lines, Santaella Yegres (1980, 1989) used a geo-historical approach to investigate regional and local dynamics in Venezuela. There have also been initiatives closely related to critical geography that discuss the territorial challenges posed in Venezuela in the light of globalization (Trinca, 2005a, 2005b). One additional element to Venezuelan critical geography is the incorporation of the "power geometry" of Doreen Massey (2005) into the country's political planning and administration. The third Chavez administration (2007-2013) adopted this tool in the implementation of territorial changes planned within its revolutionary project (Ramírez, 2010). Their proposal looked to establish a new territorial structure for people's participation in government, but it was hampered by the traditional hierarchy and centralized character of political power in Venezuela.

In Ecuador, geography as a professional field was not particularly developed until recent times. The subject has a limited presence in universities, and geography is still considered mainly a cartographic subject or a school education field. At the same time, geography maintains, to a certain extent, its military and statistical perspective. Nevertheless, during Correa's administration (2007-2017) were built up some high-level institutions for research and critical thinking about territory rights. These recent signs may show that more robust critical geographical approaches could develop. There is some evidence of a growing interest in critical geographical approaches,

involving geographers from other countries as well as Ecuador. The work of
Sarah Radcliffe (2011), which focuses on development in the post-neoliberal
era and its relationship with the proposal of "living well" in Ecuador, is
particularly important. Among geographers based in Ecuadorian geographers, a very active group has emerged, calling itself a Critical Geography
Collective of Ecuador This collective recently published a critical account
of Ecuador territorial formation and the related historical forms of capitalist accumulation(Bonilla et.al., 2016). In addition it promotes political
action to denounce mining activities and other transnational corporation
projects located in the Amazon region, and to make visible the environmental impacts of several development initiatives. Given the debates about
mining extraction activities and policies in Ecuador, the work of Alberto
Acosta and other Ecuadorian researchers and social leaders (Acosta and
Martínez, 2010) is significant. Though not geographers, they have stimulated critical thinking on the crucial topic of the relationship between oil
and life. As part of government strategy to build a national purpose and
counteract criticism, the National Center of Strategy for Territory Rights
(CENEDET) was founded in 2013. The mission of CENEDET was to foster
critical debate about the socio-spatial transformation of Ecuadorian territory, and its early directors included the geographer David Harvey and the
architect and urbanist Miguel Robles-Durán. CENEDET announced that
its main purpose was to advance radical strategies of social change, based
on socio-ecological justice and the right to the territory. Additionally, since
January of 2016, Harvey has been an Honorary Professor of the Central
University of Ecuador. In these circumstances, there are great expectations
about the fruits of CENEDET initiative in the near future, as a part of internal and national debates about the post-neoliberal governments in Latin
America (Martinez et al., 2015).

It could be claimed that despite the recognized advances and the emergence of young leaders in the field of geography, especially in Colombia
and Ecuador, critical geography in the three countries has scope for further development. A solid network of critical geographers has still not been
formed. The work is largely dispersed and relatively isolated. Under these
circumstances, the challenge is to develop more extensive research programs
with a more wide-reaching academic and social profile. The EGALES
had been the best international window to present academic and research
advances of the critical geography in Latin America, but specific discussion
spaces within critical geographers of this region are needed in order to build
a more collective agenda for the near future.

Critical geographies in the Southern Cone

In the Southern Cone, geography was historically associated with the interests of the State. Geographical societies were established in the nineteenth
century to promote the exploration and description of territories that the

state wanted to occupy.[4] The first degrees in geography began much later (in the 1950s in Argentina and Brazil, in the 1960s in Chile), but were similarly directly in the service of the State by focussing on training geography teachers, who in turn were responsible for spreading national values at the high school level. Here, the dominating image was that of the territories, which in the tradition of Vidal de La Blache were divided in regions that together formed one harmonic unit.

Within this framework, we can identify three historical moments in which the geographical institutions and the Vidalian model of geography have been challenged. The first is situated at the beginning of the 1970s, interrupted by the arrival of military governments; the second corresponds with the end of the 1970s and the early 1980s when the countries of the region experienced a transition to democracy; and the third is situated at the beginning of the twenty-first century when structural adjustment policies imposed in the 1990s showed their economic and social limitations. As we will discuss, these three moments can be seen as phases in the rise of critical geographies in the Southern Cone.

The Vidalian model started to be criticized in the 1950s by geographers who participated in processes of modernizing and professionalizing the social sciences and incorporated the analysis of social problems into geography. With regards to this incorporation of social problems, it is necessary to highlight the role of Josué de Castro (1908–1973) in Brazilian geography and Elena Chiozza (1919–2011) in Argentine geography. Josué de Castro, a civil servant in the Food and Agricultural Organization (FAO) of the United Nations from 1952 to 1956, tackled the problem of hunger as a historical-political issue in a series of texts, including *Geopolítica del Hambre* [*The geopolitics of hunger*] from 1946 (Correia de Andrade, 1997; Alves, 2007). Elena Chiozza worked as a consultant on State infrastructure projects, and coordinated two publications that spread geographical knowledge and considered the social problems that were affecting different areas of Argentina: *Mi País, Tu País* [*My country, Your country*] from 1968; and *El País de los Argentinos* [*The country of the Argentineans*] from 1974–1978. Though regional geography was not forgotten in the projects of both geographers, the regions were by then defined in the light of historical and social problems: "Not everything is harmonious now between man and nature and it appears that, to a certain level, there is a concept of relations and social conflict" (Cicalese, 2008).

It could be considered that these perspectives were the basis for the critical geography (at the time referred to as "New Geography") that began to develop in South America in the 1970s. Around this period, Latin America was experiencing political changes, such as the Cuban or Nicaraguan revolutions, and there were repercussions from student movements in places like Paris, Prague, and Mexico City. In this context, a group of geography professors and students questioned the way institutions were organized and how knowledge was produced in these institutions. This questioning occurred at two meetings. The first was held in Salto (Uruguay) in 1973, the second in Neuquén (Argentina) in 1974, where two young and committed geographers took part, German Wettstein and Milton Santos. Cicalese (2007) and Quintero et al. (2009) agree in characterizing the project of "New Geography" as a politicized form of knowledge that contemplated the situation of dependency and underdevelopment in the region. Furthermore, they concurred that "New Geography" aimed to study the reality of Latin American countries and to bring about change in this reality (particularly in the working class areas), distancing it from the European and North American proposals and international organizations (International Monetary Fund, World Bank). As Quintero et al. (2009) show, the emphasis of these meetings was on revising the practices of geography, and seeking to go beyond academia to promote a geography that, through professional activities, was more directly related to the community and more conducive to social change than the creation of new concepts.

The dictatorships and the strong social repression that accompanied the application of policies of economic adjustments in the region in the 1970s tore up many fields of society, including the universities. Coupled with the closure of some university degrees and some educational centres, the vision of geography for military purposes became hegemonic. Geographers that had started to pursue projects of critical social geography had to go into exile or hide themselves away in private research centres. Within a framework in which the military governments were trying to create internal social cohesion through exalting national values, geopolitical discourses emphasizing allusions to conflicts with neighboring countries started to dominate (Child, 1985; Hepple, 1986) and were produced and reproduced in different research and educational centres. Regional discourses went hand in hand with geopolitics. At the same time quantitative perspectives were involved in state planning activities (particularly in Chile and Brazil) (Albuquerque Bomfim, 2010; Caviedes, 1991).

In the context of transitions towards democracy in Brazil, this process started to reverse during the 1978 meeting of the Association of Brazilian Geographers (AGB) in Fortaleza. In this arena, people started to criticize the geography of professors and the general staff (following the perspective of Lacoste, 1976). Also, the association opened itself to theoretical perspectives such as Marxist standpoints, phenomenological perspectives like geography of perception, and trends that were more environmentalist in nature (Moreira, 1992). Carlos Walter Porto Gonçalves' *A Geografia está em crise. Viva a Geografia!* [*Geography is in crisis. Long live Geography!*] from 1978 is one expression of this renewed critical geography. Porto Gonçalves sought to understand the discipline in the historical context and highlighted the limits of traditional geography in understanding capitalism. He also emphasized the need to look to Marxist theory, and he proposed that the concept of social formation could help understand the uniqueness of places within the uneven historical development of capitalism. Likewise, in

Por *uma Geografía Nova* [*For a New Geography*], Milton Santos (1978/1996) found that the specificity of conventional geography had led to the "widowhood" (*viudez*) of space. By this he meant that space was analysed independently of social processes, without a dialogue with other social sciences or without integrating it with historic social dynamics. For Milton Santos, Marxist categories, together with the existentialist project of Sartre, would offer geography elements could be used to perceive and change the world.

The discussion in Brazil about the relationship between space and capitalism, mediated by the different interpretations of Marxism, resulted in different ways of understanding space: as a condition for the production and reproduction of capitalism (Moreira, 1982); as an instance of territorial appropriation in the process of capital accumulation (Moraes and Da Costa, 1982); or as an instance, means and factor for the appropriation of nature, a process at the same time global and specific to every different place on the planet (Santos, 1978/1996). These conceptual concerns impacted rural and urban analysis. Meanwhile, studies of public policies, poverty, and socio-spatial inequality, on different scales, arouse the interest of many critical geographers. This renewal of theories and concepts was also accompanied by a change in disciplinary praxis. A renewal of school geography teaching became a primary concern, and some members of the AGB directed their activities towards changing the contents of the curriculum, updating textbooks, and organizing courses for high school teachers. At the same time, the association supported social movements fighting for land and housing (Moura et al., 2008).

The transition towards democracy started later in Argentina (1983), Uruguay (1985), and Chile (1988). But also here, the process was accompanied by the emergence of a new perspective for the discipline, which gave rise to an understanding of geography as a social science and conceptions of space as a social construction. From this standpoint, problems and issues similar to those dealt with in Brazil were discussed among geographers in Argentina, Chile, and Uruguay. In the case of Argentina, this new approach is evident in the three volumes of *Aportes para el estudio del espacio socioeconómico* [*Contributions to the study of socio-economic space*], which were edited by Luis Yanes and Ana Liberalli and published in 1986, 1988, and 1989. The books analyse the development of the different sectors of the economy from a structuralist Marxist perspective and emphasize the study of the accumulation model, the role of the State, public policies, and the regional and environmental impacts of such policies (Adriani, 2007).

As the universities of these countries emerged from the dictatorships, geographers started to incorporate international geographical perspectives and, specifically, they began discussions with critical Brazilian geographers. Circulation of ideas was channeled through encounters like the EGALES, two Latin American critical geography seminars organized between the Geography Department of the University of São Paulo (1988) and the University of Buenos Aires (1990), and the 1992 *O Novo Mapa do Mundo*

[The New Map of the World] under the direction of Milton Santos. At the latter conference, the subject of globalization and the critique of this phenomenon were placed on the Southern Cone geography agenda.

In the 1990s, in a context where states moved away from the model of the welfare state in order to serve the interests of transnational capital, theoretical and epistemological reflection (which had been the main activity of critical geography up that point) gave way to studies orientated towards the historical relationship between the State, society, and space. In this framework, analyses aiming to deconstruct processes of "inventing" territories and nations became prominent (Escolar, 1996; Moraes, 1991). At the same time, a cultural perspective also started to take shape in Argentina and Brazil. The pioneering studies concerned gender analysed relations between women and labour markets, environmental questions, migration, and political participation (Veleda da Silva and Lan, 2007). In Argentina, gender geography experienced a boom between 1995 and 2000, but there are few examples of work on this subject coming out today. In contrast, this line of research is experiencing continuous and sustained interest in Brazil. This research incorporates post-structuralist perspectives into its analysis and is interested not only in the relationship between women and space, but also in the practices of queer identities in the production and transformation of space. Examples of the development of this approach can be found in the articles published in the journal *Revista Latino-americana de Geografia e Gênero* [*The Latin American Journal of Geography and Gender*] of the University of Ponta Grossa. At the same time research on cultural geography was conducted in Rio de Janeiro by the Núcleo de Estudos de Pesquisa Sobre Espaço e Cultura [The Center of Study and Research on Space and Culture], and in Góias by the Research group on Cultural Geography: territory and identities (Almeida and Ratts, 2003). These groups took up international discussions with the aim of understanding the meanings of relationships between space and culture in different social sectors. Analyses of festivities, heritage, and religious symbolism were predominant examples. In Argentina and Chile, some studies have been carried out on relationships between imaginaries, the creation of places and the establishment of urban and rural identities, and the commercial exploitation of culture and consumption. But research in this area still has a long way to go (Aliste et al., 2007; Capellá Miternique, 2009; Zusman et al., 2007).

What type of critical geography is being produced today in the Southern Cone? Although there is no singular line of work or thought defined as such, it is possible to identify a collection of studies that analyse social, economic, and environmental effects of neoliberalism and can hence be considered part of critical geography (e.g., Alessandri Carlos and Geraiges de Lemos, 2003; Arroyo, 2006; Ciccolella, 2014; Hidalgo et al., 2016; López Gallero, 2005; Ríos, 2010; Silveira, 2016; Vidal Koopman, 2005). Likewise, there are studies of relationships between space and the social movements that have taken a leading role as political agents over the last decade, for example the landless workers' movement or homeless workers' movement in Brazil, the movement of the unemployed, the *piqueteros* movement, neighbourhood assemblies, and rural and indigenous movements (Porto Gonçalves, 2006; Rizzo, 2010). These movements are characterized by the development of a new praxis and "a new value given to space-related insurgent practices" (Lopes de Souza, 2010).

In this context, the concept of territory has become a central focus. At the same time that neo-liberalists are claiming that technology has succeeded in breaking down the barriers of distance, and are declaring "the end" of territories, studies carried out in the region are recovering the concept of territory in order to demonstrate the relevance of geography as a way of understanding the dynamics of the world today. For example, transnational mining projects are creating new territories within the region's States or in their borderlands (Hevilla and Zusman, 2009). Security politics are creating new territories through controlling the displacement of people at different scales (Haesbaert, 2014). Furthermore, while capital is constantly changing through deterritorialization and reterritorialization processes (Haesbaert, 2004), various social movements find a source of empowerment in the material and symbolic appropriation of territory (Porto Gonçalves, 2001). In this sense, territory is the heritage not only of the State and the market, but also of all social actors and institutions (Silveira, 2016). Some anarchist perspectives (Lopes de Souza, 2010), and some practices of social participation, like in the case of the various social mapping projects in process (Arias, 2010; Diez Tetamanti and Escudero, 2012), are also working with this idea of territory.

In this section, we have charted some paths followed by Critical Geography in the Southern Cone. Political criticism through different epistemological perspectives guides the trends we describe. We consider that these developments could contribute to reducing the strength of the military institutions that, in countries like Chile or Paraguay, still heavily influence the national geographies in areas such as mapping. These military institutions also represent some countries in the Pan-American Institute of Geography and History (IPGH) and the International Geographical Union (IGU/UGI).

Concluding reflections

Critical Latin America geography developed from the need to question political practices, and from the necessity to go beyond disciplinary traditions rooted in classic geography, which often stem from organizations associated with the State and the liberal elites of the nineteenth century that institutionalized geography. Beyond the differences that can be observed in every national context, geographers in all the countries discussed in this chapter during the 1960s and 1970s began to search for conceptions of their field that could contribute to political projects of changing lived realities for the better. In part, this was done by offering conceptual and methodological instruments interacting with society and social science theories.

Currently, the epistemological concerns of the 1960s and 1970s are no longer central points of discussion in the discipline. But many geographers research and participate in movements that question globalization and neo-liberalism in national politics and the socioeconomic effects on local populations. These analyses are now complemented by cultural-geographical perspectives. The political praxis of new social movements are also explored through new theoretical, conceptual, and instrumental possibilities that challenge institutions in countries such as Ecuador, Paraguay, or Chile that are still strongly influenced by the military. Applying geographical knowledge to challenge the multiple forms in which neoliberalism has affected the growth and development of the continent as a whole, and specifically in every region, is one of the ways in which the next wave of Latin American critical geography will continue to generate knowledge committed to the social transformation of the continent.

The possibility of expanding critical geography in Latin America requires the formation of networks for exchange research and information between different countries. To this end, critical Latin American geographers must produce their own theoretical and substantive knowledge, whilst maintaining critical discussions with knowledge generated in other parts of the world. For this purpose, geography students have formed their own network that organizes meetings of Latin American students (Encuentro Latinoamericano de Estudiantes de Geografía-Latin American Event of Students of Geography- ELEG). This network has its own journal (RELEG). Students and professors from Colombia are organizing a Latin American network of Critical Geography with Latin American Roots: GeoRaizAl, and events such as *Geografía Crítica: territorialidad, espacio y poder en América Latina [Critical Geography: territoriality, space and power in Latin America]*, held in Bogotá in September 2011. A group of Critical Geographers – El Colectivo de Geografía Crítica Gladys Armijo – was formed in Chile. During the Regional Conference of the International Geographical Union held in Santiago at the Military School (a place where people were tortured during the dictatorship of General Augusto Pinochet) in November 2011, this group organized an alternative panel of discussion and criticism called *Geografía, movimientos sociales y territorio [Geography, social movements, and territory]*.

The formation of research and information-exchange networks could, from our point of view, be the way to address the relative isolation of Latin American geography. Although this geography has elements that unite it, Latin America stretches over such a large expanse that this impedes better interaction between countries, despite technological advances and the multiple communication possibilities presented by contemporary globalization.

Notes

1 Caldas was influenced by Alexander von Humboldt, and interacted with von Humboldt on his visit to what is now Colombia, Venezuela and Ecuador at the beginning of the nineteenth century.
2 Letter from Jamaica Kingston, September 6, 1815. Available online at http:www.analítica.com/bitbilio/bolivar/jamaica.asp
3 According to Kristin Ross (1988, 91) Reclus at one time wrote: "Geography is nothing more than history in a given space, geography is not unchanging. It is made and remade every day: in every instant it is modified by the actions of men and women". Without doubt, at that time, this was a notable example of critical thought.
4 These geographical societies were: the Historical and Geographical Institute of Río de la Plata (*Instituto Histórico y Geográfico del Río de la Plata*) created in 1843; The Argentine Geographical Institute (*Instituto Geográfico Argentino*) set up in 1879; the Argentine Geographic Society (*Sociedad Geográfica Argentina*) established in 1881; the Brazilian Historical and Geographical Institute (*Instituto Histórico Geográfico Brasilero*) created in 1822; and the Geographical Society of Río de Janeiro (*Sociedad Geográfica de Río de Janeiro*) established in 1883.

References

Acosta, A. and E. Martinez. 2010. *ITT-Yasuní: Entre el petróleo y la vida*. Quito: Abya-Yala.

Adriani, H.L. 2007. Transformaciones territoriales en los últimos veinte años. Una selección de aportes leídos desde la Geografía Económica. In, A. Camou, M.C. Torti & A. Viguera (eds.), *La Argentina democrática: los años y los libros*. Buenos Aires: Ed. Prometeo, pp. 201–224.

Alburquerque Bomfim, P. 2010. Fronteira amazónica e planejamento na época da ditadura militar: Inundar a hileia de civilização. *Boletim Goiano de Geografia* 30, 13–33.

Alessandri Carlos, A.F. and A.I. Geraiges de Lemos. 2003. *Dilemas urbanos. Novas abordagens sobre a cidade*. São Paulo: Ed. Contexto.

Aliste, E., F. Gallardo and M. Ibáñez. 2007. Fiestas religiosas de Chiloé: Patrimonio intangible y territorio. *Anales de la Sociedad Chilena de Ciencias Geográficas 2006*, 83–88.

Almeida, M.G. and A.J.P. Ratts (eds.). 2003. *Geografia. Leituras Culturais*. Goiânia: Editora Alternativa.

Alves, J.J.A. 2007. Uma leitura geográfica da fome com Josué de Castro. *Revista Norte Grande* 38, 5–20.

Arias, P.D. 2010. Mapeo autónomo y defensivo en la Zonal Pewence. *4tas Jornadas de Historia de la Patagonia (CD Rom)*. Santa Rosa, La Pampa: Universidad Nacional de la Pampa.

Arroyo, M. 2006. A vulnerabilidade dos territórios nacionais latino-americanos: O papel das finanças. In, A.I. Geraiges de Lemos, M.L. Silveira and M. Arroyo (eds.), *Questoes Territoriais na América Latina*. São Paulo: Clacso, pp. 177–190.

Barrera Lobatón, S. 2009. Reflexiones sobre Sistemas de Información Geográfica Participativos (SIGP) y cartografía. *Cuadernos de Geografía* 18, 9–23.

Bassols, A. 1967. *La división económica regional de México*. Mexico City: Universidad Nacional Autónoma de México, Instituto de Investigaciones Económicas.

Bassols, A. 2008. *No perdonar el olvido es hacer historia.* Ciudad de México: Self-published.

Bataillon, C. 1969. *Las regiones geográficas en México.* Mexico City: Siglo XXI.

Bataillon, C. 2008. *Un geógrafo francés en América Latina. Cuarenta años de recuerdos y reflexiones sobre México.* Mexico City: El Colegio de México, El Colegio de Michoacán, Centro de Estudios Mexicanos y Centroamericanos.

Berenberg, T. et al. 1988. Algunas consideraciones respecto de la situación actual en el Colegio de Geografía. *Posición* 6/7, 35–38.

Best, U. 2009. Critical Geography. In, R. Kitchin & N. Thrift (eds), *The International Encyclopedia of Human Geography* 2. Oxford: Elsevier, pp. 346–357.

Bonilla, O., P. Maldonado, M. Silveira and M. Bayan (Colectivo de Geografía Crítica del Ecuador). 2016. Nudos territoriales críticos en Ecuador. Dinámicas, Cambios y límites en la configuración territorial del Estado. *GeoGraphos, Revista Digital para Estudiantes de Geografía y Ciencias Sociales* 7(84), pp. 66–103. https://rua.ua.es/dspace/bitstream/10045/53465/1/Omar_Bonilla. pdf (13 August 2021)

Capellá Miternique, H. 2009. Por los caminos de la identidad y del desarrollo regional. *Atenea* 500, 75–90.

Caviedes, C. 1991. Contemporary geography in Chile: A story of developments and contradictions. *The Professional Geographer* 43, 359–362.

Ceballos García, B. 2001. Retos de la educación geográfica en la formación del ciudadano venezolano. *Educere* 5, 141–148. Mérida, Venezuela: Universidad de los Andes.

Chávez, M., O. González and M. del Carmen Ventura (eds.) 2009. *Geografía humana y ciencias sociales: Una relación reexaminada.* Zamora, Michoacán: El Colegio de Michoacán.

Chávez, M. and M. Checa (eds.) 2013. El Espacio en las Ciencias Sociales: Geografía, *Interdiciplinariedad y Compromiso.* Zamora, Michoacán: El Colegio de Michoacán.

Child, J. 1985. *Geopolitics and Conflict in South America: Quarrels among neighbors.* New York: Praeger.

Cicalese, G.G. 2007. Ortodoxia, ideología y compromiso político en la geografía argentina en la década de 1970. *Biblio 3 W* XII. http://www.ub.edu/geocrit/b3w-767. htm (October 2016).

Cicalese, G. 2008. La Geografía como oficio y magisterio: "Entonces, uno se queda con la satisfacción y dice: misión cumplida, la lección fue aprendida". Entrevista a la geógrafa Elena Margarita Chiozza. Notas, comentarios, recuadros y citas del entrevistador. *Geográficos*, 4. www.geograficos.com (October 2016).

Ciccollella, P. 2014. *Metrópolis latinoamericanas: más allá de la globalización.* Ciudad Autónoma de Buenos Aires: Café de las Ciudades.

Correia de Andrade, M. 1997. Josué de Castro: o homem, o cientista e seu tempo. *Estudos Avançados* 11(29), 169–194.

Cruz Báez, Á. 2007. El estado actual y perspectivas de la geografía en Puerto Rico. El Estado Actual de la Geografía en las Américas: Perspectivas sobre oportunidades de colaboración e investigación. *Panel organizado por la Asociación Americana de Geógrafos (AAG) y el Grupo de Especialidad Latinoamericanista, Reunión de la Asociación Americana de Geógrafos.* San Francisco, California. http://www.aag.org/ galleries/project-programs   files/Oportunidades_de_Colaboracion_Internacional. pdf (20 August 2021).

Delgado Mahecha, O. 2003. *Debates sobre el espacio en la geografía contemporánea*. Bogotá: Red de Estudios de Espacio y Territorio, Universidad Nacional de Colombia.

Diez Tetamanti, J.M. and B. Escudero. 2012. *Cartografía social: Investigaciones e intervención desde las ciencias sociales: métodos y experiencias de aplicación*. Comodoro Rivadavia: Universidad de la Patagonia. http://www.margen.org/Libro1.pdf (14 December 2020).

Dory, D. and D. Douzant Rosenfeld. 1995. Geografía y geógrafos en Bolivia y Cuba. Ensayo de sociología histórica comparativa. *Document's d'Anàlisi Geogràfica* 27, 57–73.

Escolar, M. 1996. *Crítica do discurso geográfico*. São Paulo: Hucitec.

Espinosa, M.A. 1997. *Región. De la teoría a la construcción social*. Ibagué: ATLAS.

Fernández Christlieb, F. 1988. La geografía que se necesita y el plan de estudios del Colegio de Geografía de la UNAM. *Posición 6/7*, 39–45.

George, P. 1969. *Geografía rural*. Barcelona: Ed. Ariel.

George, P. 1970a. *Panorama del mundo actual*. Barcelona: Ed. Ariel.

George, P. 1970b. *Geografía Económica*. Barcelona: Ed. Ariel.

George, P. 1974. *Sociología y Geografía*. Barcelona: Ediciones Península.

George, P. 1975. *Geografía Activa*. Barcelona: Ed. Ariel.

Haesbaert, R. 2004. *O mito da territorialização: Do "fim dos territórios" à multiterritorialidade*. Rio de Janeiro: Bertrand Brasil.

Haesbaert, R. 2014. *Viver no limite. Território e multi/transterritorialidade em tempos de insegurança e contenção*. Rio de Janeiro: Bertrand Brasil.

Harvey, D. 1973. *Social Justice and the City*. Oxford: Basil Blackwell.

Hepple, L. 1986. The revival of geopolitics. *Political Geography Quarterly* 5, S21–S36.

Herrera Ángel, M. 2002. *Ordenar para controlar. Ordenamiento espacial y control político en las llanuras del Caribe y en los Andes Centrales Neogranadinos. Siglo XVIII*. Bogotá: Instituto Colombiano de Antropología e Historia-Academia Colombiana de Historia.

Hevilla, C. and P. Zusman. 2009. Borders which unite and disunite: Mobilities and development of new territorialities on the Chile-Argentina frontier. *Journal of Borderland Studies* 24, 83–96.

Hidalgo, R., D. Santana, V. Alvarado, F. Arenas, A. Salazar, C. Valdebenito, L. Álvarez. 2016. *En las Costas del Neoliberalismo. Naturaleza, urbanización y producción inmobiliaria: experiencias en Chile y Argentina*. Santiago, Serie Geolibros.

Humboldt, A.v. 1848–1859. *Cosmos: A Sketch of a Physical Description of the Universe*. London: H.G. Bohn.

Humboldt, A.v. 1982. *Del Orinoco al Amazonas*. Caracas, Venezuela: Editorial labor, S.A.

Humboldt, A.v. 1991. *Viaje a las regiones equinocciales del Nuevo Continente. Segunda Edición*. Caracas, Venezuela: Monteavila Editores.

Keeding, Ekkehart. 2005. Surge la nación: la ilustración en la Audiencia de Quito (1725–1812). *Banco Central del Ecuador*. Quito.

Lacoste, Y. 1976. *La geografía: Un arma para la guerra*. Barcelona: Anagrama.

Lefebvre, H. 1973a. *El pensamiento marxista y la ciudad*. Mexico City: Editorial extemporáneos.

Lefebvre, H. 1973b. *De lo rural a lo urbano*. Barcelona: Ediciones Península.

Lefebvre, H. 1973c. *El derecho a la ciudad*. Barcelona: Ediciones Península.

Lefebvre, H. 1976. *Espacio y política*. Barcelona: Ediciones Península.

Lopez de Souza, M. 2010. Uma Geografia marginal e sua atualidade: A linhagem libertária. Opening speech at the first Coloquio Território Autônomo, Rio de Janeiro. http://territorioautonomo.wordpress.com (21 December 2020).

Lopez Gallero, A. 2005. O Uruguai em trânsito para uma transformação. In, M.L. Silveira (ed.), *Continente em Chamas. Globalização e território na América Latina*. Rio de Janeiro: Civilização Brasileira, pp. 117–143.

Martinez, E., V. Morales, C. Simbaña, J. Wilson, N. Fernández, T. Purcell and J. Rayner. 2015. Ni colonialistas ni simpáticos: una respuesta a Eduardo Gudynas. *La Línea De Fuego*. https//lalineadefuego.info/2015/10/13/ni-colonialistas-ni-simpaticos-una-respuesta-a-eduardo-gudynas (20 December 2020).

Massey, D. 2005. *For Space*. London: Sage.

McFarlane, A. 2014. *War and Independence in Spanish America*. New York: Routledge.

McKay, A.A. 2006. Cien años de geografía en Panamá (1903-2003). In, Grupo de Especialidad Latinoamericanista (ed.), El *estado actual de la Geografía en los países hispanoamericanos*, pp. 18–38. http://www.aag.org/galleries/project-programs-files/ El Estado Actual de la Geografia.pdf (20 October 2016).

Moncada, O. 1994. La geografía en México: Institucionalización académica y profesional. In, G. Aguilar and O. Moncada (eds.), *La Geografía Humana en México; institucionalización y desarrollo recientes*. Mexico City: Universidad Nacional Autónoma de México, Fondo de Cultura Económica, pp. 57–75.

Mondragón, J.M. 1988. Labor científica y social en las geografías hoy. *Posición* 6/7, pp. 30–34.

Montañez, G. 2000. Elementos de historiografía de la geografía colombiana. In, F. Leal Buitrago and G. Rey (eds.), *Discurso y razón. Una historia de las Ciencias Sociales en Colombia*. Santa Fe de Bogotá: Tercer Mundo Editores- Ediciones Uniandes, p. 53-82.

Montañez, G. 2005. Globalizações e construção do território colombiano. In: M.L. Silveira (ed.), *Continente em chamas*. Rio de Janeiro: Civilização Brasileira, pp. 85–116.

Montañez, G. 2009a. Geografía y Marxismo: lecturas y prácticas desde las obras de David Harvey, Neil Smith y Richard Peet. In, J.W. Montoya (ed.), *Lecturas en teoría de la geografía*. Bogotá: Universidad Nacional de Colombia, pp. 41–102.

Montañez, G. 2009b. Encuentros, desencuentros y reencuentros entre la geografía, las ciencias sociales y las humanidades. In, M. Chávez, O. González and M. del Carmen Ventura (eds.), *Geografía humana y ciencias sociales: Una relación reexaminada*. Zamora, Michoacán: El Colegio de Michoacán, pp. 33–72.

Moraes, A.C.R. 1991. *Bases da formação territorial do Brasil: O território colonial brasileiro no longo século XVI*. PhD thesis. São Paulo Universidade de São Pablo.

Moraes, A.C.R. and W.M. Da Costa. 1982. A geografia e o processo de valorização do espaço. In, M. Santos (ed.), *Novos rumos da geografia brasileira*. São Paulo: Hucitec, pp. 111–130.

Moreira, R. 1982. Repensando a geografia. In, M. Santos (ed.), *Novos rumos da geografia brasileira*. São Paulo: Hucitec, pp. 35–48.

Moreira, R. 1992. Assim se passaram dez anos. (A renovação da Geografia no Brasil 1978–1988). *Cadernos Prudentino* 14, 5–39.

Moura, R., D. de Olivera, H. dos Santos Lisboa, L. Martins Fontoura and J. Geraldi. 2008. Geografia Crítica: legado histórico o abordaje recurrente. *Biblio 3W* XIII. http://www.ub.es/geocrit/b3w-786.htm (14 December 2020).

Oslender, U. 2008. *Comunidades negras y espacio en el Pacífico colombiano. Hacia un giro geográfico en el estudio de los movimientos sociales.* Bogotá: Instituto Colombiano de Antropología e Historia.

Peet, R. 1969. A new left. *Antipode* 1(1), 3–5.

Peet, R. (ed.) 1977. *Radical Geography: Alternative Viewpoints on Contemporary Social Issues.* Chicago: Maaroufa Press.

Peña Reyes, L.B. 2008. Reflexiones sobre las concepciones de conflicto en la geografía humana. *Cuadernos de Geografía* 17, 89–115.

Perez, J, and A. Alberola (eds.). 1993. *España y América entre la Ilustración y el Liberalismo.* Colección de la Casa de Velázquez, Instituto de Cultura Juan Gil Albert. *Collección Seminarios, Serie mayor* 3. Alicante-Madrid: Instituto de Cultural Juan Gil Albert, Colección de la Casa de Velásquez.

Porto Gonçalves, C.W. 2001. *Geo-grafías. Movimientos sociales, nuevas territorialidades y sustentabilidad.* Mexico City: Siglo XXI.

Porto Gonçalves, C.W. 2006. A Reinvenção dos Territórios: a experiencia latino-amêricana e caribenha. In, A. Ceceña (ed.) *Los desafíos de las emancipaciones en un contexto militarizado.* Buenos Aires: Clacso, pp. 151–197.

Quintero, S., E. Dufour and V. Iut. 2009. *Los Encuentros de la Nueva Geografía y el surgimiento de la geografía crítica en Uruguay y Argentina durante los años '70.* Montevideo: 12 Encuentros de Geógrafos de América Latina. http://observatoriogeograficoamericalatina.org.mx/egal12/Teoriaymetodo/Geografiahistoricaehistoriadelageografia/02.pdf (13 August, 2021)

Radcliffe, S. 2011. Development for a postneoliberal era? Sumak kawsay, living well and the limits to decolonisation in Ecuador. *Geoforum* 43(2), 240–249.

Ramírez, B. 2007. Geographical practice in Mexico: The cultural geography project. In, R. Kitchin (ed.), *Mapping Worlds: International Perspectives on Social and Cultural Geographies.* London: Routledge, pp. 69–74.

Ramírez, B. 2010. Presentación y Mesa Redonda: Doreen Massey y las geometrías del poder, Universidad Autónoma Metropolitana-Azcapotzalco y Universidad Nacional Autónoma de México, Ciudad de México, 2, 3 y 4 de marzo 2010. *Investigaciones Geográficas* 72, 167–171.

Ramírez, B. and L. Levi. 2015. *Espacio, paisaje, región, territorio y lugar: la diversidad en el pensamiento contemporáneo.* México City: Instituto de Geografía, Universidad Autónoma Metropolitana-Xochimilco.

Ramírez, D.A. 2007. Eliseo Reclus y la Geografía de Colombia. Las Cartas de Reclus a Vergara y Velasco. https://reclus.wordpress.com/las-cartas-de -reclus-a-vergara-y-velasco/ (20 December 2020).

Ríos, D. 2010. Urbanización de áreas inundables, mediación técnica y riesgo de desastre: Una mirada crítica sobre sus relaciones. *Revista de Geografía Norte Grande* 47, 27–43.

Rizzo, P.A. 2010. El espacio público de la ciudad de Mendoza (Argentina), espacio de disputa y expresión ciudadana. *ACME* 9, 164–190.

Rodríguez, F.B. 2003. *De la huella ecológica al control territorial mediado por el abasto de alimentos a Bogotá, 1970–2003.* Bachelor thesis. Bogotá: Universidad Nacional de Colombia.

Ross, K. 1988. *The Emergence of Social Space: Rimbaud and the Paris Commune.* Minneapolis: University of Minnesota Press.

Sánchez Calderón, F.V. 2008. Elementos para una geopolítica de los megaproyectos de infraestructura en América Latina y Colombia. *Cuadernos de Geografía* 17, 7–21.

Santaella Yegres, R. 1980. *Región y localidad económica dependiente.* Caracas: Universidad Central de Venezuela.

Santaella Yegres, R. 1989. *La dinámica del espacio en la cuenca del lago de Maracaibo, 1873–1940 y su proyección hasta el presente (1980).* Caracas: Expediente Editorial José Martí, FACES UCV.

Santos, M. 1973. *Geografía y economía urbanas en los países subdesarrollados.* Barcelona: Oikos-Tau.

Santos, M. 1978/1996. *Por uma Geografia Nova.* São Paulo: Hucitec.

Silveira, M.L. 2016. *Circuitos de la economía urbana. Ensayos sobre Buenos Aires y San Pablo.* Buenos Aires. Café de las Ciudades.

Slater, D. 1978. The poverty of Modern Geographical Enquiry. In, R. Peet (ed.) *Radical Geography. Alternative Viewpoints on Contemporary Social Issues.* London: Methuen, pp. 40–50.

Tovar, R. 1986. *El enfoque geohistórico.* Caracas: Academia Nacional de Historia.

Trinca Fighera, D. 2005a. A Venezuela e os desafios territoriais do presente. In, M.L. Silveira (ed.), *Continente em chamas. Globalização e territorio na América Latina.* Rio de Janeiro: Civilização Brasileira, pp. 55–84.

Trinca Fighera, D. 2005b. Los desafíos de la globalización y la ocupación de la Amazonia. *X Encuentro de Geógrafos de América Latina.* São Paulo: Departamento de Geografía, Faculdade de Filosofia, Letras e Ciencias Humanas, Universidade de São Paulo.

Veleda da Silva, S. and D. Lan. 2007. Geography and gender studies: The situation in Brazil and Argentina. *Belgeo* 3, 371–382.

Vidal Koopman, S. 2005. La Ciudad Privada: Nuevos Actores, Nuevos Escenarios ¿Nuevas Políticas Urbanas? *Scripta Nova* 194. http://www.ub.edu/geocrit/sn/sn-194-15.htm (20 December 2020).

Zusman, P. 2002. Milton Santos. Su legado teórico y existencial. *Document's d'Anàlisi Geogràfica* 40, 205–219.

Zusman, P., H. Castro and M. Soto. 2007. Cultural and social geography in Argentina: Precedents and recent trends. *Social & Cultural Geography* 8, 775–798.

6 Critical geographies in Japan
A diverse history of critical inquiry

Koji Nakashima, Tamami Fukuda,
and Takeshi Haraguchi

Introduction

Since the Liberal Democratic Party of Japan made a comeback to power in the general election of 2012, the general situation of politics and economy in Japan has increasingly leaned right and conservative. Under neo-liberal economic policies, severe rationalization has led to forced dismissals and a rising number of temporary workers, swelling the ranks of "internet-café refugees" and homeless. Militaristic security policies continue to force Okinawa to bear the burden of the vast majority of US bases in Japan and furthermore to construct new bases in Okinawa despite Okinawan popular will of "No More Bases". In line with these disappointments, democratic reformists have gradually lost momentum while the voices of conservatives and nationalists have grown louder. To challenge these realities, contemporary critical geography in Japan needs to suggest radical ways of imagining and producing alternative realities. Focusing on recent developments in critical geography in Japan, this chapter considers how geography matters in contemporary Japanese society, while also searching for potential means of constructing alternative realities.

The history of critical geography in Japan has received much-needed attention in Mizuoka et al. (2005).[1] This path-breaking study traces the history of Japanese critical geography since the early twentieth century, suggesting that the field, as represented by the works of economic geographer Noboru Ueno (e.g. Ueno, 1972), peaked in the early 1970s. According to Mizuoka et al. (2005), the development of critical geography in Japan slowed from the late 1970s to the early 1980s, but was revived again from the middle of the 1980s. This chapter suggests an alternative reading of the history of critical geography in Japan. First, in contrast to the emphasis on the history of economic geography found in Mizuoka et al. (2005), we focus on multi-linear histories of social, cultural, and other related subfields. Second, we dedicate greater attention to more recent developments since the late 1980s, or what Mizuoka et al. (2005, 466) define as "a new era of critical geography" in Japan, spurred by critical engagement with urban space, cultural politics, neo-liberalism, and militarism and based on different theoretical

DOI: 10.4324/9781315600635-6

and ideological backgrounds. Third, we shed light on the academic and practical endeavors of individual geographers, since recent work has been advanced primarily by individuals not necessarily excluded, but often at distance, from formalized schools or groups. Fourth, we take careful notice of the theory-practice nexus in critical geographic research, particularly in research on the homeless, the feminist movement and geopolitics. By examining the theories and practices of Japanese critical geographies in heterogeneous areas, we depict how geography has engaged with the harsh realities of contemporary Japanese society.

A brief history of Japanese geography from a critical perspective

Before examining recent developments, it is helpful to set the stage by applying a critical perspective to the history of Japanese geography since the early modern era. Of particular interest are the early linkages between geography and anarchist thought, and the relationships between geography and the military during the wartime.

Anarchism and Japanese geography

In 2005, at the 4th International Conference of Critical Geography in Mexico City, a special session on "Anarchism and Geography" was held to commemorate the centenary of Elisée Reclus' death. Although the event marked the passage of a century, it is only since the late 1970s that anarchist thought has been incorporated into Anglophone geography (e.g. Dunbar, 1978; Fleming, 1979). In contrast, it was in a much earlier era that Japanese geographers discovered radical anarchist geographers like Elisée Reclus and Peter Kropotkin.

Manjiro Yamagami (1868–1946) was one such foresighted geographer. While Yamagami graduated from the Geology Department of the Imperial University of Tokyo in 1892, he had never obtained a leading position at any national university (Minamoto, 2003). He is generally known as the author of school textbooks rather than an established scholar (Nakagawa, 1978, 238–240). However, Yamagami was the first Japanese geographer to use the term "critical geography" (*hihan chirigaku*). In 1914, he published *Current Critical Geography* (*Saishin Hihan Chirigaku*), a book that introduced Elisée Reclus and explained geographical configurations of the earth by drawing from Reclus' (1868) *La Terre, Tome 1*. Although Yamagami used the word "critical geography" as a criticism of "old geography", he attached no political implications to the phrase. However, it is easy to see Reclus' philosophical influences on *Current Critical Geography*. In the last chapter, entitled "The Purpose of Critical Geography", Yamagami (1914, 395–399) raised nine tasks for the field: it should be (1) quantitative, (2) experimental, (3) respectful of historical research, (4) avoid dogmatism, (5) abandon middle-of-the-roadism, (6) abolish naïve taxonomies, (7) draw on and be philosophical, (8) logical, and (9) draw on experimental psychology. Lastly, quoting Reclus, Yamagami urged geographers "to give priority to liberal investigation, to eliminate prejudice, and to admit what you don't know as what you don't know" (Yamagami, 1914, 400).[2] Yamagami (1946) referred to Reclus again in "The past and present of Japanese geography", published in his last year. In this article, Yamagami also introduced Kropotkin as "a great geographer and a founder of communism" and declared great admiration for his achievements in geography (Yamagami, 1946, 7). As Minamoto (2003, 251) argues, it seems that Yamagami was sympathetic to the anarchist thought of Reclus and Kroptkin as well as their works of geography, although he himself never engaged in anarchist social activism.

Outside geography, Japanese anarchist and social activist Sanshiro Ishikawa (1876–1956) introduced Reclus' thought and life (Ishikawa, 1948), and translated *L'Homme et la Terre* (Reclus, 1905/1930) and "Evolution et Révolution" (Reclus, 1880/1930) into Japanese. Although Ishikawa was not a geographer, the fundamental idea of Ishikawa's anarchism – that mutual solidarity of individuals is produced in an anarchist world – has much in common with the principle of the world described in Reclus' *L'Homme et la Terre,* namely that the world consists of an organic ensemble including human beings (Nozawa, 1986, 2006, 2009). Unfortunately, the linkages forged between geography and anarchist thought in the first half of the twentieth century were not passed on to the next generation of Japanese geographers because of the involvement of Japanese geography in the total war regime. A few folklorist geographers such as Renkichi Odera and Hikoichiro Sasaki, however, inherited anarchist perspectives and conducted field surveys in rural villages in the 1930s to explore alternative social formations and mutual support systems (Mizuoka et al. 2005, 456).

Geography and the military

From the late 1930s, along with Japan's entry into the Asia-Pacific War, Japanese geographers were actively or passively enrolled into the total war regime of the Great Empire of Japan. This era is of interest not because geographers represented critical geographies, but because later critical geographers have made key contributions to the analysis of this period of the discipline – so much so that a critical perspective on the history of the discipline is at the core of critical geography in Japan. According to Takeuchi (1994, 196–200), there were three groups of geopoliticians during wartime: (1) *the Kyoto School* composed of geographers of the Imperial University of Kyoto with aspirations to reorder Japanese geopolitics (*Nihon Chiseigaku*) along the lines of traditional Tennoism[3] as a bulwark against the western world order; (2) *academics* including geographers and political scientists influenced by German Geopolitik who devoted themselves to scientific analysis and were indifferent to the actual politics of the war; and (3) *the Japanese Society for Geopolitics*, a group of geographers, journalists, military persons, and politicians that played a practical role in popularizing geographical knowledge of Asia and Pacific areas, and promoted geographical surveys in those areas.

Although the doctrines of each group have been the focus of sustained criticism (Fukushima, 1998; Okada, 2002; Shibata, 2006; Takeuchi, 1974, 1994, 2000b), the continuing confidentiality of key documents and the continuing silence of many of the individuals involved has limited our understanding of how geographers and the military were linked in practice. However, with the help of newly recovered documents, recent critical studies shed light on these linkages. Although it is generally known that the Kyoto school organized *Sogo Chiri Kenkyukai* (Society for Research of Comprehensive Geography, hereinafter SRCG) or simply *Yoshida no kai* (Group of Yoshida), an unofficial research group studying military geography under the aegis of the military authority of Japan, detailed information on the SRCG's activities was not uncovered until recently. Discovery of key documents of the SRCG and its members by critical geographers led to a revival of critical analysis of wartime geopolitics (Kobayashi et al., 2010; Mizuuchi, 2001). The documents include research reports, opinions, and letters belonging to members of the SRCG that were released by family after their deaths. Drawing on these newly found documents, Shibata (2007) and Kobayashi and Narumi (2008) bring to light the inner workings of the SRCG while critically examining the linkages between Japanese geographers and the military.

While the SRCG was mainly at work in Kyoto, The Research Committee of Military Geography (hereinafter RCMG, *Heiyo Chiri Chosa Kenkyukai*) played an active part in developing military geography in Tokyo. The RCMG was organized in early 1944 by Major Tadashi Watanabe, an officer of the General Staff Office. Even after the RCMG's dissolution, Watanabe kept the documents of the RCMG in his home and recently disclosed them to the public. Fortunately, those documents were published as "The General Staff Office and the Land Survey Department before and after the end of war: a collection of the documents possessed by Tadashi Watanabe" (Watanabe Tadashi-shi shozo siryo-shu henshu iin-kai, 2005). According to documents found in this collection, more than ten geographers of the Imperial University of Tokyo, Tokyo Bunrika University, and Tokyo Higher Normal School were engaged in organizing geographical information to prepare for the US troops' landing on the Japanese mainland. Clearly, Geography played a practical role in providing useful information for fulfilling tactics of the General Staff Office. Investigation of these newly found documents has only begun and we can expect further advances in the development of critical studies of Japanese geopolitics that may illuminate unknown relationships of complicity between geography and the military during wartime.

Post-war Marxian critical thought and social movements

Japan has a long history of Marxian critical thought that began in the early twentieth century, and continued to be deeply influential to both academics and social movements after the end of war. For example, Marxian theory played a leading role in the Miike dispute, a fierce labour dispute between the Mitsui-Miike coal-mining company and the labour union in Omuta City in the 1950s (especially in 1959–1960), which is generally known as an all-out confrontation between capital and labour all over the country. In that struggle, Marxian economist Itsuro Sakisaka (1897–1985), a professor of Kyushu University, organized private classes on Marx's *Capital* for the labourers of the Miike coal mine, instructing workers in Marxist thought in order to produce militants (Kojima, 2005). Certainly, Marxist thought played a supporting role in the Miike dispute. However, there remains a firm division of labour between intellectuals and manual labourers: Marxian critical thought has increasingly become more theoretically radical, while increasingly diverging from the reality of working class lives.

A rather different version of Marxian critical thought can be seen in the student movements of the 1960s. Japanese student movements have a long and diverse history stretching back to the early twentieth century. Although the relationship between the student movement and Marxian thought after the end of war is highly complex, it can be said that the latter provided a frame of reference for the student movements. Many leftist students protested the US-Japan military alliance and the Vietnam War by actively fighting against university authorities and the state. Particularly notable is the *Zenkyotou* movement[4] of 1968–1969 that spread all over the country, leading to numerous bloody clashes between students and riot police. However, although the issues and concerns vocalized by these students were crucial at that time, and remain highly significant even today, their movement did not succeed in altering society in any profound way (Oguma, 2009). Rather, as the movement intensified, it became divorced from the realities of everyday life and, in its final stages, a number of students turned to extremism, and their struggles eventually ended in violence and infighting. Concerning the relationship between geography and the student movement, Chiba et al. (2000, 287) note that Japanese geography in the late 1960s shared the "spirit of 1968" with Anglophone radical geography in the sense that it had become more conscious of its social relevancy. However, it is difficult to find works of geography that actually embodied this "spirit of 1968" in Japanese geography of the late 1960s. It was not until the late 1970s that Japanese geography attempted to actually become socially relevant (e.g. Takeuchi, 1976, 1979, 1980).

Mizuoka et al. (2005, 461–462) suggest that Japanese geography lost its critical orientation after the pinnacle of critical economic geography in the early 1970s. Although they ascribe the demise of this critical orientation mainly to organizational problems within economic geography in Japan, we can also point to a more theoretical explanation for its demise. The ideas of neo-Marxism (Althusser, Poulantzas, Gramsci) that flourished from the 1960s onwards among western Marxists were nearly simultaneously introduced into Japan. These works greatly influenced various fields of the humanities and social sciences in Japan, and led to exploration of new horizons in Marxian critical thought. Until recently, however, most Japanese geographers paid little attention to works of neo-Marxism. Only when Anglophone critical geographers became conscious of neo-Marxian theorists and post-Marxists, such as Laclau and Mouffe (1985), did Japanese geographers begin to examine these works from geographical perspectives (e.g. Nakashima, 1996).

Towards a new era of critical geography

During the period from the late 1980s to 2000s, attempts to organize Japanese critical geographers have been made. As described in Mizuoka et al. (2005), these attempts have manifested in two main lines of development: 1) a stream of research commissions of the Association of Japanese Geographers (AJG) on the theme of critical geography; and 2) a study group of geographical thought. The former is a series of formally organized research commissions of the AJG: "Theories and Tasks of Social Geography" (1991–1993), "Space and Society" (1994–1999), and "Critical Geography: Society, Economy and Space" (2000–2005). These research commissions have actively invited younger critical geographers to their research meetings, and organized several symposiums and conferences on critical geography (see Table 6.1). The latter study group of geographical thought has a longer history, initially forming in the late 1970s, and has maintained close relations with the Commission on the History of Geographical Thought (the present Commission on the History of Geography) of the International Geographical Union (IGU) through the intermediation of Keiichi Takeuchi (Hitotsubashi University, Komazawa University). While maintaining a focus on the history of geography and geographical thought, the group's approach gradually shifted towards critical geography after Toshio Mizuuchi (Osaka City University) became the group's organizer in the late 1990s. Another feature of this study group is the financial backing of "Grant-in-Aid for Scientific Research" from the Ministry of Education. With this financial support, the group has published several research reports, a journal (*Space, Society and Geographical Thought*) and readings in Anglophone critical geography (see Table 6.2). Members of these two groups of the AJG commissions and geographical thought study group partially overlap, and the two groups have promoted the development of critical geography in Japan in close cooperation. However, outside these broad organizational configurations, most Japanese critical geographers are not attached to a particular school or faction. They are, in most cases, individually tackling a heterogeneous mix of challenging issues from a variety of theoretical and practical approaches to critical geography.

Table 6.1 Symposiums and conferences on critical geography held in Japan

Date	Venue	Title	Conference name	Organizer
May 29–30, 1993	Tokyo	Society and Space	40th Annual Meeting of the Japan Association of Economic Geographers, 1993	The Japan Association of Economic Geographers
October 15, 1994	Nagoya	David Harvey's Theory of Spatial Configuration and Socio-Economic Geography in Japan	The General Meeting of the Association of Japanese Geographers, Autumn 1994	The Association of Japanese Geographers
September 27, 2002	Kanazawa	Critical Geography: Towards Further Contribution to the International Community of Critical Geographers	The General Meeting of the Association of Japanese Geographers, Autumn 2002	The Association of Japanese Geographers
August 5-9, 2003	Tokyo/ Osaka	Searching for alternative globalization from below	The 3rd Conference of the East Asian Regional Conference in Alternative Geography (EARCAG)	The EARCAG steering committee

Since the 2000s, some Japanese critical geographers have tackled problems in contemporary society. Yamazaki (2006) critically examined the US military presence in East Asia and the orientation of Japan's security policy from the viewpoint of critical geopolitics. Recently, Yamazaki (2011, 2014) disclosed the process of militarization of a "host" civilian society in post-war Okinawa, and local politics of the anti-US riot and place-based identities in the military base town Koza. Contemporary anti-war peace movements in Japan are faced with difficult situations under the increasing imperative of state security and militarism in East Asia. As Yamazaki (2006) notes, a new strategy for these movements will need to be created in order to mobilize the public against increasing militarism. Nakashima (2003, 2014) tackles such a task by examining how the grassroots antiwar movement in Japan

Table 6.2 Publications related to the research projects on critical geography

Year	Title	Editor	Publisher
1996	*Social Theory and Geographical Thought*	Hideki Nozawa	Kyushu University
1996– present	*Space, Society, and Geographical Thought (Annual Journal) (in Japanese)*	Toshio Mizuuchi	Osaka City University / Kyushu University / Wakayama University
1996	*Horizon in Socio-Spatial Studies: Reading Neo-Classics in Human Geography (J)*	The AJG research commission "Space and Society"	Osaka City University
1999	*Nation, Region and the Politics of Geography in East Asia*	Toshio Mizuuchi	Osaka City University
2003	*Representing Local Places and Raising Voices from Below*	Toshio Mizuuchi	Osaka City University
2006	*Critical and Radical Geographies of the Social, the Spatial and the Political*	Toshio Mizuuchi	Osaka City University

attempted to redefine the manoeuvre field of the Ground Self Defense Force of Japan as people's living environment using the concept of an "alternative production of nature" (Smith, 1998).

One of the most distinctive achievements among relatively recent critical geographies in Japan is *Tatakau Chirigku [Active Geography]* written by Japanese physical geographer Yugo Ono (2013), Professor of Hokkaido University, who has long been engaged in environmental movements against the Chitose River Water Diversion Project, Final Disposal Site for High-Level Radioactive Waste Construction Plan and many other development projects from the viewpoint of an environmental scientist and geographer. Based on his own experiences of going to court as a plaintiff, holding open forums and press conferences, and making objections in deliberative council, Ono stresses the role of geographer as an activist to intervene into and change actual politics over environmental issues. Although Ono is a physical geographer, he is also actively engaged in advocacy of Ainu indigenous people of Japan and attempts to rewrite the history and geography of "Japan" from the postcolonial view (Ono, 2013, 349–378). Ishiyama (2004, 2007) also critically examined the postcolonial history and geography of the United States' nuclear waste policy from the viewpoint of environmental justice, and disclosed its (post)colonial processes of land exploitation from indigenous people.

Other than works noted above, individual critical geographers have tended to tackle specific problems in each field. For example, political geographer Kitagawa (2007) critically examined the geopolitical and biopolitical implications of camps for "illegal immigrants" in Italy, inspired by Michel Foucault and Giorgio Agamben. In the field of cultural geography, Mori (2009) actively introduced cutting edge theories of recent Anglophone cultural geographies to Japan, and critically examined war advertisements in Japan during the Second World War from the viewpoint of "visuality" and "materiality" to reconsider the recognition of war in the present era (Mori, 2016). The following sections consider some of the challenging issues that Japanese critical geographers continue to confront, particularly in the areas of cultural geography and urban social geography.

Cultural geography

Critical trends in the geographies of culture since the Second World War

One of the strongest trends in Japanese geography in the late twentieth century has been the rise of cultural geography. Books concerning cultural geography (Nakagawa et al., 2006; Oshima et al., 1989; Takahashi et al., 1995) and annual reviews in the *Japanese Journal of Human Geography* show that the number of works related to cultural geography has rapidly increased from the late 1980s, and that research themes have diversified considerably. However, it would not be particularly helpful to merge diverse works as a singular trend of cultural geography and conclude that any specific group of studies is necessarily "critical", because these studies of culture developed under varied and changing social conditions and were conducted from a variety of perspectives.

Most studies conducted immediately after the Second World War used research frameworks established in earlier decades. It was in the mid-1910s that the concept of culture (translated as *"bunka"*) was introduced into geography in Japan. In his voluminous study of the history of cultural geography in Germany, the United States, and Japan, Hisatake (2000) separates works conducted in Japan during the pre-war period into two broad categories: (1) imperialism and ethnological studies, and (2) studies of folklore or local culture. In his explanation of the latter (Mizuoka et al., 2005), Hisatake suggests that research by folklorists was highly critical of the explanatory frameworks in use in geography at that time. After becoming aware of the detailed attention to local specificity provided by folklorists, geographers began to propose a number of new perspectives against the predominant trends influenced by German *Geopolitik*.

Post-war geographical studies related to cultural issues partly followed the pre-war studies. They applied the ethnological methodology but did not maintain the critical perspective by folklorists. They emphasized the importance of the production of geographical knowledge through field-work, and of combining this with an awareness of ethnology (Kawakita, 1956; Sasaki, 1970). Other studies, such as that of Chiba (1956), used an approach combining cultural geography with folklore studies focussed on local geographical knowledge. The authors of all of these studies tended to follow ideas about culture in the context of environmental relations as outlined in *Readings in Cultural Geography* (Wagner and Mikesell, 1962), a book that was extremely influential among Japanese cultural geographers. The idea of culture gained emphasis both in society and in the humanities and social sciences after the post-war process of democratization in Japan (Hisatake, 2000). A struggle began to preserve local forms of culture in this period of rapid social and economic change, along with a search for features of Japanese culture that could help create a new national identity. The con-tributions of cultural geography in this period were tied to this framework and would not be considered critical geography in any contemporary sense. Rather, they can be seen as playing an important role in fixing specific cul-tures in relation to particular spaces.

Since 1980, the nature of studies of culture has been dramatically trans-formed under the influence of new theoretical perspectives.[5] Studies of human-environment relationships and ways of life stressed subjectivity or inter-subjectivity, and a philosophy of humanistic geography was intro-duced into Japan from Anglophone geography (Takeuchi, 1979; Yamano, 1979). The introduction of humanistic geography encouraged Japanese cultural geographers to study the world from subjective viewpoints (e.g. Chiba, 1983). Cultural geography in this period played an important role in challenging the dominant frameworks of geography. Several geogra-phers attempted to focus on people's subjective environmental percep-tions and ways of life (Asano, 1984; Matsumoto, 1977, 1991; Sekido, 1989). Others concentrated on the symbolism of rural spaces and indigenous people's sense of place (Oshiro, 1990; Yagi, 1988). Geographers concerned with both environmental perception and symbolism shared similar per-spectives with humanistic geographers. They all wanted to protest against the predominant trend in geography of relying on nomothetic and quan-titative approaches, and instead aimed to illuminate the lived, experien-tial world largely neglected in Japanese geography. Although they never formed a fixed group, they proposed alternative perspectives whereby emphasis could be placed on local, as opposed to universal, geograph-ical knowledge. At the same time, a number of geographers, such as Naruse (1993, 1994), began to be influenced by British cultural studies and gradually shifted their focus from folk to popular culture. These studies introduced and stressed the essence of cultural studies, with its prevail-ing sense of resistance against existing hegemonies, but were not in them-selves necessarily politically motivated. However, the approach of seeing culture as inseparably related to society laid the foundation for the future development of critical cultural geography. It also provided an important perspective on the changing cultural conditions in Japanese society at the end of the twentieth century.

Japanese culture and society changed greatly in the late twentieth century. Since the late 1970s, local governments began establishing museums to represent local history and culture. This allowed for the recording and preservation of local specialties and cultures at a time of widespread social change. Such work was carried out within the framework of national and regional policies. However, as Fukuda (2005) points out, current cultural policies are not identical to earlier cultural conservation. This can be partly observed in the shift from the Basic Law of Agriculture to the Basic Law on Food, Agriculture, and Rural Areas, wherein local cultures have come to be more highly valued as active instruments for the rural economy. Thus, culture was to some extent redefined from the point of economic value, and it became a kind of commodity with additional value. As Harvey (1990, 299) noted, "we can at least begin upon the task of unraveling its [culture's] complexities under present-day conditions by recognizing that money and commodities are themselves the primary bearer of cultural codes". Such cultural commoditization has become widespread across Japanese society.

Under such social conditions, studies in cultural geography flowered, but their methods took them in many different directions. One approach regarded culture as a local resource and discussed ways in which local cultures could contribute to regional revitalization and development. This approach considers culture to be natural and taken-for-granted and supports the empowerment of people through the "objectification of culture" (Ota, 1993). However, this may conceal diversity and difference within regions or descend into essentialism (Oshiro, 1998). Another approach was to carefully discuss the socio-economic and political ways in which culture was conceived, represented, and distributed. With close connections to "new" cultural geography and the development of social theory, studies of culture flourished in Japanese geography. Research topics included not only local and ethnic cultures but also youth and street cultures (Sugiyama, 1999). The works that made a detailed examination of "culture" generally did not assert alternative positions or commit themselves directly to specific social movements. However, their elucidation of culture and the relevance of specific practices and institutions to social and political relations enabled a reexamination of the idea of culture with reference to its influence on the history of geography, as well as an examination of how culture has been utilized as a "natural" resource in modern Japanese society.

Examples of these trends in critical cultural geography include the practice of "rereading" a particular region through local people's often previously neglected memory and culture. This was proposed by Urban Research Plaza, Osaka City University, in 2008 (NPO COCOROOM, 2009), and although geographers in Japan have not often carried out such work, it is likely to occur in the future. Another example is a practical project reexamining the ideas of culture and local areas ("Kyodo" kenkyu-kai, 2003).

This was proposed and outlined in essays in the book *Kyodo* (Homeland). The project neither presupposed nor actually attempted to realize the idea of *kyodo*, but instead critically explored the process by which *kyodo* came to be taken for granted and embodied (also see Fukuda, 2014). More recently, Mori (2014) considered the production of aesthetic landscape experience through geographical theories of materiality and visually. Thus reexamination of the idea of culture could prove a very useful theme for critical geography in the future in order to avoid the essentialist understanding of culture at any scale. Influenced by Doreen Massey's influential book *For Space*, which was translated into Japanese in 2014 (Massey, 2014), reconsideration of culture with the interlinked trajectories started recently among Japanese geographers.

Gender issues in the culture of geography: Academic discipline, geographers, and their works

Cultural issues are, of course, found also in academic geography itself. Within this discipline, geographers share and inherit, or alter and oppose, the existing academic culture through their research and teaching. In recent years, some geographers have begun to reexamine the history of geography, and to focus critically on its academic and professional practices (e.g. Izumitani, 2008). Such inquiries could potentially induce other geographers to adopt critical thinking about their own positions as geographers. In terms of positionality, it is essential to consider the relations between feminism and geography. The masculinity of geographical knowledge has been a focal point of feminist geography, which was introduced to Japanese geographers through translations of books such as Gillian Rose's *Feminism and Geography* (Rose, 2001) and a review of the literature by Yoshida (1996). Several case studies were conducted from this viewpoint, focusing upon Japanese society. Yoshida (2007) paid close attention to the issue of women and labour, Kageyama (2004) explored "dwelling" and urban space from the perspective of gender, and Kimura (2008) illuminated housewives' entrepreneurial process in the suburbs. They all disclosed aspects of women's lives that had been disregarded for a long time in Japanese geography. The *Gender Atlas of Japan* (Takeda and Kinoshita, 2007) highlighted the social-geographical positions of Japanese women, for example, and has proven instructive in teaching geography.

However, in general, effective discussion about gender issues remains limited in Japanese geography. This may be partly because of the culture of geography in Japan. From 1989 to 1992, the monthly journal *Chiri* (*Geography*) ran a serial in 39 parts, titled "ethnicity and gender", but essays on gender made up less than one-third of the total. This suggests that there were few women in Japanese geography during this period (Ota, 1990). It is characteristic of critical geography in Japan that while several geographers carried out noteworthy work on power relationships and the politics of difference, gender issues were still relatively neglected. We can find a series of pioneering works by Ishizuka (e.g. 1976, 1997) that discuss issues of gender and race in the global system based on fieldwork in the Caribbean. However, the geography of gender has really only developed in Japan since the end of the twentieth century. The development of feminist geography in Japan was not directly influenced by feminism during its period as an earth-shaking social movement in the second half of the century. However, women's studies became institutionalized in Japanese academia, feminist geography was "imported" into Japan in the 1990s, and consequently, at the turn of the century feminism became implanted in the culture of geography.

Reexaminations of the culture of geography, in particular ways of producing and reproducing geographical knowledge through case studies focusing on Japanese society, has induced male geographers to reconsider their positionality. In his research on the Kamagasaki district in Osaka City, for example, a male activist social geographer Niwa (1992a) asked questions related to issues of gender and sexuality and reflected on his own positionality as a male geographer. In recent years, male geographers concerned with gender issues have also begun to publish works that pay close attention to men's positionality more generally (e.g. Kumagai, 2010; Murata, 2009). Can these works lead to a reexamination of the culture and institution of geography and positively use alternative thinking, as Niwa earlier proposed? Unfortunately, they are exceptional and do not appear to have a direct effect on changing the culture of geography as a discipline. In Japan, feminist geography is still marginal. For years, a conservative backlash against feminism has been taking place in Japanese society. Is it possible that geography will welcome alternative thinking and take a new turn by moving away from the prevailing conservative trends in Japanese society? Attitudes toward feminism will be a touchstone of Japanese geographers' ability to move forward positively in the future.

Urban social geographies

A relatively new critical trend in the field of urban social geography began in the 1990s. Three major factors can be noted as the background to this. First, the popularity of urban study increased during the 1980s and opened a new area in urban studies traversing the fields of geography, sociology, and architecture. Secondly, there was a shift in cultural theory in the 1990s. In Japan, the symposium "Dialogue with Cultural Studies" was held in 1996 with Stuart Hall as one of the guest speakers, leading to an upsurge in cultural studies. Thirdly, young geographers and sociologists were quick to take up Western-based critical geography and apply it in domestic academia.

This new critical orientation of urban social geography came to fruition in the second half of the 1990s. In 1997, the journal *10+1 [ten plus one]*, a leader in urban studies at the time, had a feature entitled "New Geography", where urban sociologists and urban geographers held exchanges of dialogue. With the publication of *From Space to Place* in 1998 (Arayama and Oshiro, 1998), young researchers presented research perspectives for critical geography, and the following year Kato (1999), from a critical perspective, introduced postmodern geography to urban studies. By critically questioning concepts of space, critical urban studies attempted a departure from the spatiality of Cartesian thought, which existing urban geography implicitly relied on. By stressing place, it spotlighted the diverse relationships among humans, society, and spaces, and it tried to decode the power relationships mediating between them. This consciousness was further developed theoretically when *The Production of Space* by Henri Lefebvre was translated into Japanese.

What were the concrete subjects of this research? As the title of Masahiro Kato's "Slums and Downtown in Osaka" (2002) indicates, urban social geography focussed on downtown areas and (inner-city) slums. The most influential study on downtown areas is *Dramaturgy of the City* by Shunya Yoshimi (1987). Kato (2002) actively integrated Yoshimi's framework into his own studies in urban social geography. Other important works of urban social geography also focusing on downtown areas include the research by Kazuaki Sugiyama and Susumu Yamaguchi, both of whom chose youth and street performers in downtown areas, while also sharing two key research interests. Firstly, they sought to uncover representational space of youths and street performers through interviews (Sugiyama, 2005; Yamaguchi, 2002). Secondly, both attempted from a critical perspective to compre- hend the power that surveys, controls, and excludes these young people (Sugiyama, 2005; Yamaguchi, 2008). These themes are based on the previ- ously mentioned re-examination of space/place. In particular, the critical reading of power is heavily influenced by a discourse on surveillance soci- ety discussed in books such as *On Liberty* by Takashi Sakai (2001) and the Japanese translation of Mike Davis' *City of Quartz*. In addition to works on downtown areas, the inner city has become another stage for urban social geography. One of the most influential essays was Mizuuchi (1994), which presented some perspectives to engage with urban question. Kato's (2002) research, referred to earlier, positions itself in compliance with Mizuuchi. Further, *Genealogy of the Modern City* by Toshio Mizuuchi, Masahiro Kato, and Naoki Oshiro (2008) is a compilation of these comprehensive studies in urban geography.

Urban social geography also concerned itself with homelessness, which during the 1990s became a grave social issue. Niwa's (1992b) research, focusing on Kamagasaki – Osaka's impoverished area – was the first of its kind to attempt an investigation of the social spaces of homelessness. Using the critical-geographical re-examination of space/place, Haraguchi (2003) investigates the geographical logics that generate homelessness. This questioning cannot be separated from the expansion of social movements involving homelessness. Since the 1980s, researchers have strongly engaged in radical social activism on the issue of homelessness. Their studies have considered the emancipation of homeless and casual workers as a practical and achievable challenge, and they have criticized the existing social system while aggressively scrutinizing themselves, not least how the "neutral" activities of researchers contribute to the reproduction of social structures. Within this trend, Niwa (1996) challenged existing geography for its objectification of social realities as a mere scientific research subject, and declared that he considered "the act of research itself an event that interrelates with a research subject" (Niwa, 1996, 10). In recent years, some researchers have been responding to this challenge. On the one hand, researchers such as Sasaki and Mizuuchi (2009) are trying to construct a geography that can actively participate in urban policies under public-private partnership, especially collaborating with Non Profit Organizations. On the other hand, researchers such as Haraguchi (2016) are searching for more radical geographies that can collaborate with global anti-capitalism struggles. The Japanese translation of Neil Smith's *The New Urban Frontier* is now accelerating the tension between these two research trends. The tension surely will become more intense especially over Tokyo Olympic Games, rescheduled for 2021.

Conclusion

As suggested in this chapter, the history of critical geography in Japan is highly complex. It cannot be adequately conveyed by a single linear history of critical geography, nor can it be easily parsed into the history of specific schools or factions of critical geographers. Particularly since the late 1970s, Japanese critical geography has increasingly diversified, individualized, and overlapped with analogous fields in the humanities and social sciences. These shifts are not simply the result of the diversity of research being conducted, but also a response to the increased complexities and heterogeneity of the hegemony confronting critical geography.

Global capitalism mobilizes every aspect of social life – from gender relationships to sexuality, cultural identity, biological resources, and relations of production – to further accumulation. Meanwhile, contemporary "empire" (Hardt and Negri, 2000) ensures its politico-economic hegemony by expanding global networks of production, consumption, and distribution. It sometimes controls existing sovereignty of the modern state, and sometimes utilizes it. As Laclau and Mouffe (1985) note, the hegemony depends on the contingency and articulative practices of society. It is not reducible to any specific instance of politics or economy, but constructed through both ideal and material practices of articulation between different elements of society. In order to confront the complicated and heterogeneous forms of hegemony of the present juncture, critical geography must be correspondingly diverse and adaptable to each individual front line.

In 1950s' and 1960s' Japan, leftist movements provided resistance against the relatively monolithic targets of monopoly capital and state power.

However, in the age of postmodernity and globalization, targets for resistance are diverse and dispersed. While global capitalism and "empire" establish their hegemony through global networks, leftists are fractionalized across the fragmented spaces of street, workplace, home, and cyberspace. In these fragmented spaces of struggle, solidarity and mutual understanding between different movements is limited. For precisely this reason, critical geography in Japan must engage the current hard realities of everyday life, while suggesting viable alternatives as well as criticizing the hegemony of global capitalism and "empire". Such tasks can only be accomplished by imagining and producing alternative spaces for solidarity and mutual understanding that can connect struggles in Japan with diverse battlefields all over the world.

Acknowledgement

We would like to thank Jay Bolthouse and Sakiko Sugawa for their reading and fruitful comments on an earlier draft.

Notes

1 An excellent account of the total history of geography and geographical thought in modern Japan is provided by Keiichi Takeuchi (2000a). Based on critical examination of a vast amount of literature by Japanese geographers, Takeuchi describes the scope and perspective of geography and geographical thought in modern Japan.
2 Yamagami did not indicate the source of his quotation from Reclus. This last phrase, "to admit what you don't know as what you don't know", is from the *Analects of Confucius.*
3 *Tenno* is the Emperor of Japan and Tennoism is the theory and thought of Emperor-worship based on Shintoism, in which *Tenno* is considered to be a descendant of an ancient mythical god.
4 *Zenkyotou* is an abbreviation of *Zengaku Kyodo Touso Kaigi* (the All-Campus Joint Struggle Committee), which was mainly organized by the non-sectional students of each university, and blockaded the university with barricades under the slogan of "demolition of university" and "self-negation".
5 Shimazu et al. (2012) review the history of geographical thought and cultural/ social geography since the 1970s.

References

Arayama, M. and N. Oshiro (eds.). 1998. *Kukan kara basho e: Chirigakuteki- souzouryoku no tankyu* [*From Space to Place: Pursuing the Geographical Imaginations*]. Tokyo: Kokon Shoin. (in Japanese).
Asano, H. 1984. Tokyo-to Miyake-jima ni okeru chikei wo shu to shita minzoku bunrui taikei [The study of geographical folk-taxonomy in Miyake Island, the Izu Islands]. *Chirigaku Hyoron* [*Geographical Review of Japan, Ser. A*] 57, 519–36. (in Japanese).
Chiba, T. 1956. *Hagemaya no kenkyu* [*A Study of Bare Hills*]. Tokyo: Nourin Kyokai. (in Japanese).

Chiba, T. 1983. *Shin chimei no kenkyu* [*A New Study of Place Names*]. Tokyo: Kokon Shoin. (in Japanese).

Chiba, T., T. Hisatake and H. Nozawa. 2000. Nihon no chirigaku ni totte no 1960 nendai [Innovations in the discipline of geography in the last years of the 1960s]. *Chirigaku Hyoron* [*Geographical Review of Japan, Ser. A*] 73, 285–7. (in Japanese).

Dunbar, G. 1978. *Elisée Reclus: Historian of Nature*. Hamden, CT: Archon Books.

Fleming, M. 1979. *The Anarchist Way to Socialism: Elisée Reclus and Nineteenth-Century European Anarchism*. London: Croom Helm.

Fukuda, T. 2005. Theorizing local culture: cultural turns in contemporary Japanese society and current studies on local culture. *Jimbun Chiri* [*Japanese Journal of Human Geography*] 57, 571–84.

Fukuda, T. 2014. Between two homes: Gentaro Tanahashi and his thoughts and practices concerning *kyodo* (homeland) and *katei* (family home). In, T. Shimazu (ed.), *Languages, Materiality, and the Construction of Geographical Modernities: Japanese Contributions to the History of Geographical Thought (10)*. Wakayama: Wakayama University, pp.71–86.

Fukushima, Y. 1998. Japanese geopolitics and its background: What is the real legacy of the past? *Political Geography* 16, 407–21.

Haraguchi, T. 2003. Yoseba no seisan katei ni okeru basyo no kochiku to seido-teki jissen: Osaka, Kamagasaki wo jirei to shite [Construction of place and institutional practice in the process of the 'production' of yoseba: the case of Kamagasaki, Osaka City]. *Jimbun Chiri* [*Japanese Journal of Human Geography*] 55, 121–43. (in Japanese).

Haraguchi, T. 2016. *Sakebi no Toshi: yoseba, Kamagasaki, ryudo-teki kaso-rodosya* [*Cry of the City: Yoseba, Kamagasaki, Fluid Lower-Class Workers*]. Kyoto: Rakuhoku Shuppan. (in Japanese).

Hardt, M. and A. Negri. 2000. *Empire*. Cambridge, Mass.: Harvard University Press.

Harvey, D. 1990. *The Condition of Postmodernity: An Enquiry into the Origins of Cultural Change*. Oxford: Blackwell.

Hisatake, T. 2000. *Bunka-chirigaku no keifu* [*Genealogies of Cultural Geography*]. Kyoto: Chijin Shobo. (in Japanese).

Ishikawa, S. 1948. *Erize Rukuryu: shiso to shogai* [*Elisée Reclus: His Thought and Life*]. Kyoto: Kokumin Kagaku-sha. (in Japanese).

Ishiyama, N. 2004. *Beikoku senjuminzoku to kaku haikibutsu: kankyo seigi wo meguru tousou* [*Native Americans and Nuclear Waste: Struggle for Environmental Justice*] Tokyo: Akashi Shoten. (in Japanese).

Ishiyama, N. 2007. Yakka Maunten keikaku to kankyo seigi: shokuminchi-syugi no chiri kukan [The Yucca Mountain project and environmental justice: the geography of colonialism]. *Meiji Daigaku Kyouyo Ronshu* [*The Bulletin of Arts and Sciences, Meiji University*] 422, 53–79. (in Japanese).

Ishizuka, M. 1976. Maruchiniku-to ni okeru kureoru shakai [Creole society in Martinique]. *Minzokugaku Kenkyu* [*The Japanese Journal of Ethnology*] 41, 155–68. (in Japanese).

Ishizuka, M. 1997. Kureoru no saikaku aruiwa henka jizai kukan no shiso [Talent of Creole or thought of the freely transmutable space). *Gendai Shiso* [*Revue de la pensée d'aujourd'hui*] 25(1), 190–9. (in Japanese).

Izumitani, Y. 2008. Basho wo meguru shiteki kankaku ni tsuite: chirigaku no kaibyaku, aruiwa katarienu mono no chirigaku [On private senses of place: the beginning of geography or geography of the inexpressible]. *Kukan, Shakai, Chirisiso* [*Space, Society and Geographical Thought*] 12, 35–49. (in Japanese).

Kageyama, H. 2004. *Toshi-kukan to jenda* [*Urban Space and Gender*]. Tokyo: Kokon Shoin. (in Japanese).

Kato, M. 1999. Postmodern chiri-gaku to modernism-teki 'toshi-e no manazashi': Harvey to Soja no hihan-teki kento wo toshite [Postmodern geography and the critique of the modernist gaze as a 'voyeuristic' way of seeing the city]. *Jimbun Chiri* [*Japanese Journal of Human Geography*] 51, 164–82. (in Japanese).

Kato, M. 2002. *Osaka no suramu to sakariba: kindai-toshi to basyo no keifugaku* [*Slums and Downtown in Osaka: Genealogy of the Modern City and Place*]. Osaka: Sogensha. (in Japanese).

Kawakita, J. 1956. Bunka no chirigaku, moshikuwa bunka no seitaigaku: Chibetto bunka no baai [Geography of culture, or ecology of culture: the case of Tibetan culture]. *Jinbun Kenkyu* [*Studies in the Humanities*] (Osaka City University) 7, 54–69. (in Japanese).

Kimura, O. 2008. Toshi kogai ni okeru jichitai no autososhingu to shufu no kigyo: Tama nyu taun Minami-osawa chiku S-sha wo jirei ni shite [Local government's outsourcing of administrative services and emerging entrepreneurship by housewives: a case study of Minami-Osawa district in Tama New Town, Tokyo]. *Jimbun Chiri* [*Japanese Journal of Human Geography*] 60, 301–22. (in Japanese).

Kitagawa, S. 2007. Gendai no chiseigaku ni okeru reigai kukan toshiteno shuyojo: Itaria no fuhoimin shuyojo e 'kantai' suru seikenryoku [The camp as a space of exception in contemporary geopolitics: biopower of 'hospitality' in camps for 'illegal immigrants' in Italy]. *Jimbun Chiri* [*Japanese Journal of Human Geography*] 59, 111–29. (in Japanese).

Kobayashi, S. and K. Narumi. 2008. Sogo chiri kenkyukai to Kosenkai [The relation between a geopolitical school and the military in Japan, 1939-1942: a critical view on the role of geographers]. *Rekishi Chirigaku* [*The Historical Geography*] 50(4), 30–47. (in Japanese).

Kobayashi, S., K. Narumi and A. Namie. 2010. *Nihon chiseigaku no soshiki to katsudo: Sogo chiri kenkyukai to Kosenkai* [*Organization and Activities of Japanese Geopolitics: Sogo chiri kenkyukai and kosenkai*]. Osaka: Institute of Human Geography, Graduate School of Letters, Osaka University. (in Japanese).

Kojima, T. 2005. *Sakisaka Itsuro, sono hito to siso* [*Itsuro Sakisaka, His Life and Thought*]. Tokyo: Erumu Shobo. (in Japanese).

Kumagai, K. 2010. Rokaru sensitibu na jenda to kaihatsu to dansei: watashi no jenda-ron [Renovating gender and development with men and local sensitivity: my trial for gender studies]. *Ochanomizu Chiri* [*Annals of Ochanomizu Geographical Society*] 50, 27–47. (in Japanese).

"Kyodo" kenkyu-kai (ed.). 2003. *Kyodo: Hyosho to jissen* [*Homeland: Representations and Practices*]. Kyoto: Sagano Shoin. (in Japanese).

Laclau, E. and C. Mouffe. 1985. *Hegemony and Socialist Strategy: Towards Radical Democratic Politics*. London and New York: Verso.

Massey, D. (translated by Mori, M.) 2014. *Kukan no tame ni* [*For Space*]. Kyoto: Getsuyo-sha. (in Japanese).

Matsumoto, H. 1977. Toresu kaikyo shoto no gyoro bunka [Fishing culture of the Torres Strait Islands]. *Minzokugaku Kenkyu* [*The Japanese Journal of Ethnology*] 41, 368–89. (in Japanese).

Matsumoto, H. 1991. Kaze no minzoku-shi, aruiwa, kaze no minzoku-shi: Toresu kaikyo shotomin no mou hitotsu no shizen [Ethnography of the winds: a Torres Strait islander view of the natural environment]. *Kokuritsu Minzokugaku Hakubutsukan Kenkyu Hokoku* [*Bulletin of the National Museum of Ethnology*] 15 (special issue), 193–235. (in Japanese).

Minamoto, S. 2003. *Kindai Nihon ni okeru chirigaku no ichi choryu* [*Beyond the Academe: Another Lineage of Modern Japanese Geography*]. Tokyo: Gakubunsya. (in Japanese).

Mizuoka, F., T. Mizuuchi, T. Hisatake, K. Tsutsumi and T. Fujita. 2005. The critical heritage of Japanese geography: Its tortured trajectory for eight decades. *Environment and Planning D: Society and Space* 23, 453–73.

Mizuuchi, T. 1994. [Kindai toshi-shi kenkyu to chiri-gaku [Geography and the studies of modern urban history]. *Keizai Chirigaku Nenpo* [*Annals of the Japan Association of Economic Geographers*] 40, 1–19. (in Japanese).

Mizuuchi, T. 2001. Tsusho 'Yoshida no kai' ni yoru chiseigaku kanren shiryo [Notes on the newly found geopolitical materials about 'Yoshida no Kai']. *Kukan, Shakai, Chirishiso* [*Space, Society and Geographical Thought*] 6, 59–63. (in Japanese).

Mizuuchi, T., M. Kato and N. Oshiro. 2008. *Modan-toshi no keifu: chizu kara yomitoku syakai to kuukan* [*Genealogy of the Modern City: Reading the Society and Space in Maps*]. Kyoto: Nakanishiya Shuppan. (in Japanese).

Mori, M. 2009. Kotoba to mono: eigoken jinbunchiri-gaku ni okeru bunkaron-teki tenkai ikou no tenkai [Tracing the discussions towards the traces of matters and the more-than-human world in Anglophone human geography]. *Jimbun Chiri* [*Japanese Journal of Human Geography*] 61, 1–22. (in Japanese).

Mori, M. 2014. The localness, materiality, and visuality of landscape in Japan. *Jimbun Chiri* [*Japanese Journal of Human Geography*] 66, 523–35.

Mori, M. 2016. *Senso to koukoku* [*War and Advertisement*] Tokyo: Kadokawa. (in Japanese).

Murata, Y. 2009. *Kukan no danseigaku* [*Men's Studies of Space*]. Kyoto: Kyoto University Press. (in Japanese).

Nakagawa, K. 1978. *Kindai chiri kyoiku no genryu* [*Origin of the Modern Geographical Education*] Tokyo: Kokon Shoin. (in Japanese).

Nakagawa, T., M. Mori and K. Kanda. 2006. *Bunka-chirigaku-gaidansu* [*Guidance to Cultural Geography*]. Kyoto: Nakanishiya Shuppan. (in Japanese).

Nakashima, K. 1996. Political geography and materialism: Towards an articulation of politics and spatiality. In, H. Nozawa (ed.), *Social Theory and Geographical Thoughts: Japanese Contributions to the History of Geographical Thought (6)*. Fukuoka: Kyushu University, pp. 29–41.

Nakashima, K. 2003. Nature as a locus of resistance: representation and appropriation of nature in the grass-roots movement against the U.S. military exercises in Hijudai, Japan. In, T. Mizuuchi (ed.), *Representing Local Places and Raising Voices from Below: Japanese Contributions to the History of Geographical Thought (8)*. Osaka: Osaka City University, pp. 91–101.

Nakashima, K. 2014. Re-appropriating the grassland: toward an alternative production of nature for changing militarized reality. *Jimbun Chiri* [*Japanese Journal of Human Geography*] 66, 565–79.

Naruse, A. 1993. Shohin to shite no machi, Daikanyama [Daikanyama: downtown as a commodity]. *Jimbun Chiri* [*Japanese Journal of Human Geography*] 45, 618–33. (in Japanese).

Naruse, A. 1994. Wagakuni no chirigaku ni okeru bunka kenkyu ni mukete [Towards cultural studies in geography in Japan]. *Chiri Kagaku [Geographical Sciences]* 49, 95–108. (in Japanese).

Niwa, H. 1992a. Dansei no shiten wa kanou ka? [Is there any possibility of men's viewpoint?] *Chiri [Geography]* 37(10), 78–80. (in Japanese).

Niwa, H. 1992b. Yoseba, Kamagasaki to nojyuku-sya: toshi-syakai chirigaku-teki kenkyu [An urban social geography of the homeless in Osaka, Japan]. *Jimbun Chiri [Japanese Journal of Human Geography]* 44, 545–64. (in Japanese).

Niwa, H. 1996. Chiri-gaku to shakai-teki genjitsu [Geography and social reality]. *Kukan, Shakai, Chirisiso [Space, Society and Geographical Thought]* 1, 2–11. (in Japanese).

Nozawa, H. 1986. Elisée Reclus no chirigaku taikei to sono siso [Geographical thought of Elisée Reclus]. *Chirigaku Hyoron [Geographical Review of Japan, Ser. A]* 59, 635–653. (in Japanese).

Nozawa, H. 2006. Elisée Reclus no chirigaku to anakizumu no siso [Anarchism and geography in Elisée Reclus]. *Kukan, Shakai, Chirisiso [Space, Society and Geographical Thought]* 10, 20–36. (in Japanese).

Nozawa, H. 2009. *Histoire de la pensée géographique en France et au Japon.* Fukuoka: Imprimerie Isseido.

NPO COCOROOM (ed.). 2009. *Kioku to chiiki o tsunagu ato purojekuto: kokoro no tane toshiteno Kamagasaki 2008 [Art Projects for Connecting Memory and Locality/ Place, "COCORO-NO-TANE" Kamagasaki 2008].* Osaka: Urban Research Plaza, Osaka City University. (in Japanese).

Oguma, E. 2009. *1968 (Vol.1 and 2).* Tokyo: Shin'yosha. (in Japanese).

Okada, T. 2002. *Chirigaku-shi: jinbutsu to ronso [History of Geography: Persons and Controversy].* Tokyo: Kokon Shoin. (in Japanese).

Ono, Y. 2013. *Tatakau Chirigku [Active Geography].* Tokyo: Kokon Shoin (in Japanese).

Oshima, J., T. Ukita and T. Sasaki. 1989. *Bunka-chirigaku [Cultural Geography].* Tokyo: Kokon Shoin. (in Japanese).

Oshiro, N. 1990. Anettai tosho no shuraku ricchi to seikatsu yoshiki: Yaeyana-gunto Kohama-jima [Geographical personality of settlement in a subtropical island: Kohama, Okinawa]. *Jimbun Chiri [Japanese Journal of Human Geography]* 42, 220–38. (in Japanese).

Oshiro, N.1998. Gendai no chiiki-hyosho to gensetsu-jokyo [Regional representation and discourse in the modern times]. In, M. Arayama and N. Oshiro (eds.), *Kukan kara basho e [From Space to Place].* Tokyo: Kokon Shoin, pp. 144–61. (in Japanese).

Ota, Y. 1990. Josei ga inai nihon no chiri-gakkai [Japanese geography without women]. *Chiri [Geography]* 35(5), 18–20. (in Japanese).

Ota, Y. 1993. Bunka no kyakutaika: kanko wo toshita bunka to aidenthithi no sozo [Objectificataion of culture: the creation of culture and identity in the tourist world]. *Minaokugaku Kenkyu [The Japanese Journal of Ethnology]* 57, 383–410. (in Japanese).

Reclus, E. 1868. *La terre: description des phénomènes de la vie du globe.* Paris: Hachette.

Reclus, E. 1880/1930. Evolution et Révolution. *Le Révolté*, 1ᵉ année, no.27 (21 février, 1880): 1-2. translated by S. Ishikawa, Shinka to kakumei [Evolution and revolution], In *Shakai Shiso Zenshu 28kan [Collected Works of Social Thoughts, Vol. 28].* Tokyo: Heibonsha, 489–514. (in Japanese).

Reclus 1905/1930. *L'Homme et la Terre, Tome 1*. Paris: Librairie Universelle. translated by S. Ihikawa, *Chijin-ron: dai 1 kan* [*Theory of Land and Human, Vol. 1*]. Tokyo: Shunjusha. (in Japanese).

Rose, G. (translated by Y. Yoshida,et al.). 2001. *Feminizumu to chirigaku* [*Feminism and Geography*]. Kyoto: Chijin shobo. (in Japanese).

Sakai, T. 2001. *Jiyu-ron: Genzaisei no keifugaku* [*On Liberty: The Genealogy of Currentness*]. Tokyo: Seidosha. (in Japanese).

Sasaki, M. and T. Mizuuchi (eds.). 2009. *Souzou-toshi to syakai-housetsu: shimin-chi, Tayousei machizukuri* [*Creative City and Social Inclusion: Civil Intellect and Developing Diverse Communities*]. Tokyo: Suiyosha. (in Japanese).

Sasaki, T. 1970. *Nettai no yakihata* [*Tropical Slash-and-Burn Agriculture*]. Tokyo: Kokon Shoin. (in Japanese).

Sekido, A. 1989. Sanson shakai no kukan kosei to chimei kara mita tochi bunrui: Nara-ken Nishiyoshino-mura Mune-gawa ryuiki wo jirei ni [The spatial organization of mountain villages and their lands as revealed in place names: a case study of Nishiyoshino-mura, Nara Prefecture]. *Jimbun Chiri* [*Japanese Journal of Human Geography*] 41, 22–43. (in Japanese).

Shibata, Y. 2006. Komaki Saneshige no "Nihon Chiseigaku" to sono shiso-teki kakuritsu: kojinshi-teki sokumen ni chakumoku shite [Saneshige Komaki's "Japanese geopolitics" and its ideological establishment]. *Jimbun Chiri* [*Japanese Journal of Human Geography*] 58, 1–19. (in Japanese).

Shibata, Y. 2007. Ajia Taiheiyo Senso-ki no senryaku kenkyu ni okeru chirigakusha no yakuwari: Sogo Chiri Kenkyukai to Sanbo honbu [The role of geographers in strategy research in the Asia-Pacific War: the Sogo Chiri Kenkyukai and the General Staff Office of the Imperial Japanese Army]. *Rekishi Chirigaku* [*The Historical Geography*] 49(5), 1–31. (in Japanese).

Shimazu, T., T. Fukuda and Oshiro, N. 2012. Imported scholarship or indigenous development?: Japanese contributions to the history of geographical thought and social/cultural geography since the late 1970s. *Jimbun Chiri* [*Japanese Journal of Human Geography*] 64, 474–96..

Smith, N. 1998. Nature at the millenium: production and re-enchantment. In, B. Braun, and N. Castree (eds.), *Remaking Reality: Nature at the Millenium*. London and New York: Routledge, pp. 271–85.

Sugiyama, K. 1999. Shakai kukan to shite no yoru no sakariba: Toyama-shi "eki-mae" chiku wo jirei to shite [The amusement quarter in the night constructed as social space: a case study of the EKIMAE district in Toyama City]. *Jimbun Chiri* [*Japanese Journal of Human Geography*] 51, 396–409. (in Japanese).

Sugiyama, K. 2005. Wakamono mondai to toshi-syakai tosei: gendai nihon no local na community policing no jirei kara [Youth problems and urban social control: evidence from a case of local community policing in contemporary Japan]. *Jimbun Chiri* [*Japanese Journal of Human Geography*] 57, 600–14. (in Japanese).

Takahashi, N., A. Tabayashi, J. Onodera and T. Nakagawa. 1995. *Bunka-chirigaku-nyumon* [*Introduction to Cultural Geography*]. Tokyo: Toho Shoin. (in Japanese).

Takeda, Y. and R. Kinoshita (eds.). 2007. *Chizu de miru nihon no josei* [*Gender Atlas of Japan*]. Tokyo: Akashi Shoten. (in Japanese).

Takeuchi, K. 1974. Nihon ni okeru Geopolitik to chirigaku [Geopolitics and geography in Japan]. *Hitotsubashi Ronso* [*The Hitotsubashi Review*] 72, 169–91. (in Japanese).

Takeuchi, K. 1976. Shakai chirigaku to daisan sekai: toshi no baai [Social geography and the third world: some cases of urban studies]. *Hitotsubashi Ronso* [*The Hitotsubashi Review*] 75, 143–61. (in Japanese).

Takeuchi, K. 1979. Shukan no chirigaku kara no gyakushosha: shakai chirigaku no iso [The position of social geography: a consideration on the so-called subjective geography]. *Hitotsubashi Ronso* [*The Hitotsubashi Review*] 81, 653–67. (in Japanese).

Takeuchi, K. 1980. Radikaru chirigaku undo to radikaru chirigaku [Ideological movement and studies in radical geography]. *Jimbun Chiri* [*Japanese Journal of Human Geography*] 32, 428–51. (in Japanese).

Takeuchi, K. 1994. The Japanese imperial tradition, western imperialism and modern Japanese geography. In, A. Godlewska and N. Smith (eds.), *Geography and Empire*. Oxford: Blackwell, pp. 188–206.

Takeuchi, K. 2000a. *Modern Japanese Geography: An Intellectual History*. Tokyo: Kokon Shoin.

Takeuchi, K. 2000b. Japanese geopolitics in the 1930s and 1940s. In, K. Dodds and D. Atkinson (eds.), *Geopolitical Traditions: A Century of Geographical Thought*. London: Routledge, pp. 72–92.

Ueno, N. 1972. *Chishigaku no genten* [The Ultimate Origin of Chorography]. Tokyo: Taimeido. (in Japanese).

Wagner, P. L. and M. W. Mikesell (eds.). 1962. *Readings in Cultural Geography*. Chicago: University of Chicago Press.

Watanabe Tadashi-shi shozo siryo-shu henshu iin-kai [Editorial committee of a collection of the documents possessed by Tadashi Watanabe] 2005. *Shusen zengo no sanbo honbu to rikuchi sokuryo-bu: Watanabe Tdashi-shi shozo siryo-shu* [The General Staff Office and the Land Survey Department Before and After the End of War: A Collection of the Documents Possessed by Tadashi Watanabe]. Osaka: Institute of Human Geography, Graduate School of Letters, Osaka University. (in Japanese).

Yagi, Y. 1988. Mura-zakai: the Japanese village boundary and its symbolic interpretation. *Asian Folklore Studies* 47(1), 137–51.

Yamagami, M. 1914. *Saishin hihan chirigaku* [Current Critical Geography]. Tokyo: Ikueishoin. (in Japanese).

Yamagami, M. 1946. Nihon chiri gakkai no konjaku [The past and present of Japanese geography]. *Kokumin Chiri* [*National Geography*] 1(7), 3–7. (in Japanese).

Yamaguchi, S. 2002. Osaka, Minami ni okeru street performers to street artists [Streets performers and street artists in Minami, Osaka]. *Jimbun Chiri* [*Japanese Journal of Human Geography*] 54, 173–89. (in Japanese).

Yamaguchi, S. 2008. Heaven Artist Program ni miru Artist no jissen to Tokyo-to no kanri [The practices of artists and control by the Tokyo Metropolitan Government in the Heaven Artist Program]. *Jimbun Chiri* [*Japanese Journal of Human Geography*] 60, 279–300. (in Japanese).

Yamano, M. 1979. Kukan kozo no jinbunshugi-teki kaidoku-ho: kon'nichi no jinbun chirigaku no shikaku [Contemporary human geography: Humanistic perspective and spatial morphology]. *Jimbun Chiri* [*Japanese Journal of Human Geography*] 31, 46–68. (in Japanese).

Yamazaki, T. 2006. The geopolitical context of "redefined" security: Japan and U.S. military presence in the post-Cold War era America. In, T. Mizuuchi (ed.), *Critical and Radical Geographies of the Social, the Spatial and the Political*. Osaka: Osaka City University Urban Research Plaza, pp. 35–50

Yamazaki, T. 2011. The US militarization of a 'host' civilian society: the case of post-war Okinawa, Japan. In, S. Kirsch and C. Flint (eds.), *Reconstructing Conflict: Integrating War and Post-War Geographies*. Surrey, UK: Ashgate, pp.253–72.

Yamazaki, T. 2014. Gunmin kyokai toshi toshiteno Koza: boudou no kioku to aid-entiti [Koza as a borderland: Memories of riot and identity]. In, T. Tani, Y. Ando and N. Noiri (eds.), *Jizoku to hen'yo no Okinawa shakai: Okinawa naru mono no genzai [Duration and Transformation of Okinawan Society: The Presence of Okinawa]*. Tokyo: Minerva Shobo, pp. 218–42. (in Japanese).

Yoshida, Y. 1996. Oubei ni okeru feminizumu chirigaku no tenkai [The development of feminist geography in Europe and North America]. *Chirigaku Hyoron [Geographical Review of Japan, Ser. A]* 69, 242–62. (in Japanese).

Yoshida, Y. 2007. *Chiiki-rodo-shijo to josei-shugyo [Regional Labor Market and Women's Employment]*. Tokyo: Kokon Shoin. (in Japanese).

Yoshimi, S. 1987. *Toshi no doramaturugi: Tokyo, sakari-ba no shakai-shi [Dramaturgy of the City: Social History of Downtowns in Tokyo]*. Tokyo: Kobundo. (in Japanese).

7 Chinese critical geography

A non-dualistic, *tongbian*-informed spatial story

Wing-Shing Tang

Introduction

This chapter discusses critical geography in China. There is a consensus in the literature that critical geography had evolved from the West in the early 1970s, fuelled on the one hand by student movements, and, on the other, by critical inquiry in social science as informed by Marx and others. Critical geography is centrally concerned with exposing the power relations and socio-spatial processes that (re)produce inequalities between people and places, including a critique of the mainstream geography discipline and the academic institutions and their rules, and advocating transformative praxis (e.g., Best, 2009; Blomley, 2006). That being the general consensus, Timár (2004, 535) once argued from the Hungarian perspective that there is hegemony even within critical geography itself, with the allegation of "Western (basically Anglo-American) theories – Eastern empirical studies". While she refers to the ideological East (within the traditional Cold War East-West divide), I would argue that it is equally true for the geographical East (in the global sense). As in Asian area studies in general (e.g., Dutton, 2005; Goss and Wesley-Smith, 2010), there is a tendency to treat China and other Asian countries as empirical cases to verify or refute theories or models developed in the West (Savage, 2003, 71). This problem cannot easily be rectified by claims that critical geography takes plural forms in many different spaces and places, as critical geographies so understood still underscore the fact that there is only one centre – the West – with one source of dynamics. I argue that developments of critical geography in China should not be understood as mere variegated critical geographies of the West (Brenner et al., 2010). Instead, critical geography in China has its own historical trajectory, and should be investigated within its own historical geography.

In order to better articulate this argument, this chapter begins by reviewing literature on the development of more critical inquiries, including critical geography, in China. Like in other parts of the non-West, there have been analyses based on random conceptual indigenization and appropriation of western concepts (see Tang, 2014a). These analyses have ignored the imperative of historical geography in any meticulous interrogation. To

DOI: 10.4324/9781315600635-7

make up this deficiency – and to situate the developments of critical geography within the development of the larger region of China and Southeast Asia – the following section argues that the East-Southeast Asian economic system, which had existed long before the formation of European capitalism, has continued to exert its influence on China after its mutual interaction with the latter. In other words, one cannot simply understand critical geography in China as following its counterparts in the West, or a case study that exhibits an independent development path, unrelated to the world. As an alternative, then, the methodology of the spatial story is introduced to unravel the continuity of the Chinese system from its past as well as its continual interaction with other parts of the world. It is based on a re-interpretation of dialectics from a Chinese perspective. As argued in the next section, this methodology has significant implications for the development of more critical inquiries in China. This is illustrated in the succeeding sections by focussing on the development of critical geography in China by stages, from the dynastic past to the twenty-first century. During each stage of this development, China's (critical) geography constantly interweaves with the wider world. The final section summarizes the arguments developed so far. Critical geography in China, or in fact in any other region in the world, is not a mere variegated form of critical geography in the West. The final section reveals that what is considered critical or not in China depends on its historical path as well as its interaction with the world, and requires an alternative methodology for its investigation. Critical geographies in the world other than the West must take a different (or should I say a critical) methodology if such analyses can be trustworthily considered critical at all.

The poverty of critical analyses on China

In the English literature, there are general works that document how the discipline of geography has developed from the colonial past, as in Hong Kong (Tang and Chan, 2008), and from the dynastic and socialist past, as in mainland China (Chai et al., 2007; Chiang, 2005; Tang, 2009; Wu, 1990; Yeung and Zhou, 1991). Obviously, in this burgeoning literature, one cannot identify an equivalent review of critical geography in China as in the West. In contrast, there is a broader literature that uses western critical theories or concepts to investigate spatial phenomena in China. It is popular nowadays for many researchers on China to invoke critical thinkers like David Harvey, Michel Foucault and Henri Lefebvre. A case in point is Teather and Chow's (2000) interpretation of *fengshui* (a Chinese philosophical system that emphasizes the harmony between people and their surrounding environments) in terms of the Lefebvrean spatial trilectics, even though Lefebvre expressed reservations about extending his insights to China (Lefebvre, 1991, 31–32). Given that Chow is an expert on *Yijing* (*I Ching, or Book of Changes*) and since they have also been perceptive in differentiating the Chinese transformational and cyclical concept of time from the West's linear one, why do Teather and

Chow still discuss *fengshui* in terms of Lefebvre (Tether and Chow, 2000)? Although Lefebvre is critical by any standard, does it mean that one must employ him uncritically? Similarly, Shin (2013) attributes the ill treatment of rural migrants in the city, and concomitantly, in the Lefebvrean sense, their rights to the city, to the urban-rural dichotomy based on household registration in China. This analysis elides the differences in the agrarian question, urbanization and politics between China and the West (Perry, 2008; see below). Besides, the exaggeration of the role of rural-urban dichotomy has exposed Shin's neglect of Lefebvre's provocative call for produced differences rather than induced ones.

China has also been scrutinized using Harvey's approach to neoliberalism. Since Harvey's treatise *A Brief History of Neoliberalism*, including his provocative coining of neoliberalism "with Chinese characteristics" (Harvey, 2005, 120–151), and Walker and Buck's (2007) diagnosis of the Chinese road as the transition to capitalism, Chinese observers have joined the bandwagon by indiscriminately simplifying the Chinese development as capitalistic, neoliberalism in particular (e.g., He and Wu, 2009). These attempts of random conceptual indigenization and random conceptual appropriation (Tang, 2014a) have neglected China's peculiar past – more recently laid down by its socialist and nationalist policies and practices in the past 50 or so years (Lin, 2006). It was the different development path in the past that led to the development of neoliberalism in the capitalist world since the 1970s. Accordingly, China's present must be interpreted differently from neoliberalism (Lo, 2016).

Similarly, Taylor (2011) insists on interpreting recent Chinese development by anti-systemic movements within Wallerstein's world-systems analysis (Wallerstein, 2002). The Chinese Communist Party has reached the core location by practicing "labour imperialism" – sucking industrial jobs from all other former "third world" countries. My query with Taylor is his insistence on explaining China's economic development with categories of the world-systems theory while ignoring facts about trade and investment (see Lo, 2016); the impressive growth attained even before opening (Cumings, 2011, 188, 191); the investment in infrastructures, hard and soft alike, that laid the backbone of the "workshop of the world"; and Chinese diaspora capital (Arrighi, 2007, 351–363). Although Wallerstein is, as Tibebu (2011, xxi–xxii) has remarked, one of the conscientious European thinkers who has tried earnestly to tackle the issue of Euro-centrism, his works have mostly focussed on the core region of the world's capitalist system, while paying only lip service to the periphery.

The crux of the problem with these Chinese observers of Lefebvre, Harvey, Wallerstein and many others is best epitomized by post-colonialist-informed researchers who have proposed alternative methodological and conceptual practices. They may not realize that, despite their invaluable contribution, they have fallen into the identical trap. To redress the imbalance caused by metrocentricity, Bunnell and Maringanti (2010), for example, proposed alternative methodological and conceptual practices to capture the diverse urban contexts of Asia. Ethnographic engagement with diverse urban spaces and lives beyond global financial centres as well as insights from feminist geographies on issues such as positionality, habitus, body and subjectivity have been identified as favoured approaches. However promising them may be, these approaches are still formulated on the basis of identical western economic and cultural processes. This is hegemonic, as Wong (1999, 210) has, in his assessment of Chinese economic historical accounts, warned us: "When we take European developments as the norm, all other experiences appear to be abnormal" and will be ignored. What he has in mind are processes not usually recognizable as Euro-American capitalism. If the processes are in fact different, so are the methodological and conceptual practices. Feminist concepts and methods may enlighten the economistic and gender-biased mind, and ethnography may add colours to more structural accounts, but they are all predicated on identical processes. To what extent can they unravel the subtleties of other processes? The issue is not so much the elaboration of, as argued by the post-structuralists too, diversity within the identical processes as, first, the recognition of some other processes more generic to China, but not necessarily found in the West. It is to these processes that we now turn.

From transcending singular modernity towards the development of the spatial story methodology

Processes related to the development of East and Southeast Asia

The processes that concern most critical geographers have usually been articulated as follows: everything in this modern world started in the fifteenth century when western capitalism first developed. The quest was, and still is, to expand capital accumulation, first in countries in the West and then to elsewhere in the world. While different scholars would elaborate the subtleties of these processes – in the case of Harvey they are spatial fix, uneven development, environmental justice and accumulation by dispossession – the conventional wisdom is to emphasize the expansion of capitalist forces over time and across space at the expense of many other forces of development.

This negligence of other forces has serious implications for how we understand China. The economic historian Takeshi Hamashita (2008) has drawn our attention to the regional economic network in East and Southeast Asia that existed even before the formation of the European capitalist system. This network has a few elements that are worthwhile highlighting. First, central to the trade with others promoted by Chinese merchants was the exchange of silver for silk, tea, porcelain and other goods. This led to the establishment of a paper currency system in capitalism (Ikeda, 1996, 68–69). Obviously, the relationships between western capitalism and our region are always both ways. Second, there had been trade among many countries in Asia (e.g., Gunn, 2011) instead of bilateral ties between any colony and its metropolis, as traditional colonial accounts would have us believe. Third, central to these multi-trade relations was the Chinese capital diaspora, generating sustained migration of Chinese to Southeast Asia (Wu, 2009) and elsewhere, and remittance-induced Chinese banking networks and overseas capital. Chinese capital diaspora is so important to the region that Hamashita (2008, 6–7) argues colonialism is of secondary importance to Chinese capital diaspora, and Studwell (2007) confirms the dominance of ethnic Chinese capital in the Southeast Asian and Hong Kong economies even nowadays. This prompts the fourth point. The East and Southeast Asian economic system that Hamashita elaborates has persisted after its incorporation into capitalism (Ikeda, 1996).

In addition to Hamashita's maritime perspective, there is the state-centric and land-centred perspective (e.g., Dardess, 2010), which, nevertheless, highlights similar distinguishing points. According to this perspective, dynastic China was an agrarian empire. Unlike European countries, however, the Chinese state relied less on rich merchant classes, taxed commerce lightly, feared the potential disruptive consequences of both concentrations of wealth and the pursuit of such wealth in the grain trade and promoted economic prosperity to gain the support of the people. The Chinese political economy also tied production and distribution, and rural industry and cash crops, together (Wong, 1999). What is of great interest is that this traditional agrarian organization is still in practice nowadays, for example, informing Beijing's late 1990s campaign to "reduce farmer burdens" (Lin, 2006, 29–30), fostering territorial urbanization (Cartier, 2015; Tang, 2006) and rejecting the practice of outright capital accumulation by dispossession (Trichur, 2012). This recognition of the past – the persistence of processes in China distinctive from those of western capitalism – is a testimony to the existence of disparate paths of modernization.

It is common in the literature to handle this problematic of recognition by the notion of path dependence (Streeck and Thelen, 2005, 6–9). One interpretation asserts that legacies of the past always affect, if not dominate, the choices and changes in the present. This simplistic, linear, repetitious, chronological account of development not only omits the complex transitions between the present and the past but also, for the present discussion, ignores the spatial linkages between various processes at particular times, which are all contradictory.

The importance of spatial linkages is best illustrated by some problematic debates in the literature. The first example relates to the issue of the rise of China: China is rising (Arrighi, 2007) or not (Cumings, 2011). Whether China is rising is a moot point in comparison with the full recognition of processes other than capitalist. There is another debate about the explanatory power of sea-bounded or land-centred processes (e.g., Perdue, 2003) and, concomitantly, whether the sea trade was part of the tributary system. These debates are problematic, since both processes are present at any time. Since Emperor Qin first built a unified kingdom in 221 BC, there had been many interactions with the outside world. Again, when China restricted trading with the outside world between the Ming and Qing dynasties, sea trades still flourished. The problem with both the sea-bounded/land-centred argument and the rise/fall argument is their failure to recognize the continued mutual embeddedness of contradictory elements: each economic system grows from the other and needs the other as partner to generate the world order that we observe nowadays. This might sound irrelevant to a critical geography of China, but without such a recognition of a distinctive past it is impossible to understand how China developed a critical geography that deviated from the West.

The spatial story methodology

This prompts us to challenge the philosophy underlying critical geography in the West and improve on it by developing a methodology that can interrogate the mutual embeddedness of contradictory elements. Dialectical thinking is central to critical geography. Its discussion usually starts with Aristotle's formal logic that "A is not not-A". This logic of identity is most closely associated with the metaphysics of Being. Hegel, praised by Lefebvre, tries to bring form and content together by introducing the third term when the two terms are in contradiction, producing a new moment of Being and of thought. But Hegel errs in reducing the truth to form itself. It was in his final period of development that, having stood Hegel's dialectic "on its head", Marx succeeds in deriving dialectical materialism by adding dialectical method to historical materialism. Instead of pure abstraction, there exists the concrete abstract (Elden, 2004, 15–64; Kipfer, 2009; Schmid, 2008). Finally, Lefebvre excels in bringing the material/mental divide by creating the third term real-and-imagined (Elden, 2004, especially 181–192) and in transcending Hegel's "thesis-antithesis-synthesis" or Marx's "affirmation-negation-negation-of-the-negation" dialectic. These developments have, undoubtedly, enlightened us on the understanding of critical geography.

I have, however, problems with this western understanding of dialectic. Despite the meticulous efforts by, for example, Marx and, even, Lefebvre, dualism still prevails. The reaction to this dualism by, for example, the post-colonial studies of the West and the non-West is, yet, another dualism of History 1 and History 2. Accordingly, China's development, as History 2, can be understood in separation from that of the West, History 1. Although Lazarus and Varma (2008) remind us that the two histories are inter-related, they still have the dualism in mind. Thus, Goswami's (2004) proposal to understand the inter-relationship as co-determination is not immune to the same criticism: there is still a dualism. This dualism feeds, for example, the aforementioned meaningless debate about the rise of China and the fall of the West.

An alternative open to us is to adopt a concept of contradiction informed by the Chinese *tongbian* thinking (Tian, 2005). *Tongbian* is, first, a strand of thought that conceives of every thing or event in the world being correlated with another. Second, it recognizes change as an embodiment of correlation in motion, or continuity between differences and varieties. Third, the most salient feature of this thought is to view the interaction and interdependence of complementary opposition as of a basic pattern as *yin* and *yang*. It is the *yin-yang* that views every thing or event in the world in constant change and movement. Fourth, unlike dialectics in the West, the unity of opposites suggests a continuity of two pairing aspects and a different interpretation of the negation of the negation. While sublation overcomes the old thing, it preserves it as a subsidiary dynamic cause. There is a continuity in it as well as a rejection of the two unrelated logics being imposed externally. Instead of co-constitution, it is mutual embeddedness and mutual transformation of quantity and quality that counts. Fifth, the becoming is due to a continuity of inner changes with external conditions. This leads to the final point that everything, as a particular focus in its own particular field, must be seen as the contextualization of its relationship with the totality. The latter comprises the full range of particular foci, each defining its own field, while a focus refers to the context within which it shapes and is shaped by that context as field. The principal contradiction is accordingly seen as a focus of correlations in a specific field (see Tang, 2014b).

To operationalize it, one needs to recognize the openness of geography and history, not one by one but mutually embedded. Western capitalism exists in contradiction with, for instance, the economic network of East and Southeast Asia. It is this mutual embeddedness between space and time that we need to unearth, and this can be achieved by the methodology of writing a spatial story. This methodology involves narrating a subject's process of becoming. It regresses from the present to the past so as to progress from the present to the future. In recognizing the importance of becoming diachronically, the methodology highlights its synchronic picture: its interdependence among its various internal components as well as with external others. All these interdependences are spatially mediated in that the resources invoked in consolidating or perpetuating the intermediation vary in spatial reaches. It is this spatio-temporality of a subject, always in the process of becoming, that a spatial story attempts to unveil (see Figure 7.1).

It is within this methodology that critical geography in China is here approached. Accordingly, critical geography is the subject in the process of becoming. Critical geography today has its historical trajectory, and has developed from its relationship with Chinese society in the past. Contradictions within this society over time help inform and mould the succeeding possible nature and contents of critical geography. As China is in constant interaction with the outside world, its critical geography also

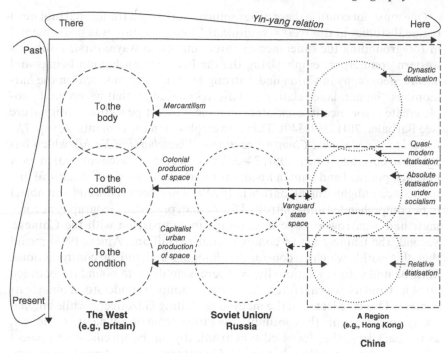

Figure 7.1 The spatial story approach.

incorporates contradictions between China and the outside world and between the geographical knowledge in China and its counterparts elsewhere. It is the spatio-temporal interdependence of the resolution of these contradictions over time that has defined critical geography. It is to the details of this Chinese critical geography as the subject in the process of becoming that we now turn.

Gazetteers and accounts of exploration in the dynastic past: Harmonious "human-land relation", contextualization and "state" governing

Geography started to flourish in China almost 2,500 years ago. During most of the dynastic period (up until 1911), there were Geographical Gazetteers (*dili zhi*), the appendices to "annals" (*zhengshi*) and local Gazetteers (*fangzhi*), and accounts of exploration. The Geographical Gazetteers represented detailed geographical studies on special topics such as the economy, the irrigation system and disasters, while the local Gazetteers provided the systematic account of any local administrative district. The exploration accounts, which were more sporadically undertaken by orders of an emperor or by missionaries and travellers, recorded the geography of the world beyond Chinese territory (Hu and Jiang, 1995; Tang, 2009).

One must not confuse these geographical works with traditional (not new) regional studies in the West. Traditional Chinese culture was instrumental in foregrounding the difference in three important ways. First, unlike its western counterpart emphasizing the duality between human beings and nature, geography in China had a strong human dimension within the harmonious "human-land relations". This acknowledges that nature and people create resonance in each other; people co-exist peacefully with nature (see Barbalet, 2011, 339–340). This is exemplified in the contents of, say, *The Sequel to the Shanghai County Gazetteers (Shanghaixian Xuzhi)*, which has a diverse coverage (Chen, 1981, 34–35) to reflect the heaven-and-earth view in which political and human history is intertwined closely with natural history. Since Enlightenment, nature in the West had been reduced to an object to be comprehended and controlled for human progress. Geography accordingly had a narrower scope in coverage in comparison with the Chinese. Second, the Chinese art of contextualizing (Hall and Ames, 1998) meant that the world, which is constantly changing and full of contradictions, must be understood relationally. As there is no discrete boundary between an object and its external forces, any object cannot be understood outside its context. This is reflected in the method of writing Gazetteers: while emphasizing diachronically the continuity over time, from one dynasty to another, as the backdrop, they focussed synchronically on the specific development of a particular region. In other words, Gazetteers conveyed one type of tempo-spatiality, the site-specific situation over the continuity of time. Time and space were, therefore, interdependent in that one could not exist without the other. Their relationship is, however, less essential than in historical materialism as practiced in the West. While Gazetteers focussed on the interdependence, developed over time, at a specific location, the art of contextualizing deterred them from searching for causal laws derived from dualistic contradictions. Third, geography was practical in orientation, and for "state"-building in particular. According to "the way" (*dao*), the emperors ruled the people on behalf of the heaven (*tian*, which itself is a broader term than nature). Knowledge was aimed not so much at seeking the truth as elaborating policies and devising implementation techniques to govern the agrarian empire. Practical knowledge about dynastic politics and the daily life of people was emphasized at the expense of the natural processes underlying the formation of the earth's surface.

Development of a quasi-discipline through interaction with western modernity: Scientification and social engineering

Since the Italian Jesuit Matteo Ricci's first visit in 1528, Renaissance geographic thought had begun to have an influence in China. Later, environmental determinism and other approaches were introduced, expediting the development of geography as a discipline independent of the cognate subject of history. The establishment of professional organizations like the Geographical Society of China and the Institute of Geography in 1934 and 1941, respectively, epitomized this development (Guo, 2000; Hu and Jiang, 1995; Peng, 1995; Tang, 2009, Yang, 1988).

This development should not be interpreted as an undisputed spatial diffusion of western Geography to the country (Guo, 2000; Peng, 1995). Environmental determinism was challenged, as the Chinese still believed that a strong human dimension always took the reins. The Chinese conception of heaven-and-earth had nevertheless started to incorporate western sciences such as cosmology, astronomy and geophysics. *Fengshui* was criticized, but the effect was not as socially penetrative as one would have expected (Guo, 2000, 49–79, 259, 267–284). Besides, at the crossroads of three systems – imperial Chinese, nation-state Chinese and the West – at the beginning of the twentieth century, the Chinese had undertaken a form of scientification different from that of the West (Wang, 2006). A totalizing scientific worldview started to prevail, reshaping intellectual agendas in the humanities and literature and, eventually, permeating people's modes of ethics and behaviour and shaping everyday life. Underlying it was the belief that scientific technology could lay the foundation for the development of a strong nation-state, the prerequisite for survival in the capitalist world system (Chen, 1997; Wang, 2002; Wang, 2006, 89). More specifically, Yang (1988, 135–136) alludes to the deployment of geographers in particular by the Republican government after 1911 in debates about new capital and provinces/regions so as to suppress localism and revolts. These attempts were nothing but the continuation of the dynastic practice that the state used to adjudicate on knowledge production for nation-building. Unlike the West, dating back to Kant, there was no clear separation of pure reason and practical reason in China. While their European counterparts had experienced the dark side of scientific and technological development, especially during World War I, and queried the associated justice issue, as Lefebvre, for example, did during his days (Elden, 2004, 145), Chinese intellectuals never pondered such a relationship. Instead they tried to prove the rationality and necessity of science, and to show how it was the foundation of an ideal political, moral and aesthetic order (Wong, 2006). In their elaboration of the moral meaning of science, the Chinese underscored "the public" at the expense of "the private". Morality as a public principle should overcome private opinions and prejudices. Science could contribute to overcoming "the private" by discovering and following natural, objective laws. Since scientific laws were universal to all human beings, science included moral laws for society, which were applicable to daily life (Wong, 2006). Thus, unlike the West, where there was a desire to protect the private, scientification in China tended to do the opposite, by, first, negating the traditional way of politics, which relied on personal perfection in the connection of heaven-and-earth, and, instead, laying down the condition for crushing daily life by mass mobilization (Chen, 2007, 262–265). Finally, having separated human culture from nature, the Chinese intellectuals concentrated on their critique of evolution within their scrutiny of science. The discussion of evolutionism was restricted to the level of science. For some, the scientific principles of evolutionism were entirely adapted to the understanding of society, history, ethics and politics. In other words, the West's query of evolutionism, which implied a modern view of history as progress, was not a matter of concern for the Chinese (Wong, 2006).

The Chinese mode of scientification had serious implications for the development of geography. Especially since the dynasties gave way to Republican China in 1911, the human world of the traditional heaven-and-earth began to be dislocated from the emperor, and the imperial authority dominated chronology started to co-exist with the western one of progress, nurturing a new concept of nature. The modern Chinese term for nature is *ziran* or *self-so-being*. Still influenced by Chinese philosophy, "the term meant both existence *and* spontaneity – features of how things worked in the organismic realm of heaven-and-earth" (Weller and Bol, 1998, 485, emphasis added). The social world was still part and parcel of *ziran*, re-confirming the traditional integration between nature and human and the process of change from the interactions of the complementary and contradictory polarities (Barbalet, 2011, 339–340). Nevertheless, although *ziran* did not refer exactly to the western concept of nature, it was now rendered concrete and objective, laid open for scientific analysis and, concomitantly, more intensive human exploitation. This led to numerous studies on the geographic environment, especially climatology and geomorphology (Guo, 2000; Hu and Jiang, 1995; Huang et al., 2012; Yang, 1988). One could also detect preliminary delineation of relationships between variables and an early search for causal laws, mostly from the perspective of environmental determinism (Hu and Jiang, 1995, 264–265). Apart from the resultant effect of distancing geography from history, these attempts laid down the societal foundation for China's later quest for social engineering.

While Chinese geographical studies increasingly incorporated other parts of the world as a by-product of the challenge to Chinese-centrism, there was also a move to investigate the trouble-stricken areas in the country like the northwest and northeast. This concern had socio-historical roots. Dating back to the early Qing dynasty, border disputes with Russia had already urged many scholars to investigate the northwestern part of the country (Guo, 2000, 160–172). The shedding of the sovereignty right of Shandong in the Treaty of Versailles in 1919 and the Japanese invasion in the northeast in 1931 were two other examples, urging Chinese geographers to better understand these areas in order to enhance the nation-building project (Peng, 1995, 50–59). These geographical regions were studied not so much for the discovery of causal laws, but rather with the functional governing of the country in mind. Nature, sovereignty, territory, region and people were nested together for social engineering.

Besides undertaking investigative work in the border areas, there was, as Peng (1995, 54–56) has enlightened us, another approach at the beginning of twentieth century, urging scholars to go to the field and be educated by the people. Originally initiated by students studying overseas, this approach started to cast doubts on the existing system. It later developed into debates about social history of China during the late 1920s and the 1930s, engaging with topics such as the nature of society, social history and the social nature of the countryside. Informed by historical materialism, these debates began to query the social causes of underdevelopment of the country and search for possible changes outside capitalism (e.g., Chen, 2010, 83–89, 96–97). It was within this context that we saw the publication of historical-materialist geographical works in the 1940s (Yang, 1988, 136), forming the backdrop to the development of Marxist-informed geography under communist China.

Geography under socialism I: "Class analysis" of development and temporal management of *ziran*

This does not imply that critical geography in China had already by the 1950s "pre-empted" the West; rather, it had basically taken a distinct path. This was attributable to the way Marxism had developed in China (e.g., Dirlik, 1997; Li, 2006; Tian, 2005). "In combining with Chinese realities (peasant revolutionary wars) and traditions (military strategist's *bianzhenfa*[1]), Marxist historical materialism and dialectics was sinicized" (Li, 2006, 34, my translation). It was the intricate intertwining of practical rationality, the quest for state governing, traditional Chinese philosophy and development experience of the past that had formulated two dominant modes of spatialization: (1) the scientific management of *ziran*, and (2) class struggle over *ziran*. These separately have serious implications for the nature and contents of critical geography during this period.

The way Marxism was introduced and then developed in China was influenced by the deep-rooted Chinese interest in practical rationality. Marxism was mainly introduced to China via Japanese translations and Russian Marxist texts. As the experience of "actual socialism" in the Soviet Union represented a successful model to emulate, selected aspects of European Marxism – in particular, Engels and Lenin rather than Marx himself – were diffused to China. Soon after historical materialism was first introduced to China in the first decade of the twentieth century, the emphasis was on class struggle. Disinterested in concrete historical analysis of the various classes in the relations of production in China, the Chinese instead focussed on their positions in distribution and consumption. The issue of class related not so much to economic foundations as to political attitude, as the main purpose of class analysis was to identify foes or friends (Li, 2006, 11–12). On the issue of dialectical materialism, drawing on Qu Qiubai's introduction from, again, the Soviet Union and his later identification of the urgency of arming the peasants, given the objective of revolution in China was to support the peasants while uprooting feudalism, Mao Zedong turned it into practices. This strategy was born of practical rationality, deduced from practical experience (Li, 2006, 22–28).

Another example is Mao's renowned tactic of encircling the towns by the countryside. It was articulated from the cumulative experiences of fighting numerous battles during the ten years of civil war, the anti-Japanese war and the liberation war between the late 1920s and 1949, when a list of efficient as well as effective tactics and strategies was devised. Mao took these particular military affairs and philosophized them, coining the renowned concept of the *maodun* [contradiction]. The peasant revolutionary war experiences in China were thus instrumental in nurturing Maoist thinking (Li, 2006, 32–33).

On the contribution of traditional Chinese thought, Mao was, according to Li (2006, 33–34), very much influenced by ancient military strategists' *bianzhenfa*. There are a few characteristics of this mode of thinking. First, the reality is at stake, and military strategic manoeuvring brings this kind of rational choice into full play. Second, observation, understanding and analysis of the phenomenon entail experience. Third, it is imperative to distil the essence of the battle and identify the contradictions. Fourth, the subject and the object are interdependent. Intimately knitted with practice, these are different from western dialectics. Tian (2005) is even more provocative in attributing this difference to *tongbian* thinking in general (as discussed above). Accordingly, Tian (2005, 150–163) has accused others of misunderstanding Maoist *bianzhenfa* as Marxist dialectics, partly due to the mis-translation of the key concepts of the former.

It was the spatial interdependence of these historical forces, both within China and beyond, that configured Maoism's different practical and socio-political emphases. As articulated in Mao's masterpiece, *On Practice*, investigative study of actual situations has a significant role to play in the comprehension of the interdependence of the phenomenon. Uninformed by historical analysis, however, this kind of study did not aim at deriving some kinds of general laws. Besides, Mao's other masterpiece, *On Contradictions*, illustrates clearly that Maoism may be one of the philosophies that has fully developed Marx's axiom that "the issue is to change the world" (Li, 2006, 37). The concern was to struggle constantly, and people were to be re-educated to contribute to the cause of revolution. The dictatorship of politics had resulted in the moralization of politics. Furthermore, instead of social justice, the tenet of western critical geography, society emphasized unselfish thought and practice; instead of "the total man" in Marx, the goal was to develop highly moralistic, ideologically pure and politically conscious party followers.

These emphases interacted with other forces to produce two prominent modes of spatialization. The development of science in the Soviet Union had generated scientific socialism. After banning debates on the New Economic Policy, Stalin launched serious challenges to sciences. For example, the "old biology" was dismissed as bourgeois for being devoted solely to understanding nature instead of changing it. A similar accusation was levelled against the field of geography (Soviet Union Institute of Geography, 1952). Physical geography was considered in total isolation from its human counterpart. While human geography was further reduced to economic geography, one faction attempted to delete the physical dimension in it from any meaningful discussion (Xu, 1985, 18–19). The Chinese, who had earlier accepted science uncritically, also approvingly embraced these Soviet developments in sciences. Harmonized with Mao's ideas about social revolution, the Soviet "new biology" was later adopted as formal Chinese Communist Party (CCP) doctrine (Marks, 2011, 112).

Ziran was increasingly accepted as something to be adapted to human needs, particularly the primacy of production. In the early 1950s, China was territorially contained by western powers and the country was poor by many standards. The concomitant quest for national independence and self-reliance rendered it imperative to adopt a rapid industrialization programme, which was funded by agricultural surpluses. Raising the income of workers and peasants co-existed comfortably with the goal of increasing the wealth of the national economy. Put differently, the growth of a commodity economy and the development of socialism were not incompatible goals, but rather consistent with the traditional Chinese interpretation of governing the agrarian empire. Thus, the Soviet model of central planning, which involved social engineering to achieve stipulated growth within a certain time frame, was welcomed, together with the new biology, economic geography and other sciences. This temporal management entailed the nationalization of all means of production, including *ziran*.

The effects on the development of geography in China were considerable. Human geography as a whole was criticized as bourgeois (Zhang, 1956). So were the sub-discipline of geopolitics and, even more particularly, Weber's industrial location theory (Chu, 1994, 28). In 1955, there was even a campaign to purge human geography (Liu, Jin and Zhou, 1999, v). Besides, Soviet geographical practices were introduced (Wu, 1958) with praise (Huang, 1958). There were attempts to introduce physical geography and economic geography to facilitate national economic development, as propagated by the concept of territorial production systems (Li, 1957). Human geography was then dominated by economic geography, which, in turn, concentrated on agricultural geography, industrial geography and transport geography only. In other words, geography was considered imperative to this new mode of spatialization based on rapid industrialization. Before any industrialization programme could proceed, knowledge about resources in the prospective region must be made available. There were then nationwide, integrative, regional surveys, spanning geography, geology, climatology, botany, zoology, soil, hydrology, agriculture, forestry and economy. Geography as a discipline excelled in these tasks because it could "bring the superiority of integrative and regional knowledge structure into full play" (Yang, 1988, 140).

After a few years of practice, it was realized that this mode of spatialization had led to many problems. As industries increasingly concentrated in cities, so did people and other resources. To secure the transfer of resources to cities, the state installed the centralized regulative mechanisms and institutions of "unified purchases and unified sales" of agricultural produces and allocation of jobs. In comparison, commodity production in the countryside was discouraged and incomes of peasants were capped at low levels. This was informed by a class analysis of the town-country relationship, emphasizing class struggles for equality at the expense of economic development and market circulation of town and country.

To comprehend the prevailing *maodun*, Mao carried out investigative surveys, including site visits and field discussions with various provinces in the latter half of 1955, work reports from various bureaux in early 1956 and, finally, reports on the formulation of the second Five Year Plan from the State Planning Commission (Wu, 2005, 50). These materials were inputs to the now renowned policy *On Ten Great Relationships* (Mao, 1999), which refers to ten *maodun*, including those between centre and local, between the coast and the interior and between China and the external world. The thinking of *tongbian* led to the formulation of spatializations that considered each pair as patterns of continuity. For example, the centre and the local were not dualities, existing in isolation of each other. They complemented each other; they were particularities on the path to socialism. The centre and the local were integrated to form a new mode of spatialization with a certain form of space and a certain mode of production. Nevertheless, given the usual emphasis on practical rationality, instead of tracing the historical causes of these *maodun*, Mao and other leaders in the higher echelon of the CCP were more concerned with practical ways to mobilize all positive factors, both within and beyond the country, for the cause of building socialism (Mao, 1999, 23).

Within China, the emphasis was on class struggle. On the issue of increasing agricultural production, it was decided to change the social relations of agricultural production by collectivization. People's communes were finally developed, after stages of collectivization, first in the countryside and, then, expanded to the city. The urban commune itself was, argued by D. Li (2006, 245–246), Mao's response to people's negative reactions, both within and beyond China, to people's communes and Great Leap Forward. It was hoped that urban communes could expedite urban construction and industrial development, thereby achieving the utopia of communism by a fast lane.

Mobilization of support by class struggle took the form of many anti-"rightist"[2] campaigns in the latter part of the 1950s (e.g., Lin, 1999). For geography, like other disciplines, the profession was "adjudicated". On the one hand, having been labelled "rightists", many human geographers were purged, causing a big gap in the development of the profession (Chu, 1994, 28). On the other, geographers were, in an annual conference on geography in 1958 (Editorial Office, 1959), requested to apply themselves to a great leap forward: the planning and design of people's communes (e.g., Beijing Normal University People's Commune Planning Group, 1959; Li, Ren and Feng, 1960; Zhang et al., 1959). In the 1960s and 1970s, economic geography increasingly shifted its focus to agricultural geography, contributing to land use planning, agricultural resource assessment and agricultural regionalization. Thus, the discipline of geography had superficially developed in those days according to political deliberation.

Mao's lens of class struggle regarded the Chinese revolution as a part of the world proletariat revolution. Again, class struggle was employed to identify friends and foes and form a united front against enemies. In the 1960s, Mao invented the "intermediate zone" theory to consolidate an anti-American united front, further expanded it into intermediate zones to cast an international united front and, finally, proposed the theory of "three worlds" to struggle against the two superpowers (Xia, 2011). While these spatializations did not substantively affect the discipline of geography at a time when the discipline was subject to serious purges under Maoism, they nevertheless form an integral part of what is now called national security geography (e.g., Shen and Lu, 2001), a sub-discipline of political geography.

Finally, Maoism affected the world in another way. Developed in China, Maoism was in interaction with Marxism, imperialism and the Cold War. It was this Maoism (Tian, 2005) that had added fuel to the student movements in the 1960s; in turn, the New Left movement as well as critical Asian studies (Lanza, 2013) and the cultural turn in Western humanities (Wolin, 2010). While the influence of the latter movement on the Chinese intellectual thought is going to be documented in the succeeding section, it is appropriate to mention the interaction between Chinese and western critical thought during this period.

Geography and socialism II: Practical rationality for the state and the spatial management of *ziran*

The downfall of the Gang of Four in 1976 led to a challenge to Maoism on the principle that "practice is the only criteria of truth", thereby prompting the emergence of various approaches. On the one hand, it is suggested that there is a continuity in the approaches represented as Marxist modernism (Gabriel, 2006). Tian (2005, 173–184) insists that the Chinese have nowadays debated economic reform, such as rationality, rules of law, democracy and rights, within *tongbian* thinking. Drawing on Daoist thought, Barbalet (2011) interprets market relations by people as *wuwei* (effortless action or non-coercive action). On the other hand, it is commonplace to attribute the recent developments in China to westernization, the United States in particular (e.g., Rofel, 2012, 447). In doing so, however, these views could not explain satisfactorily why state intervention is emphasized by the New Left in China, which must be comprehended in terms of a socialist legacy in which the state was more responsive to the needs of the population (Wang and Lu, 2012, xiv). The geopolitics of knowledge must be situated within this historical legacy as well.

The reform strategy of "groping for a way forward" by Deng Xiaoping, which provided the architecture of China's reform, aims to liberate the productive force. Besides a strong belief in the role of science and technology, enthusiastic agents are encouraged to invest their human and monetary resources, in combination with the natural ones, in the improvement of the productive force, thereby producing more resources to ameliorate shortages in the economy. The consideration of risk management is also high on the agenda of this strategy. Only those domains that, on the one hand, minimize the risk incurred, and on the other, maximize the legitimacy of reform and, accordingly, smoothen the progression, would be proposed and implemented (Qian, 2000; Wu, 2005; Zhou, 1999).

These contribute to the formulation of a distinct mode of spatialization. The aforementioned enthusiastic agents include agriculture, individual and private enterprises, foreign capital, consumers, some coastal regions, etc. These agents, who were in the periphery of the system, have their own logics of time, besides the *étatized* ones. In comparison, the latter, whose time was unintentionally slowed down in the past, due to resource hoarding or class struggle, and censored/circumvented flow of information, are still under strong state regulation as a "dominant strategic line". Given this prevalence of multiple temporalities, the state needs to invent new spatializations to enlist, enroll and mobilize enthusiastic agents and their development processes that lie outside the *étatized* network. This spatialization conveys the ontological sense of betweenness: the mutual embeddedness of imagination and practices both inside and outside of the *étatized* network. Nevertheless, it reminds us that agents and their practices are usually bounded within their respective spatial administrative hierarchy. Concomitantly, it requires a mastery of spatial rather than temporal considerations of the agents and their practices. At the risk of over-generalization, it trades time for space. Examples of new spatializations abound: open door policy, special economic zone policy, western development strategy, the "city-leading-counties" regional administrative reform, community building and the "new" socialist countryside scheme.

Accordingly, human geographers like Li Xudan and Wu Chuanjun, after decades of silence, have started to offer new and practical interpretations of the human-land relationship, energizing the development of human geography in alignment with scientific management (Wu, 1990). The publication of the specialized journals *Economic Geography* and *Human Geography*, since 1981 and 1986, respectively, marked the diversification as well as resurgence of human geography. Besides economic geography, the quest for knowledge about population, natural and tourist resources is responsible for both the establishment and the promotion of population geography, land resource geography and tourism geography. As cities are now considered active agents of growth in themselves and for their surrounding countryside, the rapid pace of urbanization alerted the national economic planners to problems like the relationship between the level of urbanization and the level of economic development, rural-urban migration, the urban-size hierarchy and the functional structure, catchment area and spatial structure of cities. The complexity of the urban question led to the birth of urban studies, with urban geography – which contributes to the regional foundations of urban development – as the dominant discipline. Urban geographers have started to investigate a whole array of issues such as land assessment, "commodity" (i.e., non-state) housing, urban transport, urban redevelopment, suburbanization, the impact of FDI, urban governance, world city formation and spatial inequalities. Lately, with the growth of the "middle-class" as well as poverty in cities, there have been attempts to explore the cultural turn of urban issues as well as urban enclaves at the fringe (e.g., Chai et al., 2007; Tang, 2009).

Amidst the quest for expedient and practical knowledge for the "Four Modernisations" programme[3], which had induced some Chinese geographers to adopt western intellectual traditions without much reservation, there were oppositions to this development, first captured in politics as the "anti-pollution" campaigns in the early 1980s, by the remnants of the "Leftists" (i.e., the "Gang of Four"). As socio-economic problems worsened in the 1990s, a loose group of critics started to voice their objections, forming an intellectual tradition generally labelled China's "New Left". The New Left refers to intellectuals, from a variety of different leftist theories, ideals and traditions, who share a fundamental concern with social inequality, justice and developmentalism under global capitalism (Gong, 2003; Li, 2015, 46–59; Mierzejewski, 2009; Wang and Lu, 2012).

According to Zhang (2013), the New Left, which should be more appropriately called nationalist socialism, is composed of two sub-groups: the New Maoist and the Nationalist. The former argues, first, that Marx's dialectical and historical materialism, with an over-emphasis on the role of productive force, fails to foresee the real possibility of communism. Besides, the "economic man" assumption of western economics, which fetishizes "man", is not only problematic but also unrelated to China. Finally, and most importantly, Maoism represents a more advanced version of Marxism, as its *bianzhenfa* assigns to consciousness a more significant role than dialectics. The Nationalist, the other sub-group, mostly consists of economists from the leftist camp, who argue, like Andre Gunder Frank more than four decades ago, that the current economic situation was dictated by the western economic powers with the objective of containing China. Unless there is a strong and corruption-free state adopting self-sufficient and self-reliant economic policy, China could not re-emerge as a great nation (Lo, 2016). Underlying this sub-group is a strong nationalist populism, a feeling energized by many developments since the 1990s, including the publication of the renowned book entitled *China Can Say No*. In sum, as Mierzejewski (2009, 17, 22) has argued, "the Chinese New Left has been mainly striving for a Chinese model of development consisting of a partly planned economy, social justice, implementation of a welfare state and nationalism based on the Chinese traditions", and, ardently criticized by others, Maoist-style campaigns of "criticism and self-criticism".

It is obvious from the above that China's New Left has no intention to uproot the regime; or to relax the regulations imposed on society and the economy by the state. Instead of the market, it calls for more comprehensive state intervention; instead of integration with the world economic system, it underscores independent nationalist development; instead of political reform, it favours some kinds of imagined state socialism. Judging from this brief overview, it is not difficult to identify the contrast in appearance between the western and Chinese New Left. As Wang and Lu (2012, xii) have observed, for example, the former is characterized by "vibrant social and political activism", while the latter tends to favour "intellectual and theoretical engagement". While there may be some truth to these observations, they are problematic if one draws the conclusion that the China New Left is not critical while its western counterpart is, merely from the absence/presence of the observable features. On the point of the absence/presence of protests, for instance, the issue is not that there are no popular protests in both town and country in China, as recent developments have proved otherwise. Neither is it that the Chinese New Left has no effective societal intervention at all, as, suggested by Mierzejewski (2009), various recent party administrations have also incorporated some of its ideas in their highly propagandized policies (as in the "Chongqing model"). The point of controversy lies in the nature and cause of the protest activities. Perry (2008) traced East-West differences in political philosophy and tradition: from the conceptions of right and of rebellion (collective livelihood *vs* inalienable individual right), the agency of rebellion (peasants *vs* urban proletariat), justification (protecting individual wellbeing *vs* checking on political tyranny) and issue (social justice *vs* moralization of politics). Other than a recognition of the modernist emphasis on development in Maoism (Perry, 2008, 41), Perry's account of the East-West dichotomy relates more to cultural difference. Our earlier account, in contrast, has emphasized the historical development from the dynastic, republican and communist past to the present, which is necessarily interwoven with the external world, from renaissance, enlightenment, Marxist intervention and Soviet development to the globalizing world. In short, it is not culture *per se* that accounts for the criticality of the China New Left and its difference with its western counterpart, but rather the mutual embeddedness between synchronic and diachronic forces in time and space.

The contrast between the West and China in the discipline of critical geography is even starker. These histories are distinct. Geography in China, as discussed above, interacted with Marxist analyses of development dating back to the 1940s. This criticality developed a particular form and content under Maoism during the first 30 years of socialism. However, the succeeding 30 years of development after reform as well as the recent intervention of China's New Left have not induced geography as a discipline to identify concerns for critical geographical issues as framed in the western tradition since the late 1960s. Undoubtedly, there are recent introductory essays drawing the attention of Chinese geographers to this critical literature, e.g., Gu (2007) on urban social geographic concepts of social polarization, new poverty, slums and unfair distribution and differences in family income (see also He and Qian, 2017); Chen, Ji and Zhang (2014) on transport equity in particular; Wang and Liu (2007) on the cultural turn in general; An and Zhu (2013) on otherness, power and the construction of place; Wan and Tang (2013) on feminist geography; Su (2013) on Gramsci's hegemony and cultural politics; and Qian (2013) on Foucault and Lefebvre. It is also interesting in passing to mention the recent resurgent interest in Marxist geography. On the one hand, this is related to various talks in China by David Harvey in June 2016 (and earlier translations of his work into Chinese). On the other, Harvey's visit may itself be a response to a call to steadfastly develop Marxism by the state; Xi Jinping, the Party Chief, made such a call in a Philosophy and Social Sciences Work Conference in celebration of the bicentennial of Marx's birth, in May of the same year (Cai et al., 2016a, 2016b). These attempts, which still have left only a shallow imprint on the field, must be deemed as a cosmetic appendix to the Chinese mainstream.

There are several reasons for this development. Many geographers in China have a background in science and engineering (Chai et al., 2007, 9). Their Soviet-inclined training has additionally rendered them less sensitive to more philosophical, theoretical and ideological debates in the humanities and social sciences, attenuating cross-over interaction between western critical ideas and geography in the first place. Their recent heavy involvement in consultancy projects, as requested by many local governments (as well as the state) on development issues, has further tilted their attention towards more practical concerns and, concomitantly, technical aspects of geography. For the more recently western-informed or trained geographers, who are mostly responsible for introducing western concepts to China, their geographical sense has, somewhat surprisingly, remained "uncritical". Their original training together with later interaction with their new peer group in the West may have unconsciously filtered the knowledge to be introduced to China. Geographers like Liu and his associates (Liu and Wang, 2013; Liu and Xu, 2008) have begun to lament the importation of western geographical theories such as those recommended in Lu, Wei and Lin (2006). This form of importation has left many critical geographical issues in China unattended (Liu and Wang, 2013, 16) – for example, migrant workers, household registration and state intervention. Although their observation is admirable, Liu and others have, mostly informed by nationalism, endorsed a post-colonialist critique of the East-West duality. This is controversial, because it has basically ignored the unequal power relations in knowledge production, on the one hand, between the West and the East and, on the other, between the authority and the expertise (Tang, 2014a). Central to these relations is the state. Many geographical ideas have been introduced with the objective of ensuring state governing, as Gu (2007) has done for promoting a harmonious society and Wang et al. (2016) have for the "new normal". The practical rationality for state governing still dominates to such an extent that the chance of developing into an expertise to proclaim a disinterested truth and promise to achieve desired results and, in turn, to crystallize into a deterring force to the domination of the state – the yardstick for western intellectuals – is still slim at this moment in history. In sum, to underscore, "critical geography" in China at this moment, which looks different from that of the West, as charted in Gregory's (1994, 5) map of intellectual landscape, is not so much due to its backwardness as due to the mutual embeddedness of contradictions within its distinct path of development and, concomitantly, different concerns and repertories.

Conclusion: Towards more critical, critical geography

This account of critical geography in China has emphatically argued that it is not a mere variegated version of its western counterpart. It is not difficult to recognize that geography in China, heavily informed by traditional *tongbian* thought, has an intellectual landscape vividly different from the West. This is also the case for critical geography in China. The objectives of critical geography in the West are to expose the power relations and socio-economic processes that (re)produce inequalities between people and places under capitalism, and to advocate transformative praxis. That said, the definition and dimension of criticality, and the methodology for deciphering this criticality, are obviously spatio-historically grounded. Central to capitalism is the capital logic and its contradictions, so compellingly summarized in Harvey (2015). There are, however, many other disparate spatio-historical pasts, and many of these are not characterized by strong private property rights, free markets and free trade. Because of this, a benchmark of criticality for critical geography that is derived from western capitalism and then uncritically deployed to other contexts, such as China, is not particularly useful or insightful. The crux of the issue is not so much the mere identification and documentation of pluralisms or diversities alongside this benchmark, but rather the need to acknowledge the existence of many more distinctive criticalities that have been derived from disparate, but inter-connected, forces and processes.

As China has undergone a historical transition from the agrarian empire situated in the context of the East and Southeast Asia economic network, and the capitalism-inclined nationalist republic to, currently, socialism, so has the prevailing status quo. During the socialist stage, for example, the status quo consists of, among others, the state ownership of means of production, the governing of the economy and society by a vanguard party, an intellectual tradition that is continuously adjudicated by the state, and a practical rationality that dominates the human-land relationship. This socialist status quo has exhibited some degrees of improvement in critical-ity over its capitalist counterpart. In the words of Marx and Engels, at least at their time, socialism represents a higher stage of development than capi-talism in that some problems related to private ownership of means of pro-'duction would have been ameliorated by the burgeoning public ownership. Due to Soviet inputs, materialist geographic works in China, however prim-itive they were, could be dated back to the 1940s, long before the Detroit Geographical Expedition led by William Bunge in the late 1960s. Besides, if critical geography in the West is concerned with exposing inherent power relations and socio-spatial processes, the Chinese under Maoism were per-haps the forerunner. Having unveiled power relations, they experimented with utopianism both in the town and in the countryside in the 1950s and the early 1960s. Furthermore, these Maoist practices influenced the New Left in the West, which informed the development of contemporary critical geography, including the political economy of the city by Harvey in the late-1960s and early 1970s. In other words, it is difficult to deny Chinese criti-cality, if the benchmark of criticality is a challenge to the capitalist system.

The discussion of the improvement in criticality, however, should not stop with state ownership of the means of production, as this is not the only dis-crepancy between capitalism and socialism. Socialism is meant to replace capitalism for the building of communism. But what Marx and Engels could not foresee almost two centuries ago is the development of socialism through the dominance of vanguard relations, i.e., a party that leads soci-ety hierarchically from a single centre. In fact, it is, according to Lebowitz (2012, 77), the vanguard party relation that brings about the property form of state ownership, not the reverse. In other words, it is the vanguard rela-tions that are the status quo, not public ownership *per se*. If the focus is only on public ownership, any discussion about criticality under socialism would have missed the point. Vanguard relations rely on the service of vanguard Marxism, the ideological apparatus, which, having rejected the perspec-tive of capital, still neglects the worker as a subject as well as his/her other relations within which the worker exists. Obviously, since criticality under socialism has been subdued by the vanguard relations, the re-installment of criticality requires, for Lebowitz (2012), going beyond vanguard Marxism.

Chinese socialism, which differs in many ways from Soviet Union social-ism, nurtures a different interpretation of criticality. On the one hand, it is difficult to deny that the Chinese vanguard party still steadfastly adheres to democratic centralism, as the latest declaration of Xi Jinping as the "core" leader has illustrated. But the planning system and regulation implanted since the 1950s has been more decentralized and fragmented. This sys-tem was further loosened in the 1960s by the attacks on bourgeois figures and ideas in the party-state. Besides, amidst the state's tight command of national territory, regional administrative units have been granted some degree of autonomy in the spatial administrative hierarchy. The recent, further release of intervention from Beijing has allowed these units to benefit, differentially, from the massive influx of foreign capital. Moreover, the ruling ideology had never guaranteed vanguard status to the peasants, who are forced to fend for themselves under the prevailing household registration system. Unlike the working class, the social contract with the peasants has never been an overwhelming hurdle to the development of socialism. To the Chinese vanguard party, then, the resulting vanguard mode of regulation is disparate. Even though the party still hierarchizes the economy and society, the looser nature of administrative-directive plan relations with enterprise managers refuse standardized regulation while the more limited social contract with the peasants has provided a source of enthusiasm for development. However, the official ideology differs considerably from vanguard Marxism. Maoism has criticized the party-state for being the source of inequalities, exploitation and alienation. Besides, due to its *tongbian* tradition, it emphasizes continuity instead of abrupt change, calling for the continuous re-interpretation of Marxism-Leninism-Mao Zedong Thought with the input of Deng Xiaoping and other leading figures, and the persistence of the Chinese-ness instead of complete westernization. Again, the old tradition of adjudication of intellectuals by the state lingers on. As a result, the Chinese state and intellectual tradition have interacted to produce a criticality that is bound to be different from its capitalist counterpart.

Accordingly, the New Left in China are not critical in that they have attributed many observable problems to the domination of a market economy and the operation of price under increasing globalization, and they have recommended an increase in state intervention. As aforementioned, it is the vanguard state, not public ownership of means of production, that is the crux of the problem. Besides, in advocating the practice of mass mobilization as during the Cultural Revolution, they have endorsed the continual leadership of the vanguard party at the expense of worker and community democratic movements from below. The class category endorsed is still limited to friends and foes as advocated by Mao. Moreover, proposals to increase "public" housing provision and other benefits have been restricted to piecemeal treatment of the labour contract to raise productivity instead of fostering the development of human capacity. Similar comments can be made in relation to the underlying logic of strengthening the vanguard state in the various regional administrative reforms of promoting cities and regions in the governing of the economy and the society. The recent call by the party to pay attention to Marxist geography is but another example of encouraging intellectual development in alignment with vanguard Marxism. In other words, although the Chinese may have done a lot in promoting criticality, critical geography in China has not advanced to a higher status, as Marx and Engels would have liked under communism.

My task in the above was to document the development of critical geography in China by arguing that the difference in critical geography between the West and the Chinese is not so much the variegated forms of one origin as the emergence of distinctive paths. I have been able to substantiate this argument by employing the methodology of a spatial story informed by the non-dualistic, *tongbian* tradition that emphasizes the mutual embeddedness between contradictions. Undoubtedly, this methodology requires further elaboration and improvement, but it has provided a pointer for a better understanding of critical geography in the world. It is in this spirit that this chapter was written.

Notes

1 The Chinese version of "dialectics".
2 "Rightists" were those who opposed utopian transformations.
3 The Four Modernisations refers to the modernisation of agriculture; industry; national defence; and science and technology.

References

An, N. and H. Zhu. 2013. Otherness, power and the construction of place: Towards a theoretical and empirical reassessment of imaginative geography. *Renwen Dili* 28, 20–47. (in Chinese)

Arrighi, G. 2007. *Adam Smith in Beijing: Lineages of the Twenty-first Century*. London: Verso.

Barbalet, J. 2011. Market relations as *wuwei*: Daoist concepts in analysis of China's post-1978 market economy. *Asian Studies Review* 35, 335–354.

Beijing Normal University People's Commune Planning Group. 1959. The role of economic geography in the planning of people's commune. *Dili Xuebao* 25(1), 40–46. (in Chinese)

Best, U. 2009. Critical geography. In, R. Kitchin and N. Thrift (eds.), *The International Encyclopedia of Human Geography* 2. Oxford: Elsevier, pp. 345–357.

Blomley, N. 2006. Uncritical critical geography? *Progress in Human Geography* 30, 87–94.

Brenner, N., J. Peck and N. Theodore. 2010. Variegated neoliberalization: Geographies, modalities, pathways. *Global Networks* 10, 182–222.

Bunnell, T. and A. Maringanti. 2010. Practising urban and regional research beyond metrocentricity. *International Journal of Urban and Regional Research* 34, 415–420.

Cai, Y. et al. 2016a. Knowledge as action: Marxist geography and its development in China. *Geographical Research* 35(7), 1205–1229. (in Chinese)

Cai, Y. et al. 2016b. Marxist geography and its development in China: Reflections on planning and practice. *Geographical Research* 35(8), 1399–1419. (in Chinese)

Cartier, C. 2015. Territorial urbanization and the party-state in China. *Territory, Politics, Governance* 3, 294–320.

Chai, Y., S. Zhou, Y. Cai, Y. Zhang, L. Wu and G. Weng. 2007. Recent progress in human geography in China. *Japanese Journal of Human Geography* 59, 2–22.

Chen, B. 2007. *Tianxia or between the Sky and the Earth: The Classical Horizon of Chinese Thoughts.* Shanghai: Shanghai Shubdian Chubanshe. (in Chinese)

Chen, C.S. 1981. *Cultural Geography of China.* Hong Kong: Joint Publishing Co. (in Chinese)

Chen, F. 2010. *The Turning Point of Republican Historiography – The Study on Social History Debates in China (1927–1937).* Jinan: Shandong Daxue Chubanshe. (in Chinese)

Chen, F., X. Ji and H. Zhang. 2014. Progress and prospect of transportation equity in urbanization process. *Renwen Dili* 29, 10–17. (in Chinese)

Chen, S. 1997. Legitimizing the state: Politics and the founding of Academia Sinica in 1927. *Papers on Chinese History* 6, 23–41.

Chiang, T.C. 2005. Historical geography in China. *Progress in Geography* 29, 148–164.

Chu, Q. 1994. An analysis of the publication of human geography literary works in our country. *Renwen Dili* 9(3), 22–31. (in Chinese)

Cumings, B. 2011. The "rise of China"? In, R.B. Marks and P.G. Pickowicz (eds.), *Radicalism, Revolution, and Reform in Modern China: Essays in Honor of Maurice Meisner.* Lanham: Lexington Books, pp. 185–207.

Dardess, J.W. 2010. *Governing China 150–1850.* Indiannapolis/Camrbidge: Hackett Publishing Company.

Dirlik, A. 1997. Mao Zedong and "Chinese Marxism". In, B. Carr and I. Mahalingam (eds.), *Companion Encyclopedia of Asian Philosophy.* London: Routledge, pp. 593–619.

Dutton, M.. 2005. The trick of words: Asian studies, translation, and the problems of knowledge. In, G. Steinmetz (ed.), *The Politics of Method in the Human Sciences: Positivism and Its Epistemological Others.* Durham: Duke University Press, pp. 89–125.

Editorial Office, The China Institute of Geography. 1959. Organize strength in preparation for the great leap forward – a record of the professional geography meeting. *Dili Xuebao* 25(1), 105–106. (in Chinese)

Elden, S. 2004. *Understanding Henri Lefebvre: Theory and the Possible.* London: Continuum.

Gabriel, S.J. 2006. *Chinese Capitalism and the Modernist Vision.* London: Routledge.

Gong, Y. (ed.) 2003. *Ideological Trends: China 'New Left' and its Effects.* Beijing: Zhongguo Shehui Chubanshe. (in Chinese)

Goss, J. and T. Wesley-Smith. 2010. Remaking area studies. In, T. Wesley-Smith and J. Goss (eds.), *Remaking Area Studies: Teaching and Learning across Asia and the Pacific.* Honolulu: University of Hawai'i Press, pp. ix–xxvii.

Goswami, M. 2004. *Producing India: From Colonial Economy to National Space.* Chicago: The University of Chicago Press.

Gregory, D. 1994. *Geographical Imaginations.* Oxford: Blackwell.

Gu, C. 2007. Study on building harmonious society and expanding social geography. *Renwen Dili* 22, 7–11. (in Chinese)

Gunn, G.C. 2011. *History without Borders: The Making of an Asian World Region, 1000–1800.* Hong Kong: Hong Kong University Press.

Guo, S. 2000. *Late-Qing Geography in the Storm Surge of the West.* Beijing: Beijing Daxue Chubanshe. (in Chinese)

Hall, D. and R. Ames. 1998. Chinese philosophy. In, E. Craig (ed.), *Routledge Encyclopedia of Philosophy.* London: Routledge. http://www.rep.routledge.com/article/G001 (21 December 2011).

Hamashita, T. 2008. *China, East Asia and the Global Economy: Regional and Historical Perspectives*. London: Routledge.

Harvey, D. 2005. *A Brief History of Neoliberalism*. Oxford: Oxford University Press.

Harvey, D. 2015. *Seventeen Contradictions and the End of Capitalism*. London: Profile Books.

He, S. and J. Qian. 2017. From an emerging market to a multifaceted urban society: Urban China studies. *Urban Studies* 54, 827–846.

He, S. and F. Wu. 2009. China's emerging neoliberal urbanism: Perspectives from urban redevelopment. *Antipode* 41, 282–304.

Hu, Y. and X. Jiang. 1995. *The History of Chinese Geography*. Taipei: Wenjian Chubanshe. (in Chinese)

Huang, B. 1958. Learning from the Soviet Union, and thanking the Soviet Union. *Dili Xuebao* 24(1), 19–23 (in Chinese)

Huang, Y., Z. Zhang, F. Yin and X. Luo (eds.) 2012. *Human Geography in Nanjing University: 1919–2012*. Nanjing: Nanjing Daxue Chubanshe. (in Chinese)

Ikeda, S. 1996. The history of the capitalist world-system vs. the history of East-Southeast Asia. *Review* 19, 49–77.

Kipfer, S. 2009. Preface to the new edition. In, H. Lefebvre (ed.), *Dialectical Materialism*. Minneapolis:The University of Minnesota, pp. xiii–xxxii.

Lanza, F. 2013. Making sense of "China" during the Cold War: Global Maoism and Asian Studies. In, J.E. Mooney and F. Lanza (eds.), *De-Centering Cold World War History: Local and Global Change*. London: Routledge, pp.147–166.

Lazarus, N. and R. Varma. 2008. Marxism and postcolonial studies. In, J. Bidet and S. Kouvelakis (eds.), *Critical Companion to Contemporary Marxism*. Leiden: Brill, pp. 309–331.

Lebowitz, M.A. 2012. *The Contradictions of "Real Socialism": The Conductor and the Conducted*. New York: Monthly Review Press.

Lefebvre, H. 1991. *The Production of Space*. Malden, MA & Oxford: Blackwell Publishing.

Li, D. 2006. *Research on the Urban People's Commune Movement*. Changsha: Hunan Renmin Chubanshe. (in Chinese)

Li, H. 2015. *Political Though and China's Transformation: Ideas Shaping Reform in Post-Mao China*. Basingstoke: Palgrave Macmillan.

Li, W. 1957. The role of physical factors and technical-economic factors in industrial allocation. *Dili Xuebao* 23(4), 399–417. (in Chinese)

Li, W., P. Ren and J. Feng. 1960. The formation of Fangshan people's commune *cum* production unit and the prospect of production allocation diversification. *Dili Xuebao* 26(2), 110–120. (in Chinese)

Li, Z. 2006. *Marxism in China*. Hong Kong: Ming Pao Publishing House. (in Chinese)

Lin, C. 2006. *The Transformation of Chinese Socialism*. Durham, NC: Duke University Press.

Lin, H. 1999. *A History of Chinese Ideological Movements*. Hong Kong: Tiandi Chubanshe. (in Chinese)

Liu, J., R. Jin and K. Zhou. 1999. *The Administrative District Geography of China*. Beijing: Kexue Chubanshe. (in Chinese)

Liu, Y. and F. Wang. 2013. Reexamining Chinese human geography: Perspective of theoretical studies in recent thirty years. *Renwen Dili* 28, 14–19. (in Chinese)

Liu, Y. and X. Xu. 2008. Duality of Chinese geography. *Dili Kexue* 28, 587–593. (in Chinese)

Lo, D. 2016. Developing or under-developing? Implications of China's "going out" for late development. *Working Papers Series* No.198, London: Department of Economics, SOAS, University of London.

Lu, L., Y. Wei and C. Lin. 2006. Urban geography of China: Views of overseas scholars. *Renwen Dili* 21, 67–71. (in Chinese)

Mao, Z. 1999. *Works of Mao Zedong*. Beijing: Zhonggong Zhongyang Wenxian Yanjiusuo. (in Chinese)

Marks, R.B. 2011. Chinese communists and the environment. In, R.B. Marks and P.G. Pickowicz (eds.), *Radicalism, Revolution, and Reform in Modern China: Essays in Honor of Maurice Meisner*. Lanham: Lexington Books, pp. 105–132.

Mierzejewski, D. 2009. "Not to oppose but to rethink": The New Left discourse on the Chinese reforms. *Journal of Contemporary Eastern Asia* 8, 15–29.

Peng, M. 1995. *Historical Geography and Modern Chinese Historiography*. Taipei: Dongda Tushu Gongshi. (in Chinese)

Perdue, P.C. 2003. A frontier view of Chineseness. In, G. Arrighi, T. Hamashita and M. Selden (eds.), *The Resurgence of East Asia: 500, 150 and 50 Year Perspectives*. London: Routledge, pp. 51–77.

Perry, E.J. 2008. Chinese conceptions of "rights": From Mencius to Mao – and now. *Perspectives on Politics* 6, 37–50.

Qian, J. 2013. Post-structuralist discourses and social theories: Michel Foucault and Henri Lefebvre. *Rewen Dili* 28, 45–52. (in Chinese)

Qian, Y. 2000. The process of China's market transition (1978–1998): The evolutionary, historical, and comparative perspectives. *Journal of Institutional and Theoretical Economics* 156, 151–179.

Rofel, L. 2012. Between *tianxia* and postsocialism: Contemporary Chinese cosmopolitanism. In, G. Delanty (ed.), *Routledge Handbook of Cosmopolitanism Studies*. London: Routledge, pp. 443–451.

Savage, V.R. 2003. Changing geographies and the geography of change: Some reflections. *Singapore Journal of Tropical Geography* 24, 61–85.

Schmid, C. 2008. Henri Lefebvre's theory of the production of space: Towards a three-dimensional dialectic. In, K. Goonewardena, S. Kipfer, R. Milgrom and C. Schmid (eds.), *Space, Difference, Everyday Life: Reading Henri Lefebvre*. London: Routledge, pp. 27–45.

Shen, W. and J. Lu. 2001. *China National Security Geography*. Beijing: Shishi Chubanshe. (in Chinese)

Shin, H.B. 2013. The right to the city and critical reflections on China's property rights activism. *Antipode* 45(5), 1167–1189.

Soviet Union Institute of Geography (ed.) 1952. *Bourgeois Geography in the Service of US Imperialists*. Beijing: Zhonghua Shudian. (in Chinese)

Streeck, W. and K. Thelen. 2005. Introduction: Institutional change in advanced political economies. In, W. Streeck and K. Thelen (eds.), *Beyond Continuity: Institutional Change in Advanced Economies*. Oxford: Oxford University Press, pp. 1–39.

Studwell, J. 2007. *Asian Godfathers: Money and Power in Hong Kong and Southeast Asia*. New York: Atlantic Monthly Press.

Su, X. 2013. Hegemony, cultural politics, and geographical ideology of Gramsci. *Renwen Dili* 28, 10–13. (in Chinese)

Tang, W.S. 2006. Planning Beijing strategically: "One world, one dream". *Town Planning Review* 77, 257–282.

Tang, W.S. 2009. Chinese-language geography. In, R. Kitchin and N. Thrift (eds.), *The International Encyclopedia of Human Geography* 2. Oxford: Elsevier, pp. 72–77.

Tang, W.S. 2014a. Governing by the state: A study of the literature on governing Chinese mega-cities. In, P.O. Berg and E. Björner (eds.), *Branding Chinese Mega-cities: Strategies, Practices and Challenges*. Cheltenham: Edward Elgar, pp. 42–63.

Tang, W.S. 2014b. When Lefebvre meets the East: Urbanisation in Hong Kong. In, A. Moravánszky, C. Schmid and L. Stanek (eds.), *After the Urban Revolution*. Surrey: Ashgate, pp. 71–91.

Tang, W.S. and K.C. Chan. 2008. Human geography in Hong Kong: A preliminary analysis. *Japanese Journal of Human Geography* 60, 36–52.

Taylor, P. 2011. Thesis on labour imperialism: How communist China used capitalist globalization to create the last great modern imperialism. *Political Geography* 30, 175–177.

Teather, E.K. and C.S. Chow. 2000. The geographer and the fengshui practitioner: So close and yet so far apart? *Australian Geographer* 31, 309–332.

Tian, C. 2005. *Chinese Dialectics: From Yijing to Marxism*. Lanham: Lexington Books.

Tibebu, T. 2011. *Hegel and the Third World: The Making of Eurocentrism in World History*. Syracuse: Syracuse University Press.

Timár, J. 2004. More than "Anglo-American", it is "Western": Hegemony in geography from a Hungarian perspective. *Geoforum* 35, 533–538.

Trichur, G.K. 2012. East Asian developmental path and land-use rights in China. *Journal of World-Systems Research* 18, 71–88.

Walker, R. and D. Buck. 2007. The Chinese road: Cities in the transition to capitalism. *New Left Review* 46, 39–66.

Wallerstein, I. 2002. New revolts against the system. *New Left Review* 18, 29–39.

Wang, B. and J. Lu. 2012. Introduction: China and New Left critique. In, B. Wang and J. Lu (eds.), *China and New Left Visions*. Lanham: Lexington Books, pp. ix–xvi.

Wang, H. 2006. Discursive community and the genealogy of scientific categories. In, M.Y. Dong and J.L. Goldstein (eds.), *Everyday Modernity in China*. Seattle: University of Washington Press, pp. 80–120.

Wang, H. and X. Tang. 2013. Review on feminist geography in the perspective of new cultural geography. *Renwen Dili* 28, 26–31. (in Chinese)

Wang, X., J. Li, X. Liu and F. Chang. 2016. The era destiny of spatial research of human geography in the new normal. *Renwen Dili* 31, 1–8 & 134. (in Chinese)

Wang, X. and Y. Liu. 2007. The development of research methodology in human geography and its schools of thought since "the cultural turn". *Renwen Dili* 23, 1–6. (in Chinese)

Wang, Z. 2002. Saving China through science: The Science Society of China, scientific nationalism, and civil society in Republican China. *Osiris (2nd Series)* 17, 291–322.

Weller, R.P. and P. Bol. 1998. From haven-and-earth to nature: Chinese concepts of the environment and their influence on policy implementation. In, M.B. McElroy, C.P. Nielson and P. Lydon (eds.), *Energizing China: Reconciling Environmental Protection and Economic Growth*. Cambridge, Mass.: Harvard University Press, pp. 473–499.

Wolin, L. 2010. *The Wind from the East: French Intellectuals, the Cultural Revolution, and the Legacy of the 1960s*. Princeton: Princeton University Press.

Wong, R.B. 1999. The political economy of agrarian empire and its modern legacy. In, T. Brook and G. Blue (eds.), *China and Historical Capitalism: Genealogies of Sinological Knowledge*. Cambridge: Cambridge University Press, pp. 210–245.

Wong, W.C. 2006. Understanding dialectical thinking from a cultural-historical perspective. *Philosophical Psychology* 19, 239–260.

Wu, C.J. 1990. The progress of human geography in China: its achievements and experiences. *Geojournal* 21, 7–12.

Wu, C. 1958. A few aspects of the situation of the geographical institutions in the Soviet Union and of the Soviet Union Geography. *Dili Xuebao* 24(4), 438–56. (in Chinese)

Wu, J. 2005. *Economic Reforms of Contemporary China*. Shanghai: Shanghai Yuandong Chubanshe. (in Chinese)

Wu, X.A. 2009. China meets Southeast Asia: A long-term historical review. In, K.L. Ho (ed.), *Connecting & Distancing: Southeast Asia and China*. Singapore: Institute of Southeast Asian Studies, pp. 3–30.

Xia, Y. 2011. Mao Zedong. In, S. Casey and J. Wright (eds.), *Mental Maps in the Early Cold War Era, 1945-68*. Basingstoke: Palgrave Macmillan, pp. 160–79.

Xu, T. 1985. The Marxist-Leninist foundation of Human Geography – the prospect of the human-land view based on the viewpoint of Marxism. In, Y. Li (ed.), *Forum on Human Geography*. Beijing: Renmin Jiaoyue Chubanshe, pp. 15–25. (in Chinese)

Yang, W. 1988. *An Outline History of Geographical Thought*. Beijing: Gaodeng Jiaoyu Chubanshe. (in Chinese)

Yeung, Y.M. and Y. Zhou. 1991. Human geography in China: Evolution, rejuvenation and prospect. *Progress in Human Geography* 15, 373–394.

Zhang, F. 2013. The picture of the contemporary China New Left. http://marxism. org.cn/detail.asp?id=2135&Channel=11&ClassID=11 (27 July 2013) (in Chinese)

Zhang, T. 1956. Human Geography serves imperialism. *Dili Xuebao* 22(1), 1–36. (in Chinese)

Zhang, T., J. Song, Y. Su, Y. Wu, S. Su and R. Hu. 1959. Preliminary experiences of the economic planning of rural people's commune. *Dili Xuebao* 25(2), 107–119. (in Chinese)

Zhou, Z. 1999. *Systemic Transformations and Economic Growth: The Chinese Experience and an Analysis of Examples*. Shanghai: Shanghai Sanlian/Shanghai Renmin. (in Chinese)

8 Francophone critical geography

Rodolphe De Koninck and Michel Bruneau

French classical geography and Élisée Reclus: The antipodes

Critical geography refers much more to a type of approach or to a point of view than to a field of knowledge per se (De Koninck, 1984). In this manner, critical science appears distinct from positivist and hermeneutic sciences. According to Habermas (1973, 143), the latter pursue "technical" and "practical" goals, while critically oriented sciences are more inclined towards "emancipation". As pointed out by Rioux (1978), these three approaches are not totally incompatible but rather complementary. However, a critical approach is both perilous, more imaginative and optimist since it "refers to value judgments to criticize what exists" and "is interested in [...] the self-creation of man [humanity?] and society" (Rioux, 1978, 9).

In Francophone geography, the origin of a truly critical approach can be found in the work of Élisée Reclus (1830–1905), whose writings spanned a period of nearly 50 years, beginning in 1860 – at least officially – and ending in 1908, some of his major publications having appeared after his death in 1905, particularly his remarkable six-volume *L'Homme et la Terre* (1905–1908), totalling over 3,500 pages (Reclus, 1869, 1876–1894, 1880, 1887, 1905–1908). Reclus was a prolific writer, to this day arguably the most prolific geographer, having, among other things, single-handedly authored the 19-volume *Nouvelle Géographie Universelle*, published between 1876 and 1894 and totalling nearly 18,000 pages! A close friend of Kropotkin (1842–1921), whose works he translated in French, Reclus was an anarchist and an activist who was jailed and even exiled from France, following his involvement with the 1871 *Commune de Paris*. He then travelled widely and eventually established himself in Belgium, where he taught at the Université libre de Bruxelles.

Elisée Reclus was both a predecessor and a contemporary of several key figures in the French school of geography, among them Paul Vidal de la Blache (1845–1918), Lucien Gallois (1857–1941) and Jean Brunhes (1869–1930). Reclus wrote on a wide variety of subjects and in some ways he was quite representative of his day, as he was most adept at broad descriptions, dwelling at length with historical issues and practising a holistic geography.

DOI: 10.4324/9781315600635-8

Yet, and more importantly, he was at the antipodes of his contemporaries, in that much of his writing was extremely critical, particularly of colonial policy, of state power and the elites, including bankers and merchants, whom he accused of manipulating the market as well as princes and kings and of making war and peace at will (Reclus (vol. 5), 1908, 332).

Reclus was one of the first geographers to convincingly condemn colonialism, largely, here again, in his *L'Homme et la Terre*. He was particularly critical of colonies where the land, resources and labour of the majority indigenous people were unashamedly controlled by a minority of conquerors, including within the French colonial empire. He demonstrated the worldwide expansion of capitalism, including the take-over by American capitalists of South American plantations, mines and local trade networks. He was also among the first to analyse the relationship between unequal development and imperialism and the latter's expansion, not only through economic and financial means but also through the manipulation of some oppressed people to oppress some others: for example the Irish throughout the British Empire. "There is no evil comparable to that of an oppressed people turning, with a vengeance, against another people which it oppresses in turn. Tyranny and trampling pile up and are hierarchized" (Reclus (vol. 1), 1908, 281).

In short, Reclus was way ahead of his times and particularly of the vast majority of geographers, the breath of his involvement and of his critical approach, as citizen, teacher, researcher and writer, having little equivalent even among contemporary geographers. What's perhaps most remarkable with Reclus' work, beyond sheer volume and breadth of issues covered, is that it illustrates the fundamental social and political pertinence of geography. This ranges from arguing for the social need to understand the origin and history of a stream or a mountain (1869 and 1875–1876), to supporting the political need for an anarchist ideal (1897) and practising a thorough analysis of the planet's geography under the impact of human occupation (1905–1908).

After his death, as the French school of geography prospered, Reclus' writings were gradually put aside – with deliberate help from some of the leaders of said school (Pelletier, 2009, 27–31) – and then largely forgotten, particularly among geographers! The period of oblivion lasted well into the 1970s.

Modern Francophone critical geography

In the post-Second World War years and throughout the 1950s and 1960s, classical human geography remained dominant, focusing on the description and "explanation" of landscapes and regions, and relying on the concepts of *genre de vie* and civilization, with Pierre Gourou (1900–1999) – the grand master of this approach in the tropical world. However, two leading geographers, Jean Dresch (1905–1994) and Pierre George (1909–2006), had begun to advocate a more critical approach, both socially and politically. Dresch was an early and quite articulate critic of colonial geography as well as of so-called tropical geography (Dresch, 1952). In advocating for social rather than human geography, George began using the Marxist concept of modes of production rather than those of *genre de vie* and civilization. And in much of his work, for example, in two books published in 1945 and 1956, he relied on concepts such as social production of space, social classes and social inequalities (George, 1945, 1956).

Students of Pierre George and Jean Dresch, such as Raymond Guglielmo, Bernard Kayser, Yves Lacoste, André Prenant, Michel Rochefort and Raymond Dugrand, were trained at the Institut de géographie de Paris. Their own families and even sometimes they themselves had been deeply involved in the anti-Nazi resistance movement and had joined the ranks of the Communist Party, which most of them fled following leakages from the supposedly secret Khrushchev report (February 1956) and the Soviet invasion of Hungary in October of the same year. These geographers all became involved in a more committed form of geography, where territorial planning played a key role. Such a vision was largely inherited from positivist Marxist thinking.

In 1965, then ideologically close to George and particularly to Dresch, Yves Lacoste published his landmark *Géographie du sous-développement*, centred on the concepts of Third World and underdevelopment and denouncing the relations of domination linking developed and underdeveloped countries. By the time the second edition of his book came out in 1976, Francophone critical geography had become increasingly influenced by the writings of Henri Lefebvre, a Marxist philosopher and sociologist, particularly *La pensée marxiste et la ville* (1972) and *La production de l'espace* (1974). The relevance of Lefebvre's approach to social and particularly to critical geography can be illustrated by this often-quoted sentence: "Social relations of production have a social existence of their own to the extent that they have a spatial existence; they are projected in space, they are engraved into it as they produce it" (Lefebvre, 1974, 152). By then, critical geography debates also became more prominent in a number of journals, particularly *Hérodote*, *L'Espace géographique*, *EspacesTemps* and the *Cahiers de géographie du Québec*.

The development of critical geography in the late 1970s

Hérodote

In 1976, Lacoste made a double hit, by publishing a well put together and highly readable book and by launching the journal *Hérodote*, which was to quickly become Francophone geography's most widely read periodical. The book, *La géographie, çà sert, d'abord, à faire la guerre*, was devoted to the demonstration of the relevance of geographical knowledge to the practice of warfare from a critical perspective (Lacoste, 1976a). Although the idea was not new, particularly to those familiar with the history of geography, the way Lacoste put it was skilful and convincing. He also argued at length that academic geography had become highly irrelevant and actually hid or at least minimized the use of geographical knowledge by the military. He deplored the fact that geography taught in schools and universities, what he called "la géographie des professeurs", stayed away from theory as well as from most key political issues, while minimizing the role of the state. After reminding his readers of the originality of Élisée Reclus' writings, Lacoste debated the relevance of Marxism to geography, particularly when applied to the analysis of urban problems, an issue he also referred to in the leading article of the first issue of *Hérodote* (Lacoste, 1976b, 31–33).

Subtitled *stratégies, géographies, ideologies*, that first issue of *Hérodote* contained several innovative pieces. These included another article by Lacoste summarizing a careful analysis of the American bombing of the Red River delta in North Vietnam in the late 1960s. By relying on highly detailed topographic maps contained in Pierre Gourou's 1936 doctoral thesis, *Les paysans du delta tonkinois* (Gourou, 1936), Lacoste was able to show that the apparently random and inaccurate bombings of the US Air Force were in fact targeted to weaken the dikes that structured the entire delta. Entitled "Enquête sur le bombardement des digues du fleuve Rouge (Vietnam, été 1972)" (Lacoste, 1976c), Lacoste's article was a summary of an elaborate report he had produced in 1972 for the Russell tribunal on war crimes, based in Stockholm. Shortly after his return from war-torn Vietnam, he had also published an article in the 16 August 1972 issue of *Le Monde*, in which he accused the US Air Force of having deliberately targeted the dikes' foundations, in order to make them so vulnerable as to eventually break open during the forthcoming rainy season, thus flooding the entire and heavily populated delta. At that time, in 1972, Lacoste's report had a major impact on public opinion and was arguably instrumental in bringing the Americans to stop their bombing before the monsoon season. He thus made two major points. First, he demonstrated that classical geographical information, such as that provided by Gourou's thesis and maps, could be used for military purposes and, second, that critical geographers should denounce and even prevent such use.

This first issue of *Hérodote* also included an interview with Michel Foucault, by then one of the intellectual heroes of the French left. With his answers, Foucault supported Lacoste's convictions concerning the strategic relevance of geographic knowledge. But he did not react clearly to his interlocutor's statement that some of Marx writings denoted "a surprising spatial sensibility" (*Hérodote*, 1976a, 1, 84).

L'Espace géographique

Under the leadership of Roger Brunet, who remained its editor for some 30 years, *L'Espace géographique* was launched in 1972, a few years earlier than *Hérodote*. This journal has become one of the most respected in the Francophone world, for a number of reasons, including for having remained open to well organized debates among its contributors. One of these debates, published in a 1977 issue, centred on the relevance of Marxism in geography. The debate was launched by Paul Claval, who argued, in an article entitled "Le marxisme et l'espace" (Claval, 1977), that Marxism was of limited use in geography because of the lack of attention it gave to space. This was challenged by a team of researchers from Bordeaux (Collectif, 1977) who counter-argued that, on the contrary, the concept of modes of production could be most useful as an analytical tool, applying it to a study of the unequal expansion of the capitalist mode of production through Thai space (Bruneau et al., 1977). Bruneau subsequently expanded his analysis in two articles published in the *Cahiers de géographie du Québec* and *Antipode* (Bruneau, 1978, 1981), while Van Beuningen joined the debate in a 1979 issue of *L'Espace géographique* (Van Beuningen, 1979).

Cahiers de géographie du Québec

In 1978, under the title of *Le matérialisme historique en géographie*, the *Cahiers de géographie du Québec* devoted an entire issue to Marxism in geography, the first Francophone journal to do so (*Cahiers de géographie du Quebec*, 1978). In the lead article, De Koninck criticized at length classical geography as well as "new geography", particularly Anglo-Saxon quantitative geography. He showed that both were fundamentally idealistic in that, each in its own way, they described their object, whether rural landscapes or American ghettos, but systematically avoided analysing the conflicts that lay behind them (De Koninck, 1978). Regarding the relevance of Marxist concepts, De Koninck (1980) joined the debate, this time in *Hérodote*. Following a thorough content analysis of articles published by Soviet geographers, whose methods were highly similar to those used by neoliberal Western geographers, he suggested that the Soviets' systematic avoidance of the use of any Marxist concept was a *contrario* demonstration of their potent analytical power.

The debates concerning the need to revive critical geography approaches in Francophone geography, whether or not through the reliance on Marxist concepts, continued for some time in the *Cahiers,* for example, in 1985 (De Koninck, 1985) involving a key contribution from Claude Raffestin (1985) (see further below).

EspacesTemps

EspacesTemps was launched in 1975, by a group of young geographers and historians, intent on making a difference among social scientists. With strong Trotskyist leanings, the journal nevertheless remained quite open to a large array of opinions and provided critical and constructive points of view on the need for theory in geography, including in the context of the debate on the relevance of Marxism in geography, to which an entire issue was devoted in 1981, under the title of *Une géographie à visage humain?* (*EspacesTemps*, 1981). In 1988, *EspacesTemps* and the *Cahiers de géographie du Québec* organized a joint conference in Paris, attended by geographers from both sides of the Atlantic. Under the common title *Géographie, état des lieux: débat transatlantique*, the publication of the Conference proceedings was divided up between the two journals, (*Cahiers de géographie du Québec*, 1988; *EspacesTemps*, 1989). The contents were a testimony to the wide spectrum of positions on the state of the discipline. Two of the most notable theoretical publications came from David Harvey and Allen Scott, in the *Cahiers de géographie du Québec*, and Claude Raffestin, in *EspacesTemps* (Harvey and Scott, 1988; Raffestin, 1989).

The originality of Claude Raffestin's contribution

Claude Raffestin's contribution to critical geography or to geography, simply and as he would prefer, is particularly original. As frequently pointed out (De Koninck, 1981; Fall, 2007; Racine, 2002), his work remains insufficiently known, perhaps because very little of it has appeared in English and because it still represents "unfinished business". In 1979, Raffestin and Mercedes Bresso co-authored a truly remarkable book entitled *Travail, espace, pouvoir* (Raffestin and Bresso, 1979). In this seminal work, by distinguishing space from territory and advocating the need to conceive territory as space in which labour (i.e. energy and information) is invested, Raffestin and Bresso proposed nothing short of an epistemological revolution in geography. From that proposal, the concepts of territoriality and territorialization could have become much more insightful than what had been advocated so far and has been since, including in Anglo-Saxon literature (Sack, 1986; Vandergeest and Peluso, 1995), where territorialization seems to be essentially a State affair. The fact that further development of Raffestin and Bresso's concepts and particularly their application to empirical analyses did not really follow suit can arguably be attributed to the publication, a year later, of Raffestin's nearly as remarkable *Pour une géographie du pouvoir* (Raffestin, 1980).

For partially understandable reasons – including its more trendy title and its publication in Paris rather than in Geneva – the latter book was given much more attention, and was picked up by political geographers, or geographers who claimed to be so. The ensuing debates focussed on the geography of power and much less on the theory of territoriality appropriation, which was the actual breakthrough suggested by Raffestin and Bresso. This appropriation, in fact, is largely achieved through the alienation of labour, as has been shown by De Koninck (1981). Not only does the study of the nuts and bolts of the alienation of territorialized labour present an extremely promising field of research for critical geographers, it is one that could make prolific use of the Marxist concepts of appropriation and alienation of labour.

Whatever the case, and to return to Raffestin, although over the years he has continued to write about territoriality and related topics, he has never really built on what he had sown in these two books. Always short and innovative, his articles have helped to broaden his readership. But it seems that neither he nor any of his readers have been able to expand meaningfully on his theoretical contribution or to test it in a broad empirical context. Raffestin has however been able to challenge those – in particular the new École française de géopolitique put together by Lacoste and the *Hérodote* group – who have been trying to reduce political geography to the study of nationalistic state power. This he clinched in yet another remarkable co-authored book, *Géopolitique et histoire*, published in 1995 (Raffestin et al., 1995). Since then, his work has remained underexploited, at least partly because Raffestin himself has been unable to render it more accessible and more operational for those intent on testing his theories (Fall, 2007).

Claiming Reclus' heritage

If the work of Élisée Reclus has been marginalized during most of the twentieth century, by the 1980s its rediscovery was well under way and it continues to this day. Much of the credit can be attributed to Yves Lacoste and one of his main followers, Béatrice Giblin, who eventually became his successor as Chief Editor of *Hérodote*. As early as 1976, under the title *Élysée Reclus, géographe, anarchiste*, the journal *Hérodote* devoted its second issue to the great geographer (*Hérodote*, 1976b) and, over the years, Lacoste himself (e.g. Lacoste, 2005) and several of the journal's contributors have kept reminding their readers of the relevance of his work. Two additional issues were devoted to that task, in 1981 under the title *Élysée Reclus, un géographe libertaire* and in 2005 as *Élysée Reclus* (*Hérodote*, 1981, 2005). Béatrice Giblin has been one of the most significant contributors to the rediscovery of Reclus's work; in 1971, she completed her PhD thesis (entitled *Élisée Reclus, géographe*) with Lacoste and, perhaps more importantly, in 1982 she assembled, commented on and published two volumes of selected writings from *L'homme et la terre* (Reclus, 1982. See also Giblin, 2005).

Another French "school" of geography, headed by Roger Brunet, also paid homage to Reclus. In 1984, Brunet founded the GIP RECLUS, which stands for *Groupe d'Intérêt Public* and *Réseau d'étude des changements dans les localisations et les unités spatiales*. The GIP RECLUS did much more than promote the study of "locations and spatial units" by developing a very broad research and publication programme, headquartered in the *Maison de la géographie* in Montpellier. One of the GIP RECLUS' signal contributions has been a ten-volume *Géographie Universelle* (GU), directed by Brunet, which appeared between 1995 and 2005 with contributions from around 100 geographers. Brunet himself, who remains one of the most influential contemporary Francophone geographers, has never been an official proponent of critical geography. But he has left a lasting imprint on French geography, through his numerous and always well-polished and very constructive publications, among which the remarkable *Le Déchiffrement du Monde* (Brunet, 2001) and numerous still blossoming initiatives, including the 14-volume Atlas de France (GIP RECLUS) and the founding or co-founding and directorship of the journals *L'Espace géographique* and *Mappemonde*.

Several geographers associated in one way or another with the RECLUS group, in particular among those involved in the GU publications, have on their own contributed to critical geography literature. This includes the authors of this chapter, but also and foremost Philippe Pelletier, the author of the Japan section of the *Chine, Japon, Corée* volume. Pelletier was instrumental in the organization of an international conference devoted to RECLUS' work, held in Lyon in September 2005 and, more importantly, in writing the excellent and already mentioned *Élysée Reclus, géographie et anarchie* (Pelletier, 2009)

Francophone tropical geography and its critique

Tropical geography has long constituted one of the most prolific fields in Francophone geography, and it remains well represented to this day, particularly in the Université de Bordeaux, which has published the journal *Cahiers d'Outre-Mer* since 1948. Tropical geography has in some ways evolved in isolation and remains a late bearer of classical and generally highly uncritical French geography. Hence, and given that French tropical geography is still influential in much of the so-called developing world, particularly Francophone Africa, its critique follows here.

The birth of critical geography in France in the days of decolonization

Tropical geography appeared in France shortly after World War II, during the early stages of the decolonization process of the British, French and Dutch empires. In some ways it can be said that it followed on from colonial geography, with Pierre Gourou's *Les Pays Tropicaux* (Gourou, 1947) playing a founding role among both Francophone and Anglophone academic circles, *The Tropical World* appearing in 1953 (Gourou, 1953). At the outset, anti-colonial activists challenged tropical geography. Such was the case with Aimé Césaire (1913–2008) who, in *Discours sur le colonialisme* (1953), pointed out that "tropicality" was presented as some sort of geographical burden justifying as well as concealing colonial exploitation and its devastations. An engaged poet and committed politician from French Martinique, Césaire qualified Gourou's basic thesis as partial and unacceptable. He questioned the position that "there has never existed a great tropical civilisation, that great civilisations have only taken root in temperate countries, that in any tropical country the seed of civilisation comes and can only come from outside the tropics and that tropical countries are weighed down either by the biological curse of the racists or, with the same consequences, by a no less efficient geographical curse" (Césaire, 1953, 32).

In France, in the aftermath of WWII, the relationship between colonial geography and burgeoning tropical geography was clearly established by Robequain who, in his 1948 review of Gourou's *Pays Tropicaux*, wrote: "However, that tropical area [...] bears a certain number of common characteristics [...]: in particular an inferiority in techniques and a lack of political cohesion which have made it the colonial area par excellence. The knowledge of tropical countries is precisely the consequence of colonisation" (Robequain, 1948, 71). This link was also clearly underlined by Césaire, when he wrote that Gourou's tropical geography was lacking in the analysis of relations of production specific to the colonial system.

As early as 1948, the Marxist geographer Suret-Canale wrote a critical review of Gourou's book, questioning the explanation of the backwardness of tropical countries as being largely attributable to poor soil quality and other environmental factors. He accused Gourou of ignoring that one of the fundamental characteristics of the countries he was studying was that they were under colonial administration and exploitation (Suret-Canale, 1948, 103).

Thus, a critical stand against tropical geography as a direct outcome of colonial geography was gradually set in motion. Dresch, in particular, criticized the methods used and the easy profits made by the monopoly positions of trading companies both in the agricultural and mining sectors, calling this a "singularly primitive and lazy economy" (Dresch, 1979, 170). Such a critical geography also looked into investment issues, labour migrations and urbanization, topics to which tropical geographers generally remained oblivious, hampered by their pessimistic interpretation of the tropical environment, an interpretation condemned by the agronomist René Dumont in a highly critical book on Africa that became a bestseller (Dumont, 1962, 16).

It was at that time that, following in Dresch and Suret-Canale's footsteps, Lacoste put together his *Geography of underdevelopment* (1965), in which he referred to the Third World, relying on Marxist concepts of relations of production, class struggle and unequal exchange and referring to the domination exerted by rich countries over poor ones. He also condemned the naturalism of "tropicalists" for giving too much importance to the so-called poverty of tropical soils in their explanation of the poverty of Third World countries. "By making it believe – so he wrote – that all Third World countries are tropical (which is false) and reciprocally that all countries located in the tropical world are poor (pretending to ignore that rich Australia is largely within the tropical belt) and owe their misery essentially to the various handicaps associated with the natural conditions specific to the tropics" (Lacoste, 1976a, 129).

However, these criticisms of Gourou's position were largely ignored in French geography's academic circles. Nevertheless, the terms of the debates, even if highly polemical and non-academic, had been clearly laid out in Césaire's 1953 book.

The debate in 1980s France on the tropicalist approach

It was only in the 1980s that an intense debate finally took place among French geographers – but still without any reference to Césaire's well-known book – coupled with an institutional crisis at the Centre d'Études de Géographie Tropicale de Bordeaux (CEGET).

The tropical geography debate centred on Gourou's thoughts which, according to its initiators, constituted a real scientific paradigm (see, for example, *Hérodote*, 1984; *L'Espace géographie*, 1984; Sautter, 1984). This paradigm was based on an approach of, first, landscape types, reconstructed by geographers at the local or regional scale, and conceived of as structures illustrating the relationships between man and natural environment; second, population distribution, generally as represented on population density maps. The concept of civilization, as defined by Gourou, acted as an explanatory principle, which posed civilization to be specific to any society on earth and based on production techniques and management techniques (techniques de production et techniques d'encadrement). Time and scale, small and large, case studies, local situations, all of these were privileged in the analysis. The intermediate scale, to which nation-states were considered to belong, was generally neglected. Geography was not seen as governed by law-like regularities, and reference to any type of model was frowned upon.

Third World geography proponents developed their criticism of this approach by relying on, first, Marxist concepts of relations of production within social formations specific to postcolonial nation-states and, second, on spatial analysis, itself based on model building and theorization. Their major criticisms were aimed at the isolation in which tropical geography found itself among social sciences. The overly descriptive approach of civilizations and techniques, even of management techniques, was said to shy away from any in-depth social or political analysis, the concept of civilization appearing totally devoid of analytical power, as it often blends or confuses landscapes with management techniques. The fact that the tropical geography approach generally neglected the role of the state and relations of domination and dependence was considered unacceptable. Finally, Gourou's neglect of the colonial factor was deplored (Suret-Canale, 1994), as was the overwhelming focus of his work on rural issues. Even in a 1982 book, in which Gourou presented himself as more hopeful about the future of the tropical world, he still shied away from the issue of differentiation and inequalities among the peasantries (Gourou, 1982). And, as Gottmann showed, he still did not take into account the process and consequences of urbanization and the role of economic and political factors (Gottmann, 1983, 41).

By deliberately sticking to a description of socio-political and cultural "encadrements", and by referring to other disciplines for more in-depth analyses, Gourou refused to integrate his approach into a broader social theory. He much preferred locking it up into an analytical black box of what he termed "encadrements", focusing instead on case studies and local characteristics and turning to "civilization" as the all-explaining variable. Gourou was fond of referring to Man, to men or sometimes to the "human group". But he never mentioned society, possibly fearing it might involve him with other social sciences, particularly sociology.

When the debate over tropical geography among French geographers died down in the late 1980s, the issue of the validity of Gourou's approach remained opened (Raison, 1989). By that time, however, British historians of the environment (e.g. D. Arnold) and geographers involved in postcolonial studies (e.g. D. Clayton, F. Driver, D.N. Livingstone, etc.) had opened a new debate on representations of the non-European world and the tropics, taking in the writings of the great nineteenth-century naturalist travellers, Humboldt, Darwin and Wallace, as well as Gourou's 1947 *Pays Tropicaux* and concept of "tropicality" (Arnold, 1996).

In other words, the critical assessment of tropical geography, particularly as conceived of by Gourou, was largely pursued outside of France, among Anglophone geographers, as illustrated in a 2005 issue of the *Singapore Journal of Tropical Geography*. With notable contributions from, among others, Bowd and Clayton (2005) (see also Bruneau, 2005; Kleinen, 2005; Raison, 2005), such assessments revealed that by putting forth western reason and science as tools essential for "development", in admiring the harmony and "intemporality" of the paddy landscapes of the Far East, Gourou's orientalism (Said, 1978) was instrumental in maintaining much of tropical geography at a safe distance from a critique of the impact of colonialism and capitalism, and it blinded tropical geography to the aspirations of Third World populations. Critical Anglophone assessments of Gourou's tropical geography have revealed the extent to which it was, in fact, a discourse of power and a way of rendering exotic a non-western, oriental and/ or tropical world. This has been developed in the recently published scientific biography of Pierre Gourou (Bowd and Clayton 2019) and the review of it by Bruneau (2020).

Recent trends in Francophone critical geography

Continuing debates in the journals

Like *Antipode* and *ACME* (*An International Journal for Critical Geographies*), a group of French geographers based at the Université de

Paris Ouest Nanterre (Paris X) founded an e-journal of critical geography. *Justice Spatiale/Spatial Justice* (JSSJ) was launched following a March 2008 Conference on *Justice et Injustice Spatiales*. This bi-annual interdisciplinary and bilingual journal (French and English) is led mainly by geographers (Bernard Bret, Frédéric Dufaux, Philippe Gervais-Lambony, Claire Hancock, Frédéric Landy), but it aims to become a forum for representatives of several social science disciplines, such as geography, planning, urbanism, urban sociology, history, philosophy and political science. As the inaugural editorial stated: "This journal aims to foster debate and therefore favours no theory or school of thought: it merely posits that the concept of justice has its place in social science and helps to make sense of places and territorialized social facts. There are several definitions of justice and several formulations of the social contract". JSSJ contends that analyses of interactions between space and society are necessary to understand social injustices and to formulate territorial policies aiming to tackle them. Over the last two years, it has published thematic issues on "Justice in the street", "Gender, sexual identities and spatial justice", "Spatial justice and environment". The word has since spread quickly: in 2009, the *Annales de géographie* published an entire double issue entitled "Justice spatiale" (Annales de géographie, 2009).

Some of the leading journals, particularly *L'Espace géographique* and *Hérodote*, have maintained the tradition of debates, many of which are very critical in nature. For example, in its January 2004 issue (*L'Espace géographique*, 2004), *L'Espace* organized a debate on postmodern geography in a much more critical fashion than had the *Cahiers de géographie du Québec* in December 1997 (*Cahiers de géographie du Québec*, 1997), whose ventures of the late 1970s, 1980s and early 1990s in critical geography seem to have since been stopped in their tracks. As for *Hérodote*, it becomes at times somewhat eclectic in its wanderings, claiming that all of geography is geopolitical, particularly since the early 1980s change in its subtitle to *Revue de géographie et de géopolitique*. But many of its (always thematic) issues represent meaningful contributions to critical geography. That was the case, for example, with the one devoted to *Menaces sur les deltas* (*Hérodote*, 2006), including an article criticizing the management of the Mississippi delta (De Koninck, 2006); and with the 2011 issue entitled *Renseignement et intelligence géographique*, literally "Intelligence and geographical intelligence" (*Hérodote*, 2011). Here, *Hérodote* seems to have somewhat returned to its critical 1976 position on the uses of geographical knowledge for war.

Gender geography

Like its Anglophone counterpart, Francophone geography has also been giving increasing voice to feminist or gender geography. In this Jacqueline Coutras has played a pioneer role in revealing the sexist foundations in the construction of urban space. "Urban design both reflects and consolidates unequal relationships between men and women" Coutras (1996) argues, for example: "It is indeed a construction, whose cohesion, and therefore solidity, is grounded in a division between the two sexes". Coutras advocated a new way of thinking the city and more specifically the "suburban crisis", from the point of view of the division between domestic labour and wage labour, with women being largely confined to the first and men having a privileged access to the latter. Such a separation is of course increasingly challenged by women's mobility resulting from their own increasing access to paid work, a transition which aggravates the urban crisis, for lack of taking into account its gendered dimensions.

For her part, Claire Hancock (2004) advocates a truly critical approach of the basic masculine fundamentals of geography. By distancing himself from his object of study through the reliance on maps, satellite images and GIS, Hancock argues, the geographer, a product of the *Lumières*, assumes a predominantly masculinist attitude: a distant and disincarnated look, which is seen as the only one capable of scientific objectivity, in contrast to a feminine approach that is said to favour subjectivity, the emotional and corporeal. Isn't the notion of territory essentially masculine? Aren't the conquest and appropriation of space related to the pleasure of penetration and possession? (Hancock, 2004). "There is probably no 'feminine' way of practising geography that can be set against a 'masculine' way", Hancock suggests; but there is "a 'masculine' construction of the object of geographical investigation that all the discipline's practitioners, men or women, should question, no doubt beneficially" (Hancock, 2004, 173).

Drawing on theories of recognition and recent work in geography, Djemila Zeinidi (2011) casts light on the different forms of injustice experienced by female Moroccan seasonal agricultural workers in Huelva (Southern Spain). Mostly married women with children, they are hired to pick fruit, strawberries in particular, in precarious conditions. Marianne Blidon (2011) examines the place of gays and lesbians in French society, suggesting that as a result of the primacy of heteronormativity, they end up as second-class citizens. Her article highlights the need for recognition in the face of violence, which renders gays and lesbians invisible and confines them to a minority status or makes them "Other" subjects. She shows how spatial confinement is predominantly based on a neo-liberal model that implies consumption, normalization and the exclusion of gay and lesbian subjects according to class and race criteria.

Critical Francophone gender geography has blossomed later than its Anglophone counterpart, from whom it originally drew its inspiration. But it has now come of age, as testified by an increasing number of challenging contributions that have appeared, for example, in a 2005 issue of the journal *Géographie et Cultures* (2005) and in a 2011 issue of *Justice Spatiale/ Spatial Justice* (2011) entitled "Gender, sexual identities and spatial justice". In addition, two international conferences on Gender and Geography have been organized in Lyon (2006) and Bordeaux (2010).

Other critical geographies

Within (or parallel to) the established schools of geography, various forms of critical geography are also emerging, such as MIGREUROP. Established in 2002, this network has since been joined by some 40 different associations as well as a large number of independent activists and researchers whose objective it is to question current European Union policies towards migrants and refugees. It has published an *Atlas des migrants en Europe,* with the subtitle *géographie critique des politiques migratoires* (MIGREUROP, 2009).

In addition, since the 1980s, the Université de Caen and a network of Universities in Western France (Rennes, Nantes, Le Mans) have rejuvenated a form of social geography inspired by Pierre George's thinking, including additionally psychological approaches of "l'espace vécu" (Frémont et al., 1984). More recently this social geography has established itself as a more socially and politically committed field and strengthened its theoretical and methodological bases (Séchet and Veschambre, 2006). In reasserting the socially produced character of space, it emphasizes "the spatial dimension of social inequalities and of relations of dominance". According to Chivallon (2003, 646–648), as a social science of space directed towards action, it has become a significant field of French geography. And as such, it represents a continuation of Marxist geography, which, however, refers more to Bourdieu or Lefebvre than to Marx or Lenin.

Much of environmental geography, including geography of hazards (Pigeon, 2005), also takes on a highly critical stance. The same can be said of urban geography, which seems to be going through some form of critical revival, in the midst of still prevailing business as usual types of publications. Examples include articles by Martin (2006), calling for a return to Henri Lefebvre's *pensée spatiale* ("spatial thinking") to criticize daily life, and Jouve (2009), calling for a renewed Marxist analysis of urban contradictions; and Milhaud's (2009) doctoral thesis which deals with the prison system in France. Overall, there is mounting evidence that the emancipatory nature of geography so dear to Reclus (Pelletier, 2009, 194–200) is finally gaining momentum among Francophone geographers.

Conclusion

Francophone critical geography has never given birth to a well-defined school of thought. Rather, it represents an approach favoured by a fluctuating number of geographers, always a small minority. Élisée Reclus was from the outset a towering figure, but one who stood on the margins of the academic world. His writings and political activism were intended for the wider audience, one deliberately ignored by most of the established members of the academic world. As a confirmed and remarkably well-articulated anarchist, Reclus considered that geography and political issues were closely interrelated. This has been Lacoste and *Hérodote*'s position, at least during the late 1960s and the 1970s, when they clearly proclaimed the political nature of geography, while questioning even challenging the nature and the role of academic geography. Now, however, *Hérodote* has in some ways become part of the establishment.

Beginning in the late 1960 and early 1970s, radical Anglophone geography referred to Marxism as an essential tool of analysis, without much interference from the political world. In France, however, the key role played by the Communist Party (PC) between 1945 and the 1980s, particularly with the decolonization wars in Indochina and Algeria, hampered the reference to Marxism as a strictly intellectual tool. Academics who proclaimed their allegiance to the PC, such as Suret-Canale (1921–2007), were rapidly marginalized, while several others, having had enough of Stalinism, had left it and remained sceptical of Marxism when not openly against it. That definitely limited its influence on critical thought, particularly among French geographers. Instead, a good number have since turned towards what became termed social geography – in this manner imitating in many ways, generally without realising it, Elisée Reclus' own approach – considered an extension of human geography, but one ideologically distinct from it, a so-called new approach within an old discipline (Frémont et al., 1984).

Bibliography

Annales de Géographie. 2009. *Justice Spatiale*. 118, 665–6.

Arnold, D. 1996. *The Problem of Nature: Environment, Culture and European Expansion*. Oxford: Blackwell.

Blidon, M. 2011. Seeking recognition: Spatial justice versus heteronormativity. *Justice Spatiale/Spatial Justice* 3. https://www.jssj.org/article/en-quete-de-reconnaissance-la-justice-spatiale-a-lepreuve-de-lheteronormativite/ (18 December 2020).

Bowd, G. and D. Clayton. 2005. French tropical geographies: Editors' introduction. *Singapore Journal of Tropical Geography* 26(3), 271–88.

Bowd G. and Clayton D. 2019. *Impure and Worldly Geography, Pierre Gourou and Tropicality*. London and New York, Routledge, 320 p.

Bruneau, M. 1978. Évolution de la formation sociale et transformations de l'espace dans le Nord de la Thaïlande (1850-1977). *Cahiers de Géographie du Québec* 22(56), 217–63.

Bruneau, M. 1981. Landscapes, social relations of production and eco-geography. *Antipode* 13(3), 26–31.

Bruneau, M. 2005. From a centred to a decentred tropicality: Francophone colonial and postcolonial geography in monsoon Asia. *Singapore Journal of Tropical Geography* 26(3), 304–22.

Bruneau M., A. Durand-Lasserve and M. Molinie. 1977. La Thaïlande, analyse d'un espace national. *L'Espace géographique* 3,179–94.

Bruneau M., 2019. Une vision postcoloniale nuancée de la tropicalité de Pierre Gourou », *l'Espace géographique*, 48 (2), p. 171-186.

Brunet, R. 2001. *Le Déchiffrement du Monde. Théorie et pratique de la géographie*. Paris, Belin.

Cahiers de géographie du Québec. 1978. Le matérialisme historique en géographie. 22(56).

Cahiers de géographie du Québec. 1988. Géographie, état des lieux: débat transatlantique. 32(87).

Cahiers de géographie du Québec. 1997. Les territoires dans l'oeil de la postmodernité. 41(114).

Césaire, A. 1953. *Discours sur le colonialisme*. Paris, rééd. Présence Africaine, 1989.

Chivallon, C. 2003. Une vision de la géographie sociale et culturelle en France. *Annales de Géographie* 634, 646–57.

Claval, P. 1977. Le Marxisme et l'Espace. *L'Espace géographique* 3, 145–64.

Collectif de chercheurs de Bordeaux. 1977. À propos de l'article de P. Claval "Le marxisme et l'espace". *L'Espace géographique* 3, 165–77.

Coutras, J. 1996. *Crise urbaine et espaces sexués*. Paris: Armand Colin.

De Koninck, R. 1978. Contre l'idéalisme en géographie. *Cahiers de géographie du Québec* 22(56), 123–45

De Koninck, R. 1980. La géographie soviétique est-elle révolutionnaire? *Hérodote* 18, 17–132.

De Koninck, R. 1981. Travail, espace, pouvoir dans les rizières du Kedah: réflexions sur la dépossession d'un territoire. *Cahiers de géographie du Québec* 25(66), 441–51.

De Koninck, R. 1984. La géographie critique. In, A. Bailly (ed.), *Les concepts de la géographie humaine*. Paris: Masson (4e édition 1998), pp. 185–98.

De Koninck, R. 1985. Idées, idéologies et débats en géographie. *Cahiers de géographie du Québec* 29(77), 175–83.

De Koninck, R. 2006. Le delta du Mississippi; une lutte à finir entre l'homme et la nature. *Hérodote* 121, 19–41

Dresch, J. 1952. L'occupation du sol en Afrique occidentale et centrale. *Symposium Intercolonial de Bordeaux* 90–6.

Dresch, J. 1979. *Un géographe au déclin des empires*. Paris: François Maspero.

Dumont, R. 1962. *L'Afrique Noire est mal partie*. Paris: Le Seuil.

EspacesTemps.1981. Une géographie à visage humain? 18/19/20.

EspacesTemps. 1989. Géographie, état des lieux: débat transatlantique. 40/41.

Fall, J.J. 2007. Lost geographers: power games and the circulation of ideas within Francophone political geographies. *Progress in Human Geography* 31(2), 195–216.

Frémont, A., J. Chevalier, R. Hérin and J. Renard. 1984. *Géographie Sociale*. Paris: Masson.

Géographie et Cutures. 2005. Le genre: construction spatiales et culturelles 54. https://doi.org/10.4000/gc.10917 (18 December 2020).

George, P. 1945. *Géographie sociale du monde*. Paris: PUF.

George, P. 1956. *Précis de géographie économique*. Paris: PUF.

Giblin, B. 2005. Élisée Reclus, un géographe d'exception. *Hérodote* 117, 11–28.

Gourou, P. 1936. *Les paysans du delta tonkinois*. Paris: Les Éditions d'Art et d'Histoire.

Gourou, P. 1947. *Les pays tropicaux, principes d'une géographie humaine et économique*. Paris: Presses Universitaires de France (nouvelle édition 1966).

Gourou, P. 1953. *The Tropical World. Its Social and Economic Conditions and Its Future Status*. London: Longmans.

Gourou, P. 1982. *Terres de bonne espérance: le monde tropical*. Paris: Plon.

Gottmann, J. 1983. The bounty of the tropics. *Times Literary Supplement*, Issue 4163, 14 January, p. 41.

Habermas, J. 1973. *La technique et la science comme idéologie*. Paris: Denoël/Gonthier.

Hancock, C. 2004. L'idéologie du territoire en géographie: incursions féminines dans une discipline masculiniste. In, C. Bard (ed.), *Le genre des territoires, féminin, masculin, neutre*, Angers: Presses de l'Université d'Angers, pp. 165–174.

Harvey, D. and A. Scott. 1988. La pratique de la géographie humaine: théorie et spécificité empirique dans le passage du fordisme à l'accumulation flexible. *Cahiers de géographie du Québec* 32(87), 291–301.

Hérodote. 1976a. Questions à Michel Foucault sur la géographie. 1.

Hérodote. 1976b. Élysée Reclus, géographe, anarchiste, 2.

Hérodote. 1981. Élysée Reclus, un géographe libertaire, 22.

Hérodote. 1984. La Géographie comme "divertissement"? Entretiens de Pierre Gourou avec Jean Malaurie, Paul Pélissier, Gilles Sautter, Yves Lacoste 33–4, 50–72.

Hérodote. 2005. Élysée Reclus 117.

Hérodote. 2006. Menaces sur les deltas 121.

Hérodote. 2011. Renseignement et intelligence géographique 140.

Jouve, B. 2009. Ville: le grand retour de la pensée critique. *Place publique* 15.

Justice Spatiale/Spatial Justice. 2011. Gender, sexual identities and spatial justice 3. https://www.jssj.org/issue/mars-2011-dossier-thematique/ (18 December 2020)

Kleinen, J. 2005. Tropicality and tropicality: Pierre Gourou and the genealogy of French colonial scholarship on rural Vietnam. *Singapore Journal of Tropical Geography* 26(3), 339–58.

Lacoste, Y. 1965. *Géographie du sous-développement*. Paris, Presses universitaires de France (2e édition 1976).

Lacoste, Y. 1976a. *La géographie, ça sert, d'abord, à faire la guerre*. Paris: François Maspero (2e édition 1982).

Lacoste, Y. 1976b. Pourquoi Hérodote? Crise de la géographie et géographie de la crise. *Hérodote* 1, 8–69.

Lacoste, Y. 1976c. Enquête sur le bombardement des villes du fleuve Rouge (Vietnam, été 1972). Méthode d'analyse et réflexions d'ensemble. *Hérodote* 1, 86–117.

Lacoste, Y. 2005. Élisée Reclus, une très large conception de la géographicité et une bienveillante géopolitique. *Hérodote* 117, 29–52.

Lefebvre, H. 1972. *La pensée marxiste et la ville*. Paris: Anthropos.

Lefebvre, H. 1974. *La production de l'espace*. Paris: Anthropos.

L'Espace géographique. 1984. Débat: géographie tropicale-géographie du Tiers Monde. 4, 305–88.

L'Espace géographique. 2004. Débat: la géographie postmoderne. 1, 1–60.

Martin, J.-Y. 2006. Une géographie critique de l'espace du quotidien. L'actualité mondialisée de la pensée spatiale d'Henri Lefebvre. *Articulo-Journal of Urban Research [Online]* 2.

Migreurop. 2009. *Atlas des Migrants en Europe: géographie critique des politiques migratoires*. Paris: Armand Colin.

Milhaud, O. 2009. Séparer et punir. Les prisons françaises: mise à distance et punition par l'espace, thèse de doctorat soutenue à l'Université de Bordeaux III.

Pelletier, P. 2009. *Élisée Reclus, géographie et anarchie*. Paris: Les Éditions du Monde libertaire.

Pigeon, P. 2005. *Géographie critique des risques*. Paris, Economica.

Racine, J.-B. 2002. La territorialité, référentiel obligé de la géographie? Une théorie encore à construire. *Cahiers Géographiques, (La Territorialité: une théorie à construire, en hommage à Claude Raffestin)* 4, 5–16.

Raffestin, C. 1980. *Pour une géographie du pouvoir*. Paris: Litec.

Raffestin, C. 1985. Marxisme et géographie politique. *Cahiers de géographie du Québec*, 29(77), 271–81.

Raffestin, C. 1989. Théorie du réel et géographicité. *EspacesTemps* 40/41, 26–31.

Raffestin, C. and M. Bresso. 1979. *Travail espace, pouvoir*. Genève: L'âge d'homme.

Raffestin, C., D. Lopreno and Y. Pasteur. 1995. *Géopolitique et histoire*. Lausanne, Paris: Payot.

Raison, J.P. 1989. Postface: Pour en finir, je l'espère, avec de fausses querelles... Si l'on parlait seulement de 'tropicalité'? In, M. Bruneau and D. Dory (eds.), *Les Enjeux de la Tropicalité*. Paris: Masson, pp.151–61.

Raison, J.P. 2005. "Tropicalism" in French Geography: reality, illusion or ideal? *Singapore Journal of Tropical Geography* 26(3), 323–38.

Reclus, É. 1869. *Histoire d'un ruisseau*. Paris: Hetzel

Reclus, É. 1876-94. *Nouvelle géographie universelle*, 19 volumes. Paris: Hachette

Reclus, É. 1880. *Histoire d'une montagne*. Paris: Hetzel

Reclus, É. 1897. *L'évolution, la révolution et l'idéal anarchiste*. Paris: Stock.

Reclus, É. 1905-08. *L'homme et la terre*, 6 volumes. Paris: Hachette.

Reclus, É. 1982. *L'homme et la terre, introduction et choix des textes par Béatrice Giblin*, 2 volumes, Paris: François Maspero/La Découverte.

Rioux, M. 1978. *Essai de sociologie critique*. Montréal: Hurtubise-HMH.

Robequain, C. 1948. Les Pays Tropicaux d'après P. Gourou. *Annales de Géographie* 305, 70–3.

Sack, R.D. 1986. *Human Territoriality: Its Theory and History*. Cambridge: Cambridge University Press.

Said, E.W. 1978. *Orientalism*. New York: Random House.

Sautter, G. 1984. La géographie tropicale de Pierre Gourou et le développement. *[débat] l'Espace géographique* 4, 335–6.

Séchet, R. and V. Veschambre (eds.). 2006. *Penser et faire la géographie sociale. Contributions à une épistémologie de la géographie sociale*. Rennes: Presses universitaires de Rennes.

Suret-Canale, J. 1948. L'exploitation coloniale est-elle une réalité géographique? *La Pensée* 16, 103–4.

Suret-Canale, J. 1994. Les géographes français face à la colonisation: l'exemple de Pierre Gourou. In, M. Bruneau and D. Dory (eds.), *Géographies des colonisations XVe-XXe siècles*, Paris: L'Harmattan, pp. 155–169.

Van Beuningen, C. 1979. Le marxisme et l'espace chez Paul Claval. Quelques réflexions critiques pour une géographie marxiste. *L'Espace géographique* 4, 263–71.

Vandergeest, P. and N. Peluso. 1995. Territorialization and State Power in Thailand. *Theory and Society* 23(3), 385–426.

Zeinidi, D. 2011. Circular migration and misrecognition. *Justice Spatiale/ Spatial Justice* 3. https://www.jssj.org/article/migrations-circulaires-et-deni-de-reconnaissance/ (18 December 2020)

9 Better late than never? Critical geography in German-speaking countries

Bernd Belina, Ulrich Best,
Matthias Naumann, and Anke Strüver

Introduction

This chapter, like others in this collection, starts with a reflection on the politics of localization of the "German-speaking countries".[1] The authors are all based in Germany. Our writing sits uneasily within the unequal relations between the countries (or parts of them) that we are writing about, with Germany being the dominant state, while Austria and parts of Switzerland are brought in by virtue of a shared language. However, the academic fields of the German-speaking areas do in fact form a strongly connected system, with people moving relatively freely between them (but not without discussions about the smaller countries being swamped with German academics and students). The areas share a central biannual convention, the German Congress of Geography, (somewhat uneasily, only referencing "German" in its title, in spite of rotating between the three countries). Moreover, publications and collaborations across state divides are manifold – including collaboration in critical geography. In this chapter, however, the focus will be on the (West) German case.

We intend to highlight a number of important moments in post-WWII German-language critical geography, adding to an international history of critical geography across scales. This will not be a complete history, but it will focus on questions of hegemony, struggle, and exclusions within German-language geography. We tell this history around four aspects that we identify as central to understanding German-language critical geography in an international context:

1 The hierarchical organization of German academia is crucial (as it also is in Austria and German-language Switzerland). Until recently, institutional power was concentrated in the hands of very few professors (almost exclusively men) of the highest rank, titled *Ordinarius*, who could and did hire and fire whomever they pleased as PhD students, faculty, and professors of lower rank, and over all of whom they had far reaching authority.

DOI: 10.4324/9781315600635-9

2 Bottom-up, grassroots movements, networks, and initiatives by under-graduate and graduate students and precariously employed young to middle-aged faculty have always been the most important source of German-language critical geographies. It is only since around 2000 that some of these initiatives became partially institutionalized within the academic discipline.

3 International connections have been a key element for German-language critical geographies, from the Weimar era (that we do not deal with here; cf. Best, 2009) to the current situation of the neo-liberalized university of "excellence".

4 Painting with a broad brush, the focus of activities within German language critical geographies since WWII moved from general emancipatory positions with a strong Marxist bent in the 1960/70s to feminist activities in the 1980/90s to more integrated new developments around class, race, and gender since around 2000.

The basis of this chapter comes from a longer-term project that has so far produced a corpus of 20+ interviews with protagonists of critical geography in Austria, Germany, and German-language Switzerland. We have also created an archive of German-language critical geography, where we have collected the mostly student-run critical geography journals that go back to the 1960s. The following also draws on a number of papers we have published in different constellations on these topics (Belina et al., 2009; Belina, 2009, 2014, 2020; Best, 2009, 2016; Strüver, 2010).

The structure of the chapter is chronological, beginning with the post-WWII situation leading to the 1960s, followed by a closer look at the struggles of the late 1960s and 1970s, a special focus on feminist geography, and finally an outlook on the current period of "international excellence" in critical and mainstream geography. We focus on key publications, events and initiatives. All translations have been made by the authors.

1945–1968: Old Nazis, old paradigms and, in the East, Marxism-Leninism

With very few exceptions (those being critical theorist Karl August Wittfogel, and Jewish landscape geographer Alfred Philippson and Jewish geographer Friedrich Leyden, all of whom were incarcerated in concentration camps, with Leyden dying in Theresienstadt), German geography and geographers were heavily involved with the Nazi regime, both ideologically and practically (Heinrich, 1990; Michel, 2014a; Rössler, 1990; Fahlbusch et al., 1989). This past was dealt with differently in West Germany (the Federal Republic of Germany, FRG) than in East Germany (the German Democratic Republic, GDR).

In West Germany, the post-WWII situation was characterized by strong continuities, both ideologically and practically. While involvement with the Nazi regime was denied, theoretical positions and empirical practices were dominated by traditional regional geography with a focus on both the notion of landscape and the determination or shaping of culture by nature. This continuity is very explicit in a 1947 paper by the well-known physical geographer Carl Troll, which opened the first issue of the new journal *Erdkunde* under the telltale title "The geographical science in Germany in the years 1933–1945: a critique and justification". In this piece, which set the tone for decades to come (cf. Michel, 2016) and which was partly translated into English in 1949, he states: "If German geographers went astray this should be regarded as part of the general intellectual and religious crisis, which also produced, as its most terrible consequence, the spirit of despotism of Hitler-Germany. German geography, however, has not been substantially deflected from its natural course of development by purely external interference." (Troll, 1949, 105). Instead, he emphasizes German geography's "noteworthy accomplishments" during that period and claims that it is only necessary to "simply [...] expurgate from the literature of this period certain falsifications" (ibid.). At the core of this redressing, all blame for involvement with the Nazi regime is put on the school of geopolitics, as developed and practiced by Karl Haushofer (1925), who himself emphasized that geopolitics emerged "from the mother liquor of geography, notably political geography" (ibid., 156). While it was indeed Haushofer who was instrumental in introducing Hitler to the thinking of traditional, environmental-determinist geography (Kost, 1988, 236–237), commentators from the 1920s and 1930s as well as later studies emphasize that there were no real theoretical differences between (political) geography and geopolitics during the 1920s, 1930s and 1940s (Kost, 1988; Lossau, 2002, 111–131; Michel, 2016).

Following the path outlined by Troll (1947, 1949) and based on a renewal of pre-war formulations of what geography ought to be (Bobek and Schmithüsen, 1949/1967), West German geographers continued to work in the tradition of environmentally determinist regional geography around the notion of landscape throughout the 1950s and 1960s.[2] At that time – the restauration period marked by US-sponsored economic recovery, ordo-liberal economic policies, anti-communism, conservative family and gender values, and a general neglect of the Nazi-past – only very few scholars attempted to renew West German human geography. Drawing on social theory, notable exceptions were Wolfgang Hartke (e.g., Hartke, 1962),[3] in Munich and Peter Schöller (e.g., Schöller, 1957). Schöller held positions in research institutions outside the university in Münster before becoming a professor at the newly founded University of Bochum in 1964.

On the other side of the iron curtain, East Germany defined itself as the "antifascist German state", which did not hinder the careers of former Nazis (Frei et al., 2019, 47–48). The "buildup of socialism" was spearheaded by the communist party, which claimed absolute leadership, following the example of and the directives from the Soviet Union. As a consequence, massive migration to West Germany, particularly of the middle classes, took place. In this context, with many geographers among those leaving for the West, a struggle over hegemony about what geography should be about took place in the post-war period. On the one hand, traditional geographers, such as the landscape ecologist Ernst Neef (Leipzig and Dresden), remained in important positions (sometimes after a brief pause following de-Nazification; cf. Schelhaas, 2004, 27). These "bourgeois" geographers, some of whom would superficially use Marxist-Leninist terminology, were accepted mainly due to the acute shortage of university lecturers and their seemingly less tainted focus on physical geography and planning. On the other hand, a new generation – usually trained in other disciplines – under the leadership of Heinz Sanke (Berlin) built what they termed economic and political geography, based on the Marxist-Leninist state ideology of the new socialist state. This school provided a poignant critique of traditional geography long before similar points were raised in West Germany in the late 1960s (Sanke, 1962). It developed a theoretical basis as well as empirical studies for the territorial planning of the socialist state (Schmidt-Renner, 1966), but it was at the same time part-and-parcel of the deterioration of Marxism from an emancipatory theory to a state ideology (cf. the reconstruction by Labica, 1986). This geography was mainly critical of what was happening in the West. The textbook by Gerhard Schmidt-Renner (1966), who maneuvered around some of the more obvious shortcomings of Stalinist Marxism-Leninism (Belina, 2014), received some positive reviews, however the work of the Marxist-Leninist GDR geographers was generally very negatively received by colleagues in West Germany, who denied it any scientific value (with the honorable exception of the aforementioned Peter Schöller; Schelhaas, 2004, 72–74).

Thus, for different reasons – Nazi continuities and anti-Marxism in the West, apolitical landscape geography, and Marxist-Leninist state ideology alongside well-hidden Nazi continuities in the East – few critical voices were heard in post-WWII German geography until well into the 1960s. All this boiled down to a common denominator: the Cold War overdetermined everything that happened in both parts of Germany in the second half of the twentieth century (and many things in Austria and Switzerland as well). Matt Hannah (2016) even argues that the Cold War repercussions are still felt in contemporary German-language geography.

West German geography's belated 1968 and its aftermath

The street protest against the visit of the Persian Shah to West Berlin and the police shooting of the protestor Benno Ohnesorg on 2 June 1967 symbolically marked the end of the West German restauration period. Following societal changes throughout the 1960s that culminated in this event, many students and other young people radicalized, asked their parents about their involvement with the Nazi regime, searched for personal and sexual freedom, and some started to read Marx and Marxist classics. Paving the way were several developments within West German society in preceding years. Beginning in 1960 and gaining pace after 1964, the West German state invested extensively into secondary and higher education to secure "resources of talent" (*Begabungsreserven*) for the revitalizing economy. This resulted in, among other things, the introduction of new applied degrees (among them the "diploma" in geography; cf. Brogiato et al., 2010, 77), a 150% increase in university students from 1965 to 1975, the funding of several new universities, some of them located in less developed regions (Wolf, 2006, 230–231), and growing participation "at all secondary and tertiary levels of education, to the benefit of rural areas, women, Catholics and the middle and lower social classes" (Mayer and Hillmert, 2003, 82). Beginning in 1963, the Frankfurt Auschwitz Trials forced a reluctant public to face the complicity of many in the atrocities of the Holocaust. Finally, West Germany's first economic crisis in 1966/67 startled the public, which had come to believe that the ordo-liberal "social market economy" would guarantee endless growth and amelioration of living conditions.

Geography was far less affected by student protests and radicalization around 1968 than other parts of academia. Still, students experimented with new lifestyles, started questioning the content and form of their studies, and some started reading Marx and writing about geography in a critical, Marx-inspired manner. The years after 1968 witnessed the emergence of several student journals such as *Geografiker* (1969–1972), *Roter Globus* [*Red Globe*] (1971–1973) and *Geographie in Ausbildung und Planung* [*Geography in Teaching and Planning*] (1973–1976) as well as book series such as *Geographische Hochschulmanuskripte* [*Geographical University Manuscripts*] (1973–1985) and *Urbs et Regio* (1976–2002) that were partly radical in content.[4] *Geografiker* was edited by students and one young faculty member at Free University Berlin (located in the Western part of the divided city). While the first two issues were devoted to technical questions of curriculum reform in geography, the third issue was a powerful critique of existing West German geography. This issue was prepared for the 37th German Congress of Geography, the central biannual conference of German language geography that took place in Kiel in 1969, and became an influential publication in West German geography. Most important, it included a position paper that tried to hammer home two central points: "1) Geography avoids to face its task and its responsibility in society. 2) In addition, as far as it adapts a self-definition as regional landscape geography [*Landschafts- und Länderkunde*] ..., it does not even meet the standard of scientific analysis." (Geografiker, 1969, 5).

At this congress the session on "The geographer [male form only!] – formation and profession" included three paper presentations, among them the position paper that was read out by four (male) students. Talking about the months prior to the congress, one of the members of the *Geografiker* group remembers: "When the establishment heard of the plans [...], they – you know, you cannot imagine this – they almost went crazy, they wet their pants" (Interview 1). This is evident in the long discussion that followed the presentation and was continued in the evening (both documented fully in the conference proceedings, Meckelein and Borcherdt, 1970, 208–232). It included comments not only by such influential figures in German geography as Carl Troll and Erich Otremba but also by several students. It becomes clear that the professors did not really understand the arguments of the students and regarded their intervention solely as an attack at the heart of geography as they understood and practiced it, and therefore as an attack of their very identity. As a student remarked in consternation, professors and students "speak different languages" (ibid., 224; cf. Belina 2020; Wardenga, 2020, 16–17). In hindsight, this appears somewhat strange, as many of the demands from the position paper are based in a neo-positivist philosophy of science and argue for a "scientific" geography that produces "applicable" knowledge. However, given the situation at the time, both the position paper and the reactions to it make perfect sense. As one member of the *Geografiker* group remembers, the group was not political in any strong sense:

> [A]ctually, the people that got together had very different biographical paths, and their habitus just happened to make them emancipatory actors. ... None of us were originally leftist, [or] came from a leftist background at home ... but certainly, if one was somehow unhappy and critical, we took this as an issue of emancipation, and also, the general political situation with the war in Vietnam etc. had become such that you had to make a decision.
>
> (Interview 1)

This kind of decision was far less important in geography than in neighbouring social sciences:

> In geography, you did not have to decide. In sociology, you had to decide, because in each class and seminar ... or at Otto Suhr Institute [at Free University Berlin] with the political scientists, you had to decide, because no seminar could start without a struggle over the content, what it was going to be about. ... In geography, this was not the case at all, up until the moment when we brought it in there.
>
> (Interview 1)

In general, the situation in geography at the time was rather grim:

> There was no theory, no empirical research that could in any way have affirmed a left political consciousness. Where should we [have been able to] read this? They [the established geographers] were all fascists,

and they were still teaching, they were still at the universities. … In the library, there were numerous books that had parts blacked out, and these were the racist statements of the professors whose lectures we took.

(Interview 1)

In this situation, the *Geografiker* position paper that mingled neo-positivism and critique, was perceived as a political attack by the old guard. Remembering a talk by Dietrich Bartels, professor in Kiel and theoretical spearhead of the German version of spatial analysis (Bartels, 1968) at the Department of Geography in Heidelberg around 1967, another interviewee remembers:

They were upset, because they … thought that their geography, what they were standing for was being destroyed. That geography generally was being destroyed if people started dealing with these new approaches. Now they had built up something for many years, something that they thought was the culmination of geography, namely regional geography, and they did call it that, and they felt personally attacked.

(Interview 2)

After "Kiel", as the 1969 German Congress of Geography is remembered and somewhat mythologized (Eisel, 2014; Helbrecht, 2014; Michel, 2014b; Sahr, 2016), it seemed as if nothing could remain the same. And indeed, change did occur. The subsequent development can be summarized as follows (Belina, 2009): The students' demands of Kiel 1969 were centred on *theory, relevance* and *critique*. As Eisel (2014, 314) points out, the demands already contained two varieties of transformation: one that was modernizing, another that was leftist. The modernizing transformation successfully became the new mainstream in the following years. *Theory* was read as methodology and, to a certain extent, as theory of science. *Relevance* was read as applicability of geographical research in (state) planning and institutions. *Critique* was seen mainly as critique of traditional regional geography, which gave rise to numerous conflicts with the older generation of "pre-modernist" geographers. The interpretations of theory and relevance created a number of attempts at an applied geography, in some cases with some guidance by theories. The second variety of transformation, the leftist critique, interpreted central terms differently. Here, theory was understood as *social theory*, critique as a *critique of society*, and relevance as *societal relevance*, meaning relevance for social movements and leftist political parties. These students read (and often criticized) geography from East Germany as well as Anglo-American radical geography (one of the journals, *Geographie in Ausbildung und Planung*, briefly cooperated with *Antipode*). The leftist main sources were Marx and Marxist classics, as well as recent theoretical and political discussions within the West German left.

Still united in Kiel by the common enemy, the split of the two varieties happened soon after. The leftist critique remained the domain of students, while young radical faculty were more or less openly forced to leave academic geography (some finding "refuge" in geographical education or other disciplines). "People were pushed to the side, there were casualties" (Helbrecht, 2014, 319; cf. Werlen, 2014, 297). The "modernist" critique founded the new neo-positivist and applied mainstream and guaranteed many of its proponents' careers within the discipline, especially at newly founded universities such as Bremen, Oldenburg, Osnabrück, Kassel, and Bochum. In these places, the power of the traditionalists was much weaker than in strongholds such as Bonn, Heidelberg or Erlangen. There was a "splintered landscape of geography departments" (Helbrecht, 2014, 320). In Erlangen, as one former student remembers, the *Ordinarius*, Eugen Wirth, "ran the department like a paramilitary organization" (Interview 3). Wirth also used his power to intervene in other departments, as in the case of Alois Kneisle, a student at Tübingen who in 1980 published a fierce Reichian-Marxist critique of Wirth's work (Kneisle, 1980; cf. Belina et al., 2009, 50–51). As a former friend of Kneisle remembers: "So Wirth positively put pressure on [name omitted, *Ordinarius* at Tübingen] to constrain [name omitted, lower ranking professor at Tübingen] so that Alois Kneisle would not write a PhD dissertation under his supervision" (Interview 4). To the best of our knowledge, this is the most extreme known example of how the careers of critical geographers in West Germany ended before they had even begun as rooted in both anti-Marxism and the power of the *Ordinarius*-professors. In most cases, the process of keeping the ranks of West German geography professors clear of radicals was more subtle.

The 1970s and onwards: New social movements and feminist geography

After 1969, the new generation of critical geographers sought to tackle topics that were closely related to the political debates and movements of the time – with the exception of the feminist movement, which only much later found its way into German-language geography (see below). In addition to isolated attempts at Marxist theory-formulation in the pages of *Geographische Hochschulmanuskripte* and *Roter Globus*, examples of important topics were urban struggles, urban inequality and urban renewal, North-South relations, "development", and international solidarity. In the literature, the focus on urban questions is dominant, with West German critical geographers connected with very active grassroots planning movements, neighbourhood committees, and criticism of urban renewal. One such example comes from Switzerland, from the pre-history of the International Network for Urban Research and Action (INURA), established in 1991. INURA is rooted in urban activism and the research of geographers from Zurich, who set up the group around 1980 when Zurich witnessed major urban upheaval (often referred to under the name of the popular documentary "Züri brännt!", i.e., "Zurich is burning!"). As one participant remembers, the "reaction of the professors was pretty much panic" (Interview 5) when they found out that their students began to read critical literature and even published the first German-language book on radical urban theory in the *Geographische Hochschulmanuskripte* series (Hartmann et al., 1985). The critical Zurich geographers also connected with fellow minded students and young faculty in Germany through the (German) federal association of geography student associations (*Bundesfachschaftentagung*). This association was also instrumental in the establishment of the *AK WISSKRI*, a multi-local discussion circle meeting throughout the 1980s that focussed on urban issues, feminism and the (Nazi) history of German geography (the latter resulting in the publication of three studies on this topic as *Urbs et Regio* (Fahlbusch et al., 1989)). Around the same time, in Austria, the Austrian Association of Critical Geographers was founded, which still exists under its new name *Kritische Geographie* (Critical Geography).[5]

It was mainly in this context that feminism began to become important in German-language critical geographical debates. Feminist geographers, many with a background in the feminist movement, have since struggled within and against the androcentric discipline of geography in the three national contexts, having criticized it as "a man's world" both in terms of researchers and subjects. They have expanded feminist geographies into *critical* geographies, having built a substantial "critical mass" for critical geography in the 1990s that now ofers a framework for feminist, gender, and queer issues.[6]

There is a tradition of networks of German-speaking feminist geographers, such as *Feministischer Georundbrief* (Feminist Geography Newsletter, since 1988, and from 2000 as the digital *Georundmail*[7]), *Geographinnentreffen* (meetings of feminist geography undergraduate-students, since 1989) and *Arbeitskreis Feministische Geographie* (Study-group Feminist Geography, since 1988), re-named *Arbeitskreis Geographie und Geschlecht* (Study-group Geography and Gender) in 2005 and back again to *Arbeitskreis Feministische Geographien* in 2019 (this time plural: Feminist Geographies). All these groups have been working with a range of different theoretical perspectives to approach questions of gender, identity, and space.

Present feminist debates, both in theory ("academia") and in practice ("politics"), are firmly rooted in the broader context of social movements that emerged in the late 1960s and early 1970s. Then, the normative position and politics of the so-called women's liberation movement was closely connected with other forms of social inequalities, including economic and class differences. Nevertheless, the overall objective of the emerging field of women's studies was to introduce gender as an analytical category, i.e., to add women to the research agendas. In German-language human geography, early work mostly dealt with discrimination against women in various social arenas and at various spatial scales, analyzing spatial structures in general and urban places in particular with respect to gender differences and discriminations (for a seminal early work, see Bock et al., 1989). Initially, the focus was on women's everyday experiences *as* women (and often also as mothers). Only in the 1990s did a German-language feminist geography come forward, not as a sub-discipline, but as an interdisciplinary and intersectional exercise that was involved with broader social conflicts and their spatial dimensions (Bühler et al., 1993; for topical summaries of feminist geographies, see Bauriedl et al., 2010; Wastl-Walter, 2010). At the same time, feminist geographers began to extend their more "politics"-oriented critique of society to that of modern science and questioned its founding philosophies, concepts, and catego-ries, especially hierarchical ways of thinking in dichotomies. The more "academic" deconstruction of these dichotomies became a crucial aspect, since dichotomies lay the foundations for treating women as supposedly essentially different from – and inferior to – men. Yet another extension of this critique was the adoption of poststructuralist thought from the social sciences in the 1990s, which was accompanied by the rejection of under-standing "women" as a unitary and homogenous category. Until then, the social category of gender – and reflections on the construction of gen-der identities in general – had been a blind spot in mainstream German-language human geography.

In terms of wider recognition and effects, the "Doreen Massey Reading Weekends" were a major turning point in both feminist and critical geog-raphy in the German-speaking context, starting with the 1998 Hettner-Lecture with Doreen Massey, organized by the geography department at the University of Heidelberg. The Hettner-Lecture were prestigious five-day visits of Anglo-American geographers – among them many critical and radical geographers such as Derek Gregory, Michael Watts, and David Harvey – consisting of two days of public lectures and three days of seminars with graduate students. During the 1998 seminars – and especially the tea breaks – a group of feminist geographers (mainly grad-uate students) spoke with Doreen Massey "identity, gender and space" and the absent topic of queer theory. Because of this desideratum, it was decided to hold another meeting among feminist geographers, focusing on queer geographies. It was called "Doreen Massey Reading Weekend", took place in January 1999 in Hamburg and was followed by five more weekends until 2004, including one in the Netherlands (University of Nijmegen). The themes discussed ranged from "queer theory" and "post-structuralist feminisms as theory and practice", in the beginning, to "cit-ies and sexualities" and "sensualities" later on, focusing on themes that were missing from geography's agenda in Germany at that time (for an extensive report, see BASSDA, 2006).

Although "gender" as an analytical category is now increasingly recognized in human geographical research and teaching, this often takes the form of "add gender and stir" rather than actual (feminist) identity theories and politics. Compared to the United States and the United Kingdom, there are still very few tenured female (let alone feminist) professors in academic geography and only recently have textbooks on feminist geographies been published. However, and partly because of the latter, the institutionalization of feminist topics into academic geography now occurs at some universities. Still, this depends on (a fortunately ever-increasing number of) feminist PhD students, postdocs and a (few) tenured professors.

The long-standing feminist critique also called attention to the significance of gender relations in both public and private spaces and to the inter-relations between the public and the private. More recent feminist works, being part of the "cultural turn" and relying on poststructuralist accounts of gender identities and relations, have therefore concentrated on the multiple constructions and the complexity of intersectional identities. These approaches increasingly reject categorical distinctions in general and class and gender as analytical categories in particular and thus have shifted the focus from the analysis of large-scale structures of inequality to small(er) scales of micro-politics and practices at the level of the embodied subject. However, a "return" to structures of inequality is becoming noticeable, especially in relation to the growing material inequalities under neo-liberalism and austerity. Still, the private sphere of home as well as increasing class inequalities between women were only recently included while dealing with questions of "care" (see Feministisches Georundmail No. 44, May 2010).

Since 2000: Acceptance through internationalization

Following German unification in 1990, the left in Germany underwent major changes: the "Second World" as a point of reference and critique had disappeared, the reality of capitalist social relations and the triumphalism of neo-liberal ideologies made any position challenging either of them almost impossible, and the biographies of both East Germans and the "1968" generation in West Germany, Austria and Switzerland were de-valued. It took around a decade for the left to regain a voice, after a government formed by Social Democrats and Greens had Germany engage in a war (in former Yugoslavia) for the first time since 1945 and introduced dramatic cuts in the welfare state.

Beginning in the late 1990s, critical perspectives became more noticeable in German-language human geography. The increasing internationalization of German academia is an integral part of opening debates for critical and sometimes explicitly left approaches (Belina et al., 2009; Hannah, 2016). Before we discuss this co-evolution of internationalization and emerging strands of critical geography in Germany, we briefly sketch the various projects and activities of German critical geography since around 2000 that, again, mainly emerged from student and young faculty initiatives:

- New groups and associations of undergraduate and graduate students committed to critical geography emerged, both as a nationwide network (established in 2005) and as numerous local groups in cities such as Berlin, Frankfurt, Hamburg, and Leipzig, often cooperating with local initiatives around, for example, urban issues. Berlin is possibly the best example of an active local group of critical geographers of different generations, consisting of students as well as precariously employed postdocs – a constellation typical for the different activities within German critical geography (Gintrac, 2016).

- Many events around critical geography were organized. These including "Antipode lectures" and "ACME lectures" at the biannual German geographers' conference, lecture series on critical geography, for example, in Berlin, Göttingen, Münster, and Potsdam, the 6th International Conference of Critical Geography (ICCG) in Frankfurt in August 2011 and, especially important in bringing together critical voices within geography and furthering exchange between critical geographers and social movements, the series Research Workshop Critical Geography (*Forschungswerkstatt Kritische Geographie*) since 2008. These low-key discussion gatherings, which usually include invited as well as submitted presentations and workshops by students and young faculty, have been held in Frankfurt, Bern, Hamburg, Bonn, Leipzig, Bremen, and Berlin.

- Numerous critical geography papers written in German were published, some in established German-language academic journals that had featured little critical and/or radical content prior to the new millennium. Two important advances occured. First, the lacunae of textbooks and other easily accessible material on critical and radical geography was filled. Notable works included an edited volume with translations of Anglo-American radical and critical geography papers (Belina and Michel, 2007), a German issue of *ACME* (2008), textbooks on feminist geography (Bauriedl et al., 2010; Fleischmann and Meyer-Hanschen, 2005; Wastl-Walter, 2010), a handbook on critical theory in spatial and urban research (Oßenbrügge and Vogelpohl, 2014) and one on critical urban geography (Belina et al., 2014). This infused classrooms with critical geography content. Second, new outlets were established, all by young, un-tenured faculty: a book series for radical geography since 2007,[8] an interdisciplinary online journal on critical urban studies with many critical geographers among the editing collective since 2013[9] and a blog on critical geography.[10]

A new and unprecedented-in-size wave of collaborations with social movements and leftist parties is observable. One focus is on urban issues and the right to the city, where successful local campaigns for affordable housing benefit from the active involvement of critical geographers,[11] and activists find accessible critical geographical publications they can actually refer to and work with (e.g., Holm and Gebhardt, 2011; Mullis, 2014; Schipper and Wiegand, 2015). Another example is the collective for critical education and creative protest, "ORANGOTANGO", which has developed critical maps for social movements in Berlin and elsewhere, building on experiences from Latin America (kollektiv orangotango+, 2018; Halder, 2018).[12]

Surprisingly, most of these activities did not cause major resistance or conflicts, unlike earlier attempts introducing critical approaches in German geography (Belina et al., 2009). We suggest that the internationalization of academia is the main driver of this. Internationalization entails different aspects. First, increasing spatial mobility of students and faculty studying or working abroad is not only supported and required for academic careers, it also exposes a new generation of German-language geographers to international debates where critical geography is important or even "hegemonic" (Berg, 2004), a process that is further enabled by new information technologies and online publications. Second, is a strong orientation towards international academic standards such as publishing in refereed journals, international research projects and paper presentations at international (i.e., Anglophone) conferences in order to meet the requirements for "excellence" and scientific competitiveness. German human geography is therefore faced with the somehow ironic situation that publishing radical/critical articles in reputed journals, such as *Antipode* and *International Journal of Urban and Regional Research*, leads to an enhanced reputation and a high significance of international critical geography in the German discipline. Radical schools of thought can be referred to and worked with more easily in the German language context these days – not because they are *critical*, but because they are *international* and *internationally successful* (Belina et al., 2009; Hannah, 2016). Marx has entered German language geography not in the original but through David Harvey, and the critical edge of the work of Foucault, Butler, or Laclau/Mouffe is, in many cases, "smuggled" in via discourse analysis and/or theorizations of the subject. Finally, we witness a deterioration of working conditions in the name of international competitiveness (Gribat et al., 2016; Sambale et al., 2008). A total of 90% of the teaching staff at German universities work on precarious, untenured fixed-term contracts. While this process mainly results in huge problems for young and middle-aged faculty, in a perverse way it also helps to increase the political consciousness of many academic geographers, both precarious and established, making critical and radical positions much more plausible.

Thus, compared to past decades, the currently very promising development of German-language critical geography is an outcome of different trends. On one hand, widespread activities of students and young faculty are leading to a rising recognition of critical and radical approaches in

human geography. On the other, the internationalization of German academia facilitates acceptance of critical geography, not because it is critical
but because it is "international" and, thus, accepted by the mainstream.
Paradoxically, emerging opportunities for critical geography go hand-in-
hand with a neoliberalization of academic work.

Conclusion and outlook

The history of radical and critical approaches within German-language
human geography is in stark contrast to that of other academic disciplines
in Germany, Austria, and German-speaking Switzerland. In many social
sciences and humanities, "1968" was followed by an influx of tenured critical and radical individuals as well as the establishment of strong radical
schools of thought, both of which were followed by a conservative rollback
during the past two decades (Brand, 2010; Bretthauer and Fromberg, 2008).
In human geography, it was only over the past two decades that openings
for critical and radical approaches truly struck roots. As a consequence,
human geography in the German-language countries is partially turning
into a "niche" where critical and leftist debates are more accepted than
elsewhere.

For this volume, however, a key issue is to situate German-language critical geographies in the context of global critical geographies. Following our
discussion of struggles of critical geographers in different periods, we will
return to the four aspects of understanding German-language critical geography, which we identified in the introduction, and discuss them in an international context. First, the hierarchical organization of German academia
can explain why German language geography is a latecomer in critical
geography, as it allowed for continuities with Nazi geography (in particular
in West Germany and Austria) and cemented the considerable conservatism
of the discipline that culminated in the struggles of 1969 at Kiel.

Second, this strong hierarchy explains why the bottom-up grassroots
initiatives of critical and radical geographers were not able to change the
institutions in the 1970s and 80s, as was the case in Anglophone critical
geography. Former student-led journals from the United States, such as
Antipode, became internationally highly ranked academic outlets with
major publishers, but not many people have ever heard of German journals such as *Roter Globus* – even at the department where it was produced
40 years ago. Pioneering North American and British critical geographers
who became tenured professors and are now retiring have educated students, who have themselves become established in the discipline. But Few
have heard of their German-language counterparts, like Alois Kneisle, who
were basically kicked out of human geography.

Third, it is interesting that German-language critical geographies always
have oriented towards other critical geographies. Examples include the
cooperation between *Antipode* and *Geographie in Ausbildung und Planung*,

early translations of English-language publications from radical geography (often by students or young faculty), and the international orientation of some German-speaking critical geographers since the 1990s. Following the internationalization of mainstream geography and academia in the German-speaking world, however, we would argue that this orientation has become even more important. The fast growth of international mobility, the importance placed on international experiences and connections, the "excellence" of international visibility – all these are aspects of mainstream academia's growing interest in situating itself globally, which somewhat ironically have benefitted German-speaking critical geographers.

Fourth, after different focuses in the past, German-language critical and radical geography today can best be described as a movement within the discipline producing and disseminating knowledge around injustices based on class, race and gender, often in cooperation with social movements. However, such positions are still taken by only few geographers with permanent positions. By unearthing histories of past struggles and the roles they play in contemporary German-language geography, we hope to build a consciousness that can support more to pursue critical and radical geography.

Interviews

Interview 1: conducted on 4 July 2008 by Ulrich Best and Bernd Belina in Mainz.

Interview 2: conducted on 2 August 2008 by Ulrich Best and Bernd Belina in Göttingen.

Interview 3: conducted on 5 July 2008 by Ulrich Best and Bernd Belina in Mainz.

Interview 4: conducted on 9 February 2010 by Bernd Belina in Frankfurt.

Interview 5: conducted on 13 July 2008 by Ulrich Best and Thomas Bürk in Berlin.

Notes

1 The authors would like to thank Boris Michel and Daniel Mullis for helpful comments on an earlier version of this chapter.

2 Michel (2014b, 302) reminds us that neither physical geography nor economic geography or the rudimentary attempts at social geography (mentioned at the end of this paragraph) practised regional geography based on the notion of landscape; according to Sahr (2016, 80), this kind of geography had "largely collapsed already" prior to 1969. Still, it was the only theoretical foundation of geography accepted and acceptable in West Germany in the 1950s and 60s.

3 Werlen (2010, 111) argues that Hartke's innovative position, met with hostility within West German geography in the 1950s and 60s, enters theoretical discussion in the FRG only in the 1970s via the "detour" of the reception of Anglo-American behavioral geography, where Hartke's work was positively received. We will make a similar argument below concerning more radical positions in the 2000s.

4 We have made many of these journals available in an online archive at http://
   kritische-geographie.de/textarchiv-kritische-geographie.
5 See www.kritische-geographie.at/kg.htm.
6 A detailed introduction to the development of German-speaking feminist
   geography can be found in Fleischmann and Meyer-Hanschen (2005, 29–53);
   for summaries in English, see Bäschlin (2002) and Bauriedl (2008).
7 See archive on https://ak-feministische-geographien.org/rundmail.
8 www.dampfboot-verlag.de/shop/kategorie/raumproduktionen-theorie-
   und-gesellschaftliche-praxis.
9 www.zeitschrift-suburban.de.
10 www.kritische-geographie.de.
11 In Berlin: mietenvolksentscheidberlin.de; in Frankfurt: www.stadt-fuer-alle.
    net.
12 orangotango.info.

References

ACME (ed.). 2008. Theme issue: German critical geographies. *ACME* 7(3).
Bartels, D. 1968. *Zur wissenschaftstheoretischen Grundlegung einer Geographie des Menschen*. Stuttgart: Franz Steiner.
Bäschlin, E. 2002. Feminist geography in the German-speaking academy: History and movement. In, P. Moss (ed.), *Feminist Geography in Practice: Research and Methods*. Oxford: Blackwell, pp. 25–9.
BASSDA. 2006. A kind of queer geography/räume durchqueeren: The Doreen Massey reading weekends. *Gender, Place & Culture* 13(2), 173–86.
Bauriedl, S. 2008. Still gender trouble in German-speaking feminist geography. In, P. Moss and K. F. Al-Hindi (eds.), *Feminisms in Geography: Rethinking Space, Place, and Knowledges*. Lanham: Rowman & Littlefield, pp. 130–9.
Bauriedl, S., M. Schier and A. Strüver (eds.). 2010. *Geschlechterverhältnisse, Raumstrukturen, Ortsbeziehungen. Erkundungen von Vielfalt und Differenz im spatial turn*. Münster: Westfälisches Dampfboot.
Belina, B. 2009. Theorie, Kritik und Relevanz in der deutschsprachigen sozialwissen-schaftlichen Geographie 40 Jahre nach Kiel, mit einigen bescheidenen Vorschlägen letztgenannte im Arbeitsalltag als gesellschaftliche zu füllen. *Rundbrief Geographie* 221, 18–20.
Belina, B. 2014. Was der Mythos der modernen Geographie nach Kiel ausschließt. *Geographica Helvetica* 69, 305–7.
Belina, B. 2020. Nach Kiel. Geschichtsphilosophisch inspirierter Kommentar zu Ute Wardengas GZ Journal Lecture 2019. *Geographische Zeitschrift* 108(1), 23–31.
Belina, B., U. Best and M. Naumann. 2009. Critical geography in Germany: From exclusion to inclusion via internationalization. *Social Geography* 4(1), 47–58.
Belina, B., M. Naumann and A. Strüver (eds.). 2014. *Handbuch Kritische Stadtgeographie*. Münster: Westfälisches Dampfboot.
Belina, B. and B. Michel (eds.). 2007. *Raumproduktionen. Beiträge der Radical Geography. Eine Zwischenbilanz*. Münster: Westfälisches Dampfboot (Raumproduktionen, 1).
Berg, L.D. 2004. Scaling knowledge: Towards a critical geography of critical geographies. *Geoforum* 35(5), 553–8.
Best, U. 2009. Critical Geography. In, R. Kitchin and N. Thrift (eds.), *International Encyclopedia of Human Geography*. Oxford: Elsevier, pp. 345–57.

Best, U. 2016. Competitive internationalisation or grassroots practises of internationalism? The changing international practises of German-language critical geography. *Social & Cultural Geography* 17(1), 23–38.

Bobek, H. and J. Schmithüsen.1949/1967. Die Landschaft im logischen System der Geographie. In, W. Storkebaum (ed.), *Zum Gegenstand und zur Methode der Geographie.* Darmstadt: Wissenschaftliche Buchgesellschaft, pp. 257–76.

Bock, S., U. Hünlein and H. Klamp (eds.). 1989. *Frauen(t)räume in der Geographie. Beiträge zur feministischen Geographie.* Kassel: Gesamthochschule Kassel (Urbs et Regio, 52).

Brand, U. 2010. Bedingungen und Möglichkeiten kritischer Wissenschaft. spw *Zeitschrift für sozialistische Politik und Wirtschaft* 181, 36–43.

Bretthauer, L. and D. Fromberg. 2008. Prekarisierung und Marginalisierung der Kritik. In, Torsten Bultmann (ed.), *Prekarisierung der Wissenschaft.* Berlin: Karl Dietz, pp. 23–40.

Brogiato, H.P., D. Hänsgen, N. Henniges, B. Schelhaas and U. Wardenga. 2010. "Ich kann sie nicht mehr gebrauchen, die Geographen, wie sie heute sind". Zur Gründungsgeschichte des DVAG. *Standort* 34, 74–9.

Bühler, E., H. Meyer, D. Reichert and A. Scheller (eds.). 1993. *Ortssuche. Zur Geographie der Geschlechterdifferenz.* Zürich: eFeF.

Eisel, U. 2014. Alte Zeiten, neue Zeiten – Ein Bericht, verbunden mit einigen Gedanken über neugierige Identitätssuche. *Geographica Helvetica* 69, 313–7.

Fahlbusch, M, Rössler, M, and D. Siegrist. 1989. *Geographie und Nationalsozialismus (Urbs et Regio 51).* Kassel: Gesamthochschule Kassel.

Fleischmann, K. and U. Meyer-Hanschen. 2005. *Stadt Land Gender. Einführung in Feministische Geographien.* Königstein/Taunus: Ulrike Helmer.

Frei, N., F. Maubach, C. Morina and T. Maik. 2019. *Zur rechten Zeit: Wider die Rückkehr des Nationalismus.* Berlin: Ullstein.

Gintrac, C. (2016): Kritische Stadtgeographie – ein Archipel epistemischer Gemeinschaften. *sub\urban. zeitschrift für kritische stadtforschung* 4(2/3), 59–82.

Geografiker. 1969. Sonderheft zum 37. Deutschen Geographentag. [Special Issue prepared for the 37th German Geographical Conference] Vol. 3.

Gribat, N., S. Hoehne, B. Michel and N. Schuster. 2016. Kritische Stadtforschungen. Ein Gespräch über Geschichte und Produktionsbedingungen, Disziplinen und Interdisziplinarität. *sub\urban. zeitschrift für kritische stadtforschung* 4(2/3), 11–36.

Halder, S. 2018. *Gemeinsam die Hände dreckig machen. Aktionsforschungen im aktivistischen Kontext urbaner Gärten und kollektiver Kartierungen.* Bielefeld: transcript.

Hannah, M.G. 2016. Innovations in the afterlife of the Cold War: German-language human geography. *Social & Cultural Geography* 17(1), 71–80.

Hartke, W. 1962. Die Bedeutung der geographischen Wissenschaft in der Gegenwart. In, Wolfgang Hartke and Friedrich Wilhelm (eds.), *Tagungsberichte und Abhandlungen des 33. Deutschen Geographentages in Köln 1961.* Wiesbaden: Steiner, pp. 113–31.

Hartmann, R., H. Hitz, C. Schmid and R. Wolff. 1985. Theorien zur Stadtentwicklung. *Geographische Hochschulmanuskripte* 12. Oldenburg.

Haushofer, K. 1925/1977. Politische Erdkunde und Geopolitik. In, Josef Matznetter (ed.), *Politische Geographie.* Darmstadt: Wissenschaftliche Buchgesellschaft, pp. 138–61.

Heinrich, H.-A. 1990. Der politische Gehalt des fachlichen Diskurses in der Geographie Deutschlands zwischen 1920 und 1945 und dessen Affinität zum Faschismus. *Geographische Zeitschrift* 78(4), 209–26.

Helbrecht, I. 2014. Der Kieler Geographentag 1969: Wunden und Wunder. *Geographica Helvetica* 69, 319–20.

Holm, A. and D. Gebhardt (eds.). 2011. *Initiativen für ein Recht auf Stadt. Theorie und Praxis städtischer Aneignungen.* Hamburg: VSA.

Kneisle, A. 1980. *"Offene" Wissenschaftstheorie oder Anbiederung an die Forschergemeinde.* Karlsruhe: Universität Karlsruhe.

kollektiv orangotango+ (ed.). 2018. *This Is Not an Atlas. A Global Collection of Counter-Cartographies.* Bielefeld: transcript.

Kost, K. 1988. *Die Einflüsse der Geopolitik auf Forschung und Theorie der Politischen Geographie von ihren Anfängen bis 1945.* Bonn: Dümmler.

Labica, G. 1986. *Der Marxismus-Leninismus. Elemente einer Kritik.* West-Berlin: Argument.

Lossau, J. 2002. *Die Politik der Verortung. Eine postkoloniale Reise zu einer >anderen< Geographie der Welt.* Bielefeld: transcript.

Mayer, K.U. and S. Hillmert. 2003. New ways of life or old rigidities? Changes in social structures and life courses and their political impact. *West European Politics* 26(4), 79–100.

Meckelein, W. and C. Borcherdt (eds.). 1970. *Deutscher Geographentag Kiel, 21. bis 26. Juli 1969. Tagungsbericht und wissenschaftliche Abhandlungen.* Stuttgart: Frank Steiner.

Michel, B. 2016. "With almost clean or at most slightly dirty hands". On the self-denazification of German geography after 1945 and its rebranding as a science of peace. *Political Geography* 60, 135–43.

Michel, B. 2014a. Antisemitismus, Großstadtfeindlichkeit und reaktionäre Kapitalismuskritik in der deutschsprachigen Geographie vor 1945. *Geographica Helvetica* 69, 193–202.

Michel, B. 2014b. Wir sind nie revolutionär gewesen – Zum Mythos des Kieler Geographentags als der Geburtsstunde einer neuen Geographie. *Geographica Helvetica* 69, 301–3.

Mullis, D. 2014. *Recht auf die Stadt. Von Selbstverwaltung und radikaler Demokratie.* Münster: Unrast.

Oßenbrügge, J. and A. Vogelpohl (eds.). 2014. *Theorien in der Raum- und Stadtforschung. Einführungen.* Münster: Westfälisches Dampfboot.

Rössler, M. 1990. *"Wissenschaft und Lebensraum". Geographische Ostforschung im Nationalsozialismus.* Berlin/Hamburg: Dietrich Reimer.

Sahr, W.-D. 2016. KIEL 1969 – eine *Mythanalyse* zur Epistemologiegeschichte der deutschen Nachkriegsgeographie. *Geographica Helvetica* 71, 77–85.

Sambale, J., V. Eick and H. Walk. 2008. *Das Elend der Universitäten. Neoliberalisierung deutscher Hochschulpolitik.* Münster: Westfälisches Dampfboot.

Sanke, H. 1962. *Entwicklung und gegenwärtige Probleme der politischen und ökonomischen Geographie in der Deutschen Demokratischen Republik. (Sitzungsbericht der Deutschen Akademie der Wissenschaften zu Berlin, Klasse für Philosophie, Geschichte, Rechts- und Wirtschaftswissenschaften. 4.)* Berlin: Akademie.

Schelhaas, B. 2004. *Institutionelle Geographie auf dem Weg in die wissenschaftspolitische Systemspaltung: die Geographische Gesellschaft der DDR bis zur III. Hochschul- und Akademiereform 1968/69.* Leipzig: Leibniz-Institut für Länderkunde.

Schipper, S. and F. Wiegand. 2015. Neubau-Gentrifizierung und globale Finanzkrise. Der Stadtteil Gallus in Frankfurt am Main zwischen immobilienwirtschaftlichen Verwertungszyklen, stadtpolitischen Aufwertungsstrategien und sozialer Verdrängung. *sub\urban. zeitschrift für kritische stadtforschung* 3(3), 7–32.

Schmidt-Renner, G. 1966. *Elementare Theorie der ökonomischen Geographie. 2. durchgesehene und erweiterte Auflage*. Gotha: Haack.

Schöller, P. 1957. Wege und Irrwege der Politischen Geographie und Geopolitik. *Erdkunde* 11(1), 1–2.

Strüver, A. 2010. KörperMachtRaum und RaumMachtKörper: Bedeutungsverflechtungen von Körpern und Räumen. In, S. Bauriedl, M. Schier and A. Strüver (eds.), *Geschlechterverhältnisse, Raumstrukturen, Ortsbeziehungen*. Münster: Westfälisches Dampfboot, pp. 217–237.

Troll, C. 1947. Die geographische Wissenschaft in Deutschland in den Jahren 1933 bis 1945. Eine Kritik und Rechtfertigung. *Erdkunde* 1(1), 3–48.

Troll, C. 1949. Geographic Science in Germany during the Period 1933-1945: A Critique and Justification. *Annals of the Association of American Geographers* 39(2), 99–137.

Wardenga, U. 2020. Vergangene Zukünfte – oder: Die Verhandlung neuer Möglichkeitsräume in der Geographie. *Geographische Zeitschrift* 108(1), 4–22.

Wastl-Walter, D. 2010. *Gender Geographien. Geschlecht und Raum als soziale Konstruktionen*. Stuttgart: Franz Steiner.

Werlen, B. 2010. Wolfgang Hartke – Begründer der sozialwissenschaftlichen Geographie. In, *Gesellschaftliche Räumlichkeit 1. Orte der Geographie*. Franz Steiner: Stuttgart, pp. 102–125.

Werlen, B. 2014. Kiel 1969 – Leuchtturm oder Irrlicht? *Geographica Helvetica* 69, pp. 293–299.

Wolf, F. 2006. Bildungspolitik: Föderale Vielfalt und gesamtstaatliche Vermittlung. In, M.G. Schmidt and R. Zohlnhöfer (eds.), *Regieren in der Bundesrepublik*. Wiesbaden: VS Verlag für Sozialwissenschaften, pp. 221–241.

10 Presence and absence
Ireland and critical geographies

Mary Gilmartin

Introduction

Ireland occupies an uneasy place within contemporary academic geography. Attempts to name and critique the hegemony of 'Anglo-American' geography often seem unsure of where Ireland fits. If 'Anglo-American' geography is defined linguistically, then Ireland – where English is the *de facto* first language – is certainly part of the core. At previous international critical geography conferences, for example Mexico City, the association of disciplinary hegemony with the English language meant that Ireland was clearly classified as 'Anglo-American'. Yet, definitions based on linguistic competence occlude the long and deeply problematic colonial and postcolonial relationship between Britain and Ireland. English is spoken in Ireland as a direct consequence of colonialism, and the experience of colonialism continues to shape everyday life in Ireland, not least through the continuing geopolitical divide on the island of Ireland, where the northeast remains an integral part of the United Kingdom. This postcolonial reality challenges the supremacy of linguistic competency as the key marker of 'Anglo-American' geography.

However, the uneasy place of Ireland also raises questions about the geographies of knowledge production. In relation to critical geography, important questions have been raised by people who work outside English-speaking contexts. Kirsten Simonsen, writing from Denmark, highlights the power-knowledge system at work in 'Anglo-American' or 'Anglophone' critical geography (she uses the two terms interchangeably). For Simonsen, 'knowledge produced in the UK and USA [is seen as] 'unlimited', 'universal' or at least 'transferable', while knowledge produced in almost all other places ... is constructed as 'limited', 'local' and deeply 'embedded' in its local/national context' (2004, 526). For the Spanish geographer Maria-Dolors García-Ramon, Anglophone critical geography 'has constructed a privileged position in which it is the theoretical centre and other scholarly geographical traditions are considered peripheral and a-theoretical, having to conform to the centre's theoretical framework and agenda' (García-Ramon, 2004). Ireland may well be Anglophone, but it is far from the centre

DOI: 10.4324/9781315600635-10

of geographic knowledge production. Indeed, geographers working in and on Ireland are not unused to hearing their research described as local, parochial or provincial by colleagues – in Ireland and further afield.

Yet, to describe Ireland solely in terms of its colonial past, its postcolonial present and its marginal position within contemporary geography is just one part of a more complex and complicated story. Academic geography in Ireland has long been shaped by the closeness of Britain. Many of the first lecturers and professors of geography were hired from British universities: for example, Estyn Evans (appointed lecturer at Queen's University Belfast in 1928) and Tom Jones-Hughes (appointed professor at University College Dublin in 1960) were graduates of the University of Wales. This pattern continues today, where three of the five chairs of geography in the Republic of Ireland, as well as a majority of permanent staff, are graduates of UK universities. In addition to the ongoing movement of people and ideas between Britain and Ireland, academic geography in Ireland is also influenced by measures of 'excellence' emerging from the United Kingdom, and by the regular use of geographers from the United Kingdom as external examiners for degree programmes and postgraduate theses. Less obviously, however, the influence of UK-trained geographers in Ireland meant that the discipline was less shaped by the Catholic Church than was the case for other social sciences in universities in Ireland, particularly sociology and philosophy. This was certainly the case at University College Dublin and, latterly, at the National University of Ireland Maynooth, where the Department of Geography was the first department to appoint a professor from a non-clerical or non-religious background.

In all these ways, therefore, Ireland raises questions that challenge any simplistic understanding of place and knowledge production. What does it mean to write about Ireland, given its colonial past? Does that include just the Republic of Ireland, or also Northern Ireland? Does it include geographers working in and/or on Ireland, regardless of their location and/or nationality? Does it include members of the Irish diaspora, regardless of their areas of interest? And what does it mean to write of critical geography in the context of Ireland? Is this critical geography as defined by 'the symbols of bourgeois prestige' (Blomley, 2007, 53), with a specific form of international reach? Or is it critical geography in a broader sense, concerned with socio-geographical change and with combating exploitation and oppression at a variety of scales? What might a situated understanding of 'critical' mean in the specific context of Ireland?

In this chapter, I explore these questions by focusing on three main areas. The first interrogates the situatedness of knowledge production. The second highlights the role of colonialism and its aftermath in the production and circulation of critical geography in, on and from Ireland. Third, and by way of conclusion, I consider the future of critical geography from the perspective of Ireland. Throughout these three broad areas of focus, tensions around the meaning of place and knowledge continually emerge. Rather than bend them into shape, these tensions are an integral part of the chapter.

The situatedness of knowledge production: Being critical in Ireland

In March 1978, a group of geographers organized a conference in Dublin. The topic – 'Radical approaches to Irish geography: historical and contemporary' – marked a new development in academic geography in Ireland; a fact marked by publication of a special issue on Ireland in *Antipode* in Summer 1980, following on from the earlier conference (Mac Laughlin et al., 1980). The growing influence of a political economy approach to Irish geography is evident in many of these papers (in particular, see Byrne, 1980; Mac Laughlin, 1980; Parson, 1980; Perrons, 1980; Regan, 1980; Walsh, 1980). However, several of the papers also touch on a key issue in Ireland in the 1970s and 1980s: that of Northern Ireland, both from political economy and geopolitical perspectives (see Anderson, 1980; Boal, 1980; Pringle, 1980).

The energy of the conference and the subsequent publication of the papers appeared to herald a new beginning for the discipline of geography in Ireland. The editorial collective commented on what they saw as 'almost a complete lack of concern with major social and economic issues' (Mac Laughlin et al., 1980, iii) among professional geographers. Despite this early flourishing, however, there is limited evidence of the further development of this particular radical strain in Irish academic geography, as well as suggestions of a rather muted response to these radical approaches. This is not to suggest that a radical geographical perspective did not develop in Ireland. Indeed, some of the contributors to the *Antipode* special issue continued to approach their research from such a perspective, most notably Francis Walsh (who now writes as Proinnsias Breathnach) and Jim Mac Laughlin (see, for example, Breathnach, 2007, 2010; Mac Laughlin, 1994). Rather, the extent to which radical geographers struggled to get academic appointments in Ireland suggests the academy did not prove a welcoming environment for such approaches. Though radical perspectives and beliefs remain, their articulation is less apparent in the official chronicles of Irish geography, while more visible in other sites, for example, in the work of a range of non-governmental and quasi-governmental organizations where radical geographers later found employment. The flowering of radical geography in Ireland was stifled, to an extent, by a dominant version of historical geography that was often explicitly apolitical and atheoretical. The hegemonic form of historical geography that developed in Ireland was consensual and non-confrontational, and shied away from direct engagement with political topics. The influence of historical geography on emerging departments of geography meant that geography was more likely to ally itself with the humanities rather than the social sciences. And, while historical geographers in Ireland paid attention to mapping colonialism, they were significantly less attuned to the need to decolonize the discipline of Irish geography – the ongoing patriarchal nature of academic geography in Ireland is a clear case in point (Ní Laoire and Linehan, 2002). The dominance of historical geography at a time when academic geography was developing in Ireland placed the limits to growth on other, more radical approaches to the study of Irish geography. As a consequence, explicitly radical or critical approaches often appear marginal in an Irish context. It is, it seems, difficult to identify an Irish school of critical geography.

Yet, the apparent lack of critical geography in Ireland is misleading. Geography in Ireland has a very different tradition and trajectory to the United Kingdom or the United States, particularly because of its sustained focus on grounded, empirical work. This work is often locally engaged and embedded, and very often demonstrates a strong commitment to the politics of place. A different form of scholarship is thus evident in Irish geography, one that pays less attention to the vagaries of theoretical fashion, and that is more concerned with developing a deep knowledge of place. This commitment to place, and also to a particular version of spatial justice, marks the work of academic and activist geographers in Ireland, though it may not be explicitly framed as critical geography. This point was already made by the *Antipode* editorial collective in 1980, rather apologetically, when they wrote that some of the articles in the collection:

> May not be considered 'radical' by many readers of *Antipode*, but they do constitute genuine attempts by concerned individuals to begin to develop a real understanding of certain crucial issues, and form important building blocks upon which further analysis can be based.
>
> (Mac Laughlin et al., 1980: iii)

However, this comment rests on a particular understanding of 'radical': that of a Marxist perspective. A more expansive understanding of radical offers up an alternative, critical perspective: one that recognizes the importance of documenting inequality, difference and injustice as a means of effecting change.

Seen from this perspective, geographers in Ireland pay significant attention to specific rural and urban areas, and to their socio-spatial marginalization. A key publication was *Poor People, Poor Places* (Pringle et al., 1999), which emerged from a conference on 'Geography and poverty' held in 1996. The organizers of the conference included the Combat Poverty Agency, a quasi-governmental agency with a focus on anti-poverty strategies and policy interventions. One of the organizers, Jim Walsh from the Combat Poverty Agency, trained as a geographer, and he was an advocate for area-based approaches to poverty reduction in the context of Ireland. The edited volume addressed broad areas of concern: there were chapters on both rural and urban poverty, on particular places (for example, Dublin and Limerick) and on a cross-border analysis of poverty on the island of Ireland. There is also an explicit political economy perspective underpinning some recent work on urban sites in Ireland (MacLaran and Kelly, 2014). As an example, Michael Punch's analysis of problem drug use in Dublin discusses uneven development as a consequence of neo-liberal urban restructuring (Punch, 2005). His focus on heroin use in inner-city Dublin communities, emerging as a significant issue from the early 1980s onwards, directly relates the drug problem to the unevenness of economic restructuring. As Punch describes, urban restructuring under neo-liberal globalization led to 'mass unemployment, low incomes, limited job prospects, educational disadvantage, and the effective marginalization of sectors of the population...physical decay, isolation, limited services and an uncertain housing situation' (Punch, 2005, 771). Ireland, as Punch pointed out so starkly, 'remains one of the most unequal countries in Europe' (Punch, 2005, 761). In addition to a strong political economy base, more recent work also integrates cultural studies and post-structuralist thought in its analysis of cities in Ireland. This is particularly the case for a growing body of work on Cork which, while framed within a broader analysis of urban change, offers critical commentaries on the commodification of culture and on the role of representation at times of crisis (see O'Callaghan, 2012; O'Callaghan and Linehan, 2007). This work is not just confined to the current situation. Jacinta Prunty's research on Dublin slums draws on a wide range of original, empirical sources to carefully recraft the urban geographies of exclusion and marginalization in the city (Prunty, 1998), while Stephen Royle charts the growth and development of Belfast as an industrial city with the concomitant rise in levels of poverty, disease and conflict (Royle, 2011). As well as more explicitly theoretical work, designed for academic audiences, geographers in Ireland also have a long tradition of working with state and civil society organizations, using geographical analysis to address issues of urban inclusion, exclusion and marginalization. For example, recent work in collaboration with local authorities focusses on urban change and housing in Limerick, on social exclusion in Cork and on social inclusion in Dublin (Gleeson et al., 2009; Linehan and Edwards, 2005; McCafferty, 2005; McCafferty and Canny, 2005), while recent work in association with the Combat Poverty Agency focusses on greater community participation in urban planning, with particular reference to areas of disadvantage (MacLaran et al., 2007).

Geographers working in and on Ireland also have a long and sustained engagement with rural areas, from both historical and contemporary perspectives. The comment of the editors of the *Antipode* special issue in many ways serves as a critique of the pre-eminence of historical geography: thus the observation that major social and economic issues were ignored. Yet, it could be argued that historical geography in Ireland, with its strong focus on rural areas, did deal with the social, cultural and economic consequences of colonialism, particularly in terms of landholding and settlement. Here, the influence of Tom Jones Hughes, the first professor of geography at University College Dublin, is worth highlighting. Jones Hughes supervised a wide range of postgraduate theses on aspects of rural life in Ireland (Jones Hughes, 2010, 347), and the men whose PhDs he supervised became, in turn, professors of geography at National University of Ireland Maynooth, University College Cork and University College Dublin (Smyth, 2010, xviii–xx). On one level, this approach to historical geography is firmly rooted in an empiricist and, some argue, descriptive tradition, drawing heavily on documentary sources and on fieldwork. On another level, this approach also insisted on the importance of local knowledge and on rural ways of life, coming under threat in the postcolonial rush to modernization. This was not an attempt to preserve tradition, but rather a desire to understand 'how variations in the economic, social and cultural needs of communities have either bonded or separated people at a variety of territorial levels' (Smyth, 2010, xxiii).[1] In addition, Jones Hughes' interest in rural areas, and the encouragement he provided for students from rural areas, marked a moment of critical engagement for the discipline of geography in Ireland. With those 'country people', he provided a:

> Sympathetic sounding board in what for some was an uncongenial university environment, stiffening their appreciation of the validity of their own culture, giving them a new pride in their localities but above all making them probe deeper into the intertwined meanings of society and place.
>
> (quoted in Smyth, 2010, xvii)

The spirit of this particular approach to historical geography is best captured by the county *History and Society* series, edited by William Nolan. The aim of this series is to produce an edited volume on each of the 32 counties in Ireland: between 1985 and 2020, 27 volumes have been published. This represents the sustained commitment to place that characterizes the work of many geographers in Ireland, and the kind of long-term thinking that is rarely rewarded in the current academic climate, concerned with immediate outputs in limited formats.

While the study of rural areas was dominated by historical geographers, it was not just their preserve. A smaller number of social and economic geographers, whose interests lay in contemporary rural issues, worked contemporaneously with their counterparts, charting the ways in which rural Ireland was undergoing significant and rapid transformation. The Department of Geography at University College Galway (now the School of Geography and Archaeology at NUI Galway) has had a long emphasis on rural studies, and work emanating from NUI Galway has been joined by that of other geographers, concerned with key issues affecting landscapes and lives in changing rural Ireland. One longstanding area of focus is rural development, which has charted changes in rural economies and highlighted alternative means of fostering development (see, for example, Cawley, 1983, 2010; Gillmor, 1987; McDonagh, 2001; McDonagh et al., 2010;

Storey, 2004). Population change in rural Ireland is a second area of sustained focus, whether charting the sustained depopulation of rural Ireland (Cawley, 1994), or focusing on the impacts of migration from rural areas on those who stayed behind (Ní Laoire, 2000, 2001). In recent years, the changing relationship between urban and rural has received more attention, both in terms of the built environment (particularly apt in the context of the short-lived housing boom in Ireland) and in terms of mobility, whether on a temporary or a permanent basis (Gkartzios and Scott, 2010; Horner, 1999; Mahon, 2007; Meredith et al., 2007). Meanwhile, issues that transcend the rural/urban divide in the practise of critical geography in Ireland include health, sustainability and migration. Studies of health, from a critical perspective, highlight spatial inequalities in the distribution of illness and in access to services and resources (Houghton, 2005; Kalogirou and Foley, 2006; Morrissey et al., 2008). A growing body of work on sustainability and the environment focusses on rural areas, on waste and waste management, on environmental governance, and on foodscapes (Davies, 2008; 2012; Fahy and Ó Cinnéide, 2009; McDonagh et al., 2009; Sage, 2010). For geographers interested in migration, both the Irish diaspora (Boyle, 2011; Jenkins, 2013; Nash, 2008) as well as recent migrants to Ireland have received attention from critical geographers. In particular, the work of the Migrant Children's research project team, based at University College Cork, provided new insights into children's migration in Ireland and more generally (Ní Laoire et al., 2011).

From its inception, therefore, academic geography in Ireland has been deeply embedded in place. This means that the work of academic geographers has not always been translated to other contexts, leading to claims of parochialism. Yet, as this overview shows, the commitment to place and context means that academic geographers, as well as those trained as academic geographers and employed in other settings, have a sustained engagement with Ireland, from both historical and contemporary perspectives. Much of this locally embedded work is critical in form and focus, concerned with documenting inequalities, influencing policies, understanding and indeed valorizing the everyday life-worlds of people in Ireland, and charting the myriad ways in which the space of Ireland is shaped by broader global processes. One of those global processes – colonialism – is the focus of the next section of this chapter.

Colonialism and critical geography

Ireland is a 'neo-colony' (Walsh, 1980, 67). So wrote Francis Walsh, as part the *Antipode* special issue in 1980. Walsh marshalled a range of evidence to support his assertion, focusing particularly on the level of dependence on foreign investment. Here, and elsewhere, Walsh uses the case of Irish mining to highlight a 'classical form of neo-colonialism' (Walsh, 1980, 70; see also Regan and Walsh, 1976). These are among a small number of attempts to use neo-colonial theory to explain the particular economic circumstances of Ireland. Walsh and Regan were among a group of geography students from Ireland who had studied at Simon Fraser University in Canada in the early 1970s. Phil O'Keefe (O'Keefe, n.d.), editor of *Antipode* from 1978–1980, remarked that Simon Fraser, with its 'strong Irish flavour', was central to radical and socialist geography in the period: it had, according to Peet, the largest membership of the Union of Socialist Geographers (Peet, n.d.). The influence of radical, socialist and Marxist thought on the work of geographers in Ireland continues, in both academic and activist contexts. However, despite Walsh and Regan's early attempts to interrogate Ireland from a neo-colonial perspective, few geographers followed their lead. Instead, postcolonial theory, interpreted through the cultural turn in geography, became more influential. As a consequence, Ireland has been, and continues to be, an important site for the elaboration of postcolonial approaches within geography more generally, and within critical geography.

Two areas of focus are particularly important. The first offers a critique of the practise of colonialism in Ireland, while the second uses a postcolonial perspective to interrogate the contemporary situation in Ireland, as a partially or problematically postcolonial state. The focus on colonial Ireland has been most common, with a wide range of geographical work attempting to reclaim and rewrite the stories of Ireland's experience of and with the British colonial endeavour. William J. Smyth's comprehensive engagement with the early colonization of Ireland exemplifies this tradition (Smyth, 2006). His insistence on the need to get 'behind and beyond the more richly documented worlds of the ruling elites' is significant, both because of how it highlights the 'lives and localities of ...ordinary men and women' (Smyth, 2006, xx), and because of how it mirrors other engagements with the practise of British colonialism, most notably that of the Subaltern Studies group. More recently, David Nally's discussion of the Great Irish Famine – where around one million died and another two million emigrated – skilfully interweaves analyses of colonialism, capitalism and biopolitics together in its insistence on the Famine as a 'colonial experience' (Nally, 2011, 23), while the *Atlas of the Great Irish Famine* shows the effects of the Famine in vivid detail (Crowley et al., 2012). These substantial recent engagements with colonial Ireland are part of a much broader body of research, carried out by geographers based in and beyond Ireland, on the practise of colonialism in Ireland and on anti-colonial movements (see, for example, Kearns, 2003, 2004, 2006; Morrissey, 2003, 2005). While this work explicitly engages with colonialism from a postcolonial perspective, other work charts the observable impacts of colonialism on aspects of life in colonized Ireland, such as landownership, land use, mapping, environmental transformation, population change and movement and changing landscapes (see, for example, Andrews, 1975; Duffy, 2007; Whelan, 2003). John Andrews' meticulous study of the Ordnance Survey in Ireland (Andrews, 1975) focusses on 'those who wrote things down' (Friel et al., 1983, 119). Yet, his work was used to dramatic effect by playwright Brian Friel in his play 'Translations'. Friel described his encounter with *A Paper Landscape* in this way:

> And suddenly here was the confluence – the aggregate – of all those notions that had been visiting me over the previous years: the first half of the nineteenth century; an aspect of colonialism; the death of the Irish language and the acquisition of English. Here were all the elements I had been dallying with, all synthesised in one very comprehensive and precise text. Here was the perfect metaphor to accommodate and realise all those shadowy notions – map-making. Now, it seemed to me, all I had to do was dramatise *A Paper Landscape*.
>
> (Friel et al., 1983, 123)

Friel acknowledged that, in the end, his attempts to dramatize Andrews' work were less than successful, but he certainly drew from the text in charting the relationships that developed between colonizer and colonized, between people and place and between language and identity. In short, the geographical representation of one of the key tools of the colonial project in Ireland led to a highly influential cultural representation of colonialism, from a postcolonial perspective.

The second area of focus is the contemporary situation in Ireland. Again, this work at times draws explicitly on postcolonial theory, but at other times focusses on the social, cultural and political expressions of a society struggling to make sense of its colonial past. There are two specific areas of focus here. The first, and most explicit, is work on Northern Ireland as a site and space of conflict. Among geographers, perhaps the most influential early work was by Frederick Boal, whose research on Belfast from the 1960s onwards highlighted the extent of segregation and territoriality, based on religious affiliation, among working-class neighbourhoods in that city (Boal, 1969). Reflecting on the 1969 paper from the vantage point of 2008, Boal commented modestly that it 'lacked a discussion of the broader historical and political contexts ... was insufficiently theorised [and] did not offer any predictive insights' (Boal, 2008, 334). Yet, Boal's work set the stage for attempts by social, political and cultural geographers in Ireland to make sense of the Troubles, a 'pernicious and persistent conflict' (Shirlow, 2008, 337) that, though alleviated by ceasefires and political resolutions from 2004 onwards, continues in other forms today. A second, though significantly less substantial, body of work focusses on contemporary Ireland as a postcolonial site. While this is gaining some momentum, the contribution made by geographers is markedly less than that made by literary scholars in particular, whose use of postcolonial theory to interrogate contemporary Ireland has in general been more nuanced.

Ireland, North and South, has been marked by the Troubles, and the discipline of geography reflects the ways in which this conflict has been expressed and experienced. Political geographers have focussed on the border between Northern Ireland and the Republic of Ireland, often from an historical perspective (Anderson and O'Dowd, 2007; Rankin, 2007), while more recently others – including cultural and social geographers – have focussed on the border as a landscape, and as a site of everyday lives (Nash et al., 2010; Nash and Reid, 2010). Of particular relevance is the Borderlands project, funded by the Arts and Humanities Research Council in the United Kingdom, which sought to explore contemporary meanings and experiences of the border as well as its historical geographies.[2] Others have addressed the intersections between political and social geographies, particularly in urban areas of Northern Ireland. The empirical focus of Boal and, later, Paul Doherty (Boal, 1969; 1976; Boal et al., 1974; Doherty, 1990) is taken up by Peter Shirlow. Shirlow's work draws on insights from political economy as well as from social and cultural studies, and engages with ongoing violence and related inequalities in Northern Ireland (Shirlow, 2006; Shirlow and Gallaher, 2006; Shirlow and Murtagh, 2006). In recent years, a strong focus on questions of identity and belonging has also emerged. This focus, inspired by the work of Brian Graham (2007), draws on the cultural turn within geography, and has two distinct forms of expression. The first comes from a position of scepticism around the 'postcolonial' status of Ireland/Northern Ireland. Cultural geographer Bryonie Reid expresses this ambivalence well when she says that 'the portrayal of Ireland as a postcolonial state is one which I find difficult, implying as it can that Northern Ireland remains a colonial society: in which case it is suggested I think of myself as a colonist' (Reid, 2011). As a consequence, some work in this vein is concerned with reasserting the legitimacy of the unionist presence in Ireland (Switzer, 2007), while other work focusses on the complex and imbricated relationship between Ireland and Britain (Johnson, 2003). As concepts, memory and/or place have become a means of reflecting on the Troubles from positions of spatial and temporal proximity (Graham and McDowell, 2007; McDowell, 2009; McDowell and Switzer, 2009; Reid, 2007). The second focusses on the practises of nationalism and loyalism (see Dowler, 1998, 2001, 2002; Gallaher, 2007), both during and after the Troubles, but often comes from the work of academic geographers based outside Ireland, albeit following extensive fieldwork. It is possible to argue that the contemporary focus on memory mirrors an earlier focus on historical geography: it allows for an indirect engagement with topics that remain raw and contested. The past thus offers a less fraught route to reflections on the geographies of the present.

However, the Border has proven to be an intellectual as well as political divide for geographers in Ireland. While geographers based in Northern Ireland and outside, particularly in the United States and Britain, have researched and written about the causes and consequences of the Troubles, geographers based in the Republic of Ireland have struggled with the issue.

Beyond the 1980 special issue of *Antipode*, there are limited examples of direct engagement with Northern Ireland by southern geographers. However, it is interesting to speculate on the extent to which growing up in a divided place shaped the research interests and careers of a range of well-known political geographers, whose early university education was in Ireland, but who later moved to the United States and the United Kingdom. Gearóid Ó Tuathail/Gerard Toal's experience of growing up near the border is an implicit presence in *Critical Geopolitics* (Ó Tuathail, 1996), and his experience of Ireland is explicitly discussed in a forum on that book (Toal, 2000). Others who work as political geographers, but whose early training was in Ireland, include Simon Dalby, John O'Loughlin and David Storey. While their research rarely engages directly with Ireland (with the exception of Storey, 2001), their concentration in the subdiscipline of political geography raises questions about the relationship between place autobiographies and academic trajectories. Now, academic geographers in the Republic of Ireland are beginning to engage with Northern Ireland. However, much of this work focusses on effectively reshaping the territory of Ireland. For example, AIRO (the All-Ireland Research Observatory) is an explicitly cross-border spatial data project.[3] On one level, the work of AIRO clearly highlights the differences in everyday life that emerge based on which side of the border a person lives or works, thus building on a long tradition of geographical analysis in Ireland that points to spatial injustices. However, its representations of the island of Ireland, no longer with empty spaces to denote the North or the South, are helping to craft a new image of Ireland, no longer rendered separate by a line on a map.

Others use postcolonial approaches to reflect on contemporary Ireland, and use Ireland – broadly defined – to contribute to the development of a spatial imaginary within postcolonial theory. Catherine Nash has made extensive use of postcolonial theory to reflect on identity in the context of Ireland, on issues ranging from gender and landscape in Ireland as expressed in the work of artist Kathy Prendergast (Nash, 1993), to a postcolonial analysis of place-based identities in Ireland (Nash, 1999), to more recent work on Irish identities as expressed through genealogy and genetics (Nash, 2008). In his study of Irish Catholics in Scotland, Mark Boyle makes extensive use of Jean Paul Sartre's theory of colonialism, arguing that the migration and settlement of Irish Catholics in Scotland represents a form of colonial encounter, with long-lasting consequences (Boyle, 2011). More modest engagements with postcolonial theory involve reflections on pipeline politics in the West of Ireland (Gilmartin, 2009), on 'contrapuntal urbanisms' (O'Callaghan, 2012), and on the ways in which attention to the specific experience Ireland fundamentally challenges the ways in which postcolonialism is theorized in the discipline of geography (Gilmartin and Berg, 2007).

Critical geography: Insights from Ireland

The story of critical geography in and beyond Ireland is highly influenced by migration and mobility. The official narrative of academic geography in Ireland highlights the extent to which departments of geography were shaped by migrant geographers from Britain: Estyn Evans in Belfast, Tom Jones Hughes in Dublin (Kearns, 2013). This movement continues, with UK-trained geographers having a significant influence on the practise of geography in contemporary Ireland. Equally relevant, however, is the importance of student migration: the postgraduate students who travelled to Vancouver, Toronto and New York, for example, and who infused geography in Ireland with new ideas and new approaches. Many of those did not necessarily work in academic geography, but instead brought geographic perspectives to the study of issues of poverty in Ireland, or of development in the Global South. This long history of movement and mobility has led to a certain porosity about disciplinary borders in Ireland, and a fluidity in thinking about the practise of geography.

Despite these international links and connections, geography in Ireland is often very focussed on the local. This has led to charges of insularity and parochialism, and nascent attempts to introduce research assessment strategies in Ireland prioritize research that is seen as 'international' over research that is seen as 'national'. Yet, as I have argued here, the nature and form of much geographic research in Ireland has a strong commitment to spatial justice, whether through its attempts to describe and highlight inequalities, or through applied work that attempts to directly or indirectly influence policy. This commitment to place is a component of the best work by critical geographers in and on Ireland (for a recent example, see Kearns et al., 2014). Yet, much of this work receives a limited audience outside Ireland. Indeed, critical geographers from Ireland are best known for work that has no direct relationship with Ireland (for example, Kitchin and Dodge, 2011). However, this insistence on the importance of place has helped contribute to the strength of academic geography in Ireland. The subject remains remarkably popular at school and university level, and geographers are important public intellectuals within Ireland, offering critical perspectives on contemporary Irish society. Thus, Ireland offers an alternative way of thinking about critical geography: as local practise, where the local is understood as a site of decoloniality, marked by its connections to other local places rather than by its global designs (Gilmartin, 2009; Mignolo, 2000; 2007). This is particularly important when we consider the specific context of contemporary Ireland, shaped by its experience of colonialism in the heart of Europe, by its experience of conflict in the form of the Troubles, and by its new troubles as a state experiencing structural adjustment as a consequence of the 'great recession' (Fraser et al., 2013). Geographers in Ireland continue to emphasize the need for critical engagement with the local, in a variety of forms (Kitchin et al., 2013; see also Ireland after Nama[4]). Our work shows the importance of understanding critical geographies as 'situated knowledges' (Larner, 2011), and of recognizing alternative ways to practise critical geographies that respond to scale and context, and not just to globalizing theory.

Acknowledgements

Thanks to Proinnsias Breathnach, Stephen McCarron and Peter Shirlow for their helpful comments on an earlier draft of this chapter.

Notes

1 This is Smyth's description of the work of Tom Jones Hughes, but it is an equally apt description of the work of many of his postgraduate students, such as Smyth himself, William Nolan and Patrick Duffy.
2 See the project website at http://www.irishborderlands.com/project/index. html (accessed 24 August 2021).
3 See http://airo.maynoothuniversity.ie/ for more information about AIRO.
4 See irelandafternama.wordpress.com, an explicit response by geographers in Ireland to the country's current and devastating economic crisis.

References

Anderson, J. 1980. Regions and religions in Ireland: A short critique of the 'two nations' theory. *Antipode* 12(1), 44–52.

Anderson, J. and L. O'Dowd. 2007. Imperialism and nationalism: The Home Rule struggle and border creation in Ireland, 1885-1925. *Political Geography* 26(8), 934–50.

Andrews, J.H. 1975. *A Paper Landscape: The Ordnance Survey in Nineteenth-Century Ireland*. Oxford: The Clarendon Press.

Blomley, N. 2007. Critical geography: Anger and hope. *Progress in Human Geography* 31(1), 53–65.

Boal, F.W. 1969. Territoriality on the Shankill-Falls divide, Belfast. *Irish Geography* 6, 30–50.

Boal, F.W. 1976. Ethnic residential segregation. In, D.T. Herbert and R.J. Johnson (eds.), *Social Areas in Cities, Vol. 1*. Chicester: John Wiley, pp. 235–51.

Boal, F.W. 1980. Two nations in Ireland. *Antipode* 12(1), 38–43.

Boal, F.W. 2008. Territoriality on the Shankill-Falls divide: Being wise after the event. *Irish Geography* 41(3), 329–35.

Boal, F.W., P. Doherty and D. Pringle. 1974. *The Spatial Distribution of Some Social Problems in the Belfast Urban Area*. Belfast: Northern Ireland Community Relations Project.

Boyle, M. 2011. *Metropolitan Anxieties: On the meaning of the Irish Catholic Adventure in Scotland*. Farnham: Ashgate.

Breathnach, P. 2007. Occupational change and social polarisation in Ireland: Further evidence. *Irish Journal of Sociology* 16, 22–42.

Breathnach, P. 2010. From spatial Keynesianism to post-Fordist neoliberalism: Emerging contradictions in the spatiality of the Irish state. *Antipode* 42(5), 1180–99.

Byrne, D. 1980. Spatial underdevelopment: The strategy of accumulation in Northern Ireland. *Antipode* 12(1), 87–96.

Cawley, M. 1983. Part time farming in rural development: Evidence from western Ireland. *Sociologia Ruralis* 23, 63–75.

Cawley, M. 1994. Desertification: Measuring population decline in rural Ireland. *Journal of Rural Studies* 10(4), 395–407.

Cawley, M. 2010. Adding value locally through integrated rural tourism: Lessons from Ireland. In, G. Halseth, S. Markey and D. Bruce (eds.), *The Next Rural Economies: Constructing Rural Place in a Global Economy.* Wallingford: CABI, pp. 89–101.

Crowley, J., W.J. Smyth and M. Murphy (eds.). 2012. *The Atlas of the Great Irish Famine.* Cork: Cork University Press.

Davies, A. 2008. *Geographies of Garbage Governance: Interventions, Interactions and Outcomes.* Aldershot: Ashgate.

Davies, A. 2012. Geography and the matter of waste mobilities. *Transactions of the Institute of British Geographers* 37(2), 191–6.

Doherty, P. 1990. Residential segregation, social isolation and integrated education in Belfast. *Journal of Ethnic and Migration Studies* 3, 391–401.

Dowler, L. 1998. 'And they think I'm just a nice old lady': Gender identities and war, in Belfast, Northern Ireland. *Gender Place and Culture* (5)2, 159–76.

Dowler, L. 2001. No man's land: Transgressing the boundaries of West Belfast, Northern Ireland. *Geopolitics* 6(3), 158–76.

Dowler, L. 2002. Till death do us part: Masculinity, friendship and nationalism in Belfast Northern Ireland. *Environment and Planning D, Society and Space* 20(1), 53–71.

Duffy, P. 2007. *Exploring the History and Heritage of Irish Landscapes.* Dublin: Four Courts Press.

Fahy, F. and M. Ó Cinnéide. 2009. Re-Mapping the urban landscape: Community mapping – an attractive prospect for sustainability? *Area* 41(2), 167–75.

Fraser, A., E. Murphy and S. Kelly. 2013. Deepening neoliberalism via austerity and 'reform': The case of Ireland. *Human Geography* 6, 38–53.

Friel, B., J. Andrews and K. Barry. 1983. Translations and a paper landscape: Between fiction and history. *The Crane Bag* 7(2), 118–24.

Gallaher, C. 2007. *After the Peace: Loyalist Paramilitaries in Post-Accord Northern Ireland.* Ithaca: Cornell University Press.

García-Ramon, M.-D. 2004. The spaces of critical geography: An introduction. *Geoforum* 35(5), 523–4.

Gillmor, D.J. 1987. Concentration of enterprises and spatial change in the agriculture of the Republic of Ireland. *Transactions of the Institute of British Geographers NS* 12(2), 204–16.

Gilmartin, M. 2009. Border thinking: Rossport, Shell and the political geographies of a gas pipeline. *Political Geography* 28(5), 274–82.

Gilmartin, M. and L. Berg. 2007. Locating postcolonialism. *Area* 39(1), 120–4.

Gkartzios, M. and M. Scott. 2010. Residential mobilities and house-building in rural Ireland: evidence from three case studies. *Sociologia Ruralis* 50(1), 64–84.

Gleeson, J., R. Kitchin, B. Bartley and C. Treacy. 2009. *New Ways of Mapping Social Inclusion in Dublin City.* Dublin City Development Board/NIRSA.

Graham, B. (ed.). 2007. *In Search of Ireland: A Cultural Geography.* London: Routledge.

Graham, B. and S. McDowell. 2007. Meaning in the Maze: The heritage of Long Kesh. *Cultural Geographies* 14(3), 343–68.

Horner, A. 1999. The tiger stirring: Aspects of commuting in the Republic of Ireland 1981-1996. *Irish Geography* 32(2), 99–111.

Houghton, F. 2005. Hiding the evidence, the State and spatial inequalities in health in Ireland. *Irish Geography* 37(2), 96–106.

Jenkins, W. 2013. *Between Raid and Rebellion: The Irish in Buffalo and Toronto, 1867-1916*. Montreal: McGill-Queens University Press.

Johnson, N.C. 2003. *Ireland, the Great War and the Geography of Remembrance*. Cambridge: Cambridge University Press.

Jones Hughes, T. 2010. *Landholding, Society and Settlement in Nineteenth-Century Ireland: A Historical Geographer's Perspective*. Dublin: Geography Publications.

Kalogirou, S. and R. Foley. 2006. Health, place & Hanly: Modelling accessibility to hospitals in Ireland. *Irish Geography* 39(1), 52–68.

Kearns, G. 2003. Nation, empire, cosmopolis: Ireland and the break with Britain. In, D. Gilbert, D. Matless and B. Short (eds.), *Geographies of British Modernity: Space and Society in the Twentieth Century*. Oxford: Blackwell, pp. 204–28.

Kearns, G. 2004. Mother Ireland and the revolutionary sisters. *Cultural Geographies* 11, 459–83.

Kearns, G. 2006. Bare life, political violence and the territorial structure of Britain and Ireland. In, D. Gregory and A. Pred (eds.), *Violent Geographies: Fear, Terror and Political Violence*. New York: Routledge, pp. 9–34.

Kearns, G. 2013. Introduction to special issue: Historical geographies of Ireland: Colonial contexts and postcolonial legacies. *Historical Geography* 41, 22–34.

Kearns, G., D. Meredith and J. Morrissey (eds.). 2014. *Spatial Justice and the Irish Crisis*. Dublin: Royal Irish Academy.

Kitchin, R. and M. Dodge. 2011. *Code/Space: Software and Everyday Life*. Massachusetts: The MIT Press.

Kitchin, R., D. Linehan, C. O'Callaghan and P. Lawton. 2013. Public geographies through social media. *Dialogues in Human Geography* 3(1): 56–72.

Larner, W. 2011. Economic geographies as situated knowledges. In, J. Pollard, C. McEwan and A. Hughes (eds.), *Postcolonial Economies: Rethinking Material Lives*. London: Zed Press, pp. 81–106.

Linehan, D. and C. Edwards. 2005. *City of Difference: Mapping Social Exclusion*. Cork: Cork City Council.

MacLaran, A., V. Clayton and P. Brudell. 2007. *Empowering Communities in Disadvantaged Areas: Towards Greater Community Participation in Irish Urban Planning?* Dublin: Combat Poverty Agency.

MacLaran, A. and S. Kelly (eds.). 2014. *Neoliberal Urban Policy and the Transformation of the City: Reshaping Dublin*. Basingstoke: Palgrave Macmillan.

Mac Laughlin, J. 1980. Industrial capitalism, ulster unionism and orangeism: An historical reappraisal. *Antipode* 12(1), 15–27.

Mac Laughlin, J., D. Pringle, C. Regan and F. Walsh (eds). 1980. Special Issue on Ireland. *Antipode* 12(1), 1–115.

Mac Laughlin, J. 1994. *Ireland: The Emigrant Nursery and the World Economy*. Cork: Cork University Press.

Mahon, M. 2007. New populations, shifting expectations: The changing experience of rural space and place. *Journal of Rural Studies* 23(3), 345–56.

McCafferty, D. 2005. *Limerick: Profile of a Changing City*. Limerick: Limerick City Development Board.

McCafferty, D. and A. Canny. 2005. *Public Housing in Limerick: A Profile of Tenants and Estates*. Limerick: Limerick City Council.

McDonagh, J. 2001. *Renegotiating Rural Development in Ireland*. Aldershot: Ashgate.

McDonagh, J., M. Farrell, M. Mahon and M. Ryan. 2010. New opportunities and cautionary steps? Farmers, forestry and rural development in Ireland. *European Countryside* 2(4), 236–51.

McDonagh, J., T. Varley and S. Shortall (eds.). 2009. *A Living Countryside? The Politics of Sustainable Development in Rural Ireland*. Aldershot: Ashgate.

McDowell, S. 2009. Negotiating places of pain in post-conflict Northern Ireland: Debating the future of the Maze prison/Long Kesh. In, W. Logan and K. Reeves (eds.), *Places of Pain and Shame: Dealing with Difficult Heritage*. London: Routledge, pp. 215–30.

McDowell, S. and C. Switzer. 2009. Redrawing cognitive maps of conflict: Lost spaces and forgetting in the centre of Belfast. *Memory Studies*, 2(3), 337–53.

Meredith, D., M. Charlton, R. Foley and J. Walsh. 2007. *Modelling Commuting Catchments in Ireland: A Hierarchical Approach using GIS*. The Rural Economy Research Centre Working Paper Series, Working Paper 07-WP-RE-12, Teagasc.

Mignolo, W. 2000. *Local Histories/Global Designs: Coloniality, Subaltern Knowledges, and Border Thinking*. Princeton: Princeton University Press.

Mignolo, W. 2007. Delinking: the rhetoric of modernity, the logic of coloniality and the grammar of de-coloniality. *Cultural Studies* 21(2-3), 449–514.

Morrissey, J. 2003. *Negotiating Colonialism*. London: Royal Geographical Society (HGRG Research Series).

Morrissey, J. 2005. Cultural geographies of the contact zone: Gaels, Galls and overlapping territories in late medieval Ireland. *Social and Cultural Geography* 6(4), 551–66.

Morrissey, K., G. Clarke, D. Ballas, S. Hynes and C. O'Donoghue. 2008. Examining access to GP services in rural Ireland using microsimulation analysis. *Area* 40(3), 354–64.

Nally, D.P. 2011. *Human Encumbrances: Political Violence and the Great Irish Famine*. Notre Dame, Indiana: University of Notre Dame Press.

Nash, C. 1993. Remapping and renaming: New cartographies of identity, gender and landscape in Ireland. *Feminist Review* 44, 39–57.

Nash, C. 1999. Irish placenames: Post-colonial locations. *Transactions of the Institute of British Geographers* 24, 457–80.

Nash, C. 2008. *Of Irish Descent: Origin Stories, Genealogy and the Politics of Belonging*. Syracuse, NY: Syracuse University Press.

Nash, C., L. Dennis and B. Graham. 2010. Putting the border in place: Customs regulation in the making of the Irish border, 1921–1945. *Journal of Historical Geography* 36, 421–31.

Nash, C. and B. Reid. 2010. Border crossings: New approaches to the Irish border. *Irish Studies Review* 18(3), 265–84.

Ní Laoire, C. 2000. Conceptualising Irish rural youth migration: A biographical approach. *International Journal of Population Geography* 6, 229–43.

Ní Laoire, C. 2001. A matter of life and death? Men, masculinities and staying 'behind' in rural Ireland. *Sociologia Ruralis* 41, 220–36.

Ní Laoire, C., F. Carpena-Mendez, N. Tyrrell and A. White. 2011. *Childhood and Migration in Europe: Portraits of Contemporary Ireland*. Farnham: Ashgate.

Ní Laoire, C. and D. Linehan. 2002. Engendering the human geographies of Ireland: A thematic section. *Irish Geography* 35(1), 1–5.

O'Callaghan, C. 2012. Lightness and weight: (Re)reading urban potentialities through photographs. *Area* 44(2), 200–7.

O'Callaghan, C. and D. Linehan. 2007. Identity, politics and conflict in dockland development in Cork, Ireland: European Capital of Culture 2005. *Cities* 24(4), 311–23.

O'Keefe, P. n.d. Editing Antipode. http://onlinelibrary.wiley.com/journal/10.1111/(ISSN)1467-8330/homepage/editor_s_past_reflections.htm (5 January 2021).

Ó Tuathail, G. 1996. *Critical Geopolitics.* London: Routledge.

Parson, D. 1980. Spatial underdevelopment: The strategy of accumulation in Northern Ireland. *Antipode* 12(1), 73–86.

Peet, Richard. n.d. Reminiscing the early Antipode. http://onlinelibrary.wiley.com/journal/10.1111/(ISSN)1467-8330/homepage/editor_s_past_reflections.htm (5 January 2021).

Perrons, D. 1980. Ireland and the break-up of Britain. *Antipode* 12(1), 53–65.

Pringle, D.G. 1980. The Northern Ireland conflict: a framework for discussion. *Antipode* 12(1), 28–37.

Pringle, D.G., J. Walsh and M. Hennessy (eds.). 1999. *Poor People, Poor Places: A Geography of Poverty and Deprivation in Ireland.* Dublin: Oak Tree Press.

Prunty, J. 1998. *Dublin Slums, 1800-1925, A Study in Urban Geography.* Dublin: Irish Academic Press.

Punch, M. 2005. Problem drug use and the political economy of urban restructuring: Heroin, class and governance in Dublin. *Antipode* 37(4), 754–74.

Rankin, K. 2007. Deducing rationales and political tactics in the partitioning of Ireland, 1912-1925. *Political Geography* 26(8), 909–33.

Regan, C. 1980. Economic development in Ireland: The historical dimension. *Antipode* 12(1), 1–14.

Regan, C. and F. Walsh. 1976. Dependence and underdevelopment: The case of mineral resources and the Irish Republic. *Antipode* 8(3), 46–59.

Reid, B. 2007. Creating counterspaces: Identity and the home in Ireland and Northern Ireland. *Environment and Planning D* 25, 933–50.

Reid, B. 2011. Reflecting on Troubling Ireland: a cultural geographer's perspective. http://troublingireland.com/essay/essay (5 January 2021).

Royle, S. 2011. *Portrait of an Industrial City: 'Changing Belfast', 1750-1914.* Belfast: Ulster Historical Foundation.

Sage, C. 2010. Re-imagining the Irish foodscape. *Irish Geography* 43(2), 93–104

Shirlow, P. 2006. Belfast: The post-conflict city. *Space and Polity* 10(2), 99–107.

Shirlow, P. 2008. Sympathies, apathies and antipathies: The Falls-Shankill divide. *Irish Geography* 41(3), 337–40.

Shirlow, P. and C. Gallaher. 2006. The geography of Loyalist paramilitary feuding in Belfast, Northern Ireland. *Space and Polity* 10(2), 149–69.

Shirlow, P. and B. Murtagh. 2006. *Belfast: Segregation, Violence and the City.* London: Pluto Press.

Simonsen, K. 2004. Differential spaces of critical geography. *Geoforum* 35(5), 525–8.

Smyth, W.J. 2006. *Map-making, Landscapes and Memory: A Geography of Colonial and Early Modern Ireland c.1530-1750.* Cork: Cork University Press.

Smyth, W.J. 2010. Tom Jones Hughes and historical geography: An introduction. In, T. Jones Hughes, *Landholding, Society and Settlement in Nineteenth-Century Ireland: A Historical Geographer's Perspective*. Dublin: Geography Publications, pp. x–xxviii

Storey, D. 2001. *Territory: The Claiming of Space*. Harlow: Prentice Hall.

Storey, D. 2004. A sense of place: Rural development, tourism and place promotion in the Republic of Ireland. In, L. Holloway and M. Kneafsey (eds.), *Geographies of Rural Cultures and Societies: Perspectives on Rural Policy and Planning*. Aldershot: Ashgate, pp. 197–13.

Switzer, C. 2007. *Unionists and Great War Commemoration in the North of Ireland 1914–1939: People, Places and Politics*. Dublin, Irish Academic Press.

Toal, G. 2000. Dis/placing the geo-politics which one cannot not want. *Political Geography* 19(3), 385–96.

Walsh, F. 1980. The structure of neo-colonialism: The case of the Irish Republic. *Antipode* 12(1), 66–72.

Whelan, Y. 2003. *Reinventing Modern Dublin: Streetscape, Iconography and the Politics of Identity*. Dublin: University College Dublin Press.

11 Italian critical geographies

A historical perspective

*Elena dell'Agnese, Claudio Minca, and
Marcella Schmidt di Friedberg*

Critical glimpses

Today, it may be difficult to draw a tentative map of an *Italian* critical geography. Like many other academic fields, in the last decades Italian geography has undergone a process of profound internationalization. This process has seen, on the one hand, the departure (and sometimes the return) of many academics; on the other, an increased tendency to spend extensive study and work periods abroad on the part of most early career geographers. What is more, publishing in international peer-reviewed journals and with publishing houses with a global reach has become a common practice especially for the newer generations of geographers, a tendency that is now positively evaluated in terms of career progression.

Having said that, Italian geography still suffers from some longstanding problems, some of which characterize other disciplines in the academic national context as well. One major issue is the gender gap, which is still remarkable if one observes the demographics of the apical academic positions in geography, where the number of male full professors is double that of women, and in the leading roles within the most relevant geography associations and societies. For example, the Italian Geographical Society and the Association of Italian Geographers (AGeI) have never appointed a woman as a President. A second, more general problem has to do with the disproportion between the relatively large number of PhD students and post-doctoral positions and the relatively few academic posts available, something that has triggered a significant 'brain drain' among the ranks of geographers. Thirdly, many research areas and topics that have emerged and become well-established in geographical debates internationally remain somewhat niche or even marginalized, including, among others, critical animal geographies, more-than-human geographies, vegan geographies, LGBTQIA+ geographies. In our view, this may represent another reason why some early career geographers seek work opportunities abroad.

Despite all these difficulties, noteworthy critical work has been done by Italian geographers working either in Italy or in other countries, in the past decade or so. This includes, to name a few, Federico Ferretti's investigation of

DOI: 10.4324/9781315600635-11

anarchic geographies (Ferretti, 2011, 2013, 2017) and his organization of the first International Conference of Anarchist Geographies and Geographers (ICAGG) in Reggio Emilia (2017) and the work of Chiara Brambilla (2015) on borderscapes; Filippo Celata and Raffaella Coletti (2019) and Chiara Giubilaro (2017) on migration and border geographies in Italy and the Mediterranean and Emanuela Casti's (2013) critical cartographies; but also Teresa Isenburg's (2000) writings on the relationship between social space and the question of illegality; Cesare Di Feliciantonio's (2017), Ugo Rossi's (2017) and Alberto Vanolo's (2015) critical analysis of neoliberal urban practices; Annalisa Colombino and Paolo Giaccaria's (2016) interventions on biocapitalism; Egidio Dansero's team working on food and Alternative Food Networks (Barbera et al., 2014); Claudio Cerreti's Franco-Italian seminars on social geography (Cerreti, Dumont and Tabusi, 2012); Paolo Bonora's (2015) research on land consumption; Marco Grasso's (2020); on the oil and gas industry's impact on climate change; Andrea Zinzani's (2020) political ecologies; and Salvo Torre's (Torre, Benegiamo and Dal Gobbo, 2020) and Rachele Borghi's writings on decolonial theory and practice. Rachele Borghi (2020), now based at the University of Paris-Sorbonne, has published a critical essay on *Decolonialità e privilegio: Pratiche femministe e critica al sistema-mondo* [*Decoloniality and Privilege: Feminist practices and a critique of the world-system*], which represents an attempt to propose new spaces for reflection on 'the decolonial' by trying to reconcile theoretical work with militant practices and activism in the Italian academic context.

As may be glimpsed from the aforementioned examples, or from an overview of the topics discussed at the most recent Italian Geographical Congress (held in Rome in 2017), critical research by contemporary Italian geographers today covers a variety of topics and has entered a fruitful and productive dialogue with the existing international critical debates in the discipline. However exciting and promising these developments appear, it remains somewhat difficult to draw a cognitive mapping of Italian critical geography; that is, to identify a body of work marked by a specific set of critical discourse and practices, and a related coherent set of methodological and theoretical approaches. Italian critical geographies, despite the rich and diverse work of the past decade, remain a relatively fragmented field, something that is possibly linked to the complicated history of critical thought in Italian geography to which this chapter is dedicated.

Consequently, writing a historical account of Italian critical geography is not an easy task. Every attempt at mapping out the threads and the 'moments' of critical geographical practice in Italy is in fact confronted by the imbroglio of defining gestures and political moves that characterize that very history; a history of fragmented but important critical 'positionings' accompanied, sometimes on the part of the same scholars, by institutional and at times rather conservative academic practices. The bumpy terrain of critical work in Italian geography is indeed marked by a series of excellent episodes of radical interventions, in a context shaped by a record of loose networks and lost opportunities; sometimes even by practices of 'strategic forgetting'.

However, this is no reason to claim that past critical work in geography has not been relevant and influential, quite the contrary. Some of the protagonists of what may be defined as the most explicit season of Italian critical geography – that is, the geographers affiliated in the most diverse and personalized ways to a 'movement' described in the 1970s as 'Geografia Democratica' – have later become key players in the development of the discipline. It would thus be unfair to deny the existence of a few key moments in which the opportunity to develop an Italian critical approach to things geographical was present, and also that these moments represented a real possibility (then) to intervene in and influence international debates, also in the light of the broader political events in Italy. This has in fact happened in other disciplines, like political philosophy, in which the Italian debate initiated in the late 1960s and early 1970s has become key to international theoretical discussions (see Minca, 2012a). In geography, such a possibility was arguably missed. Still, it would be similarly unfair to ignore the fact that several Italian geographers continued to produce important critical work, despite the unfavourable academic context and the equally unfavourable changing political climate in Italy.

This explains why, when faced with the task of narrating a putative critical geography in Italy, we had to selectively operate a series of choices about which roots-and-routes to describe and recover under that label, and which ones to exclude from this account, in the full awareness that the picture provided by the present review will not only be incomplete, but also not fully coherent and genealogically reliable. However, this fuzziness is the result of a disciplinary history that is difficult if not impossible to describe in any discreet terms; a typical history of a 'critical periphery' (Minca, 2000, 2003), marked by many 'local' and individual interventions of great quality and originality – again something typical of a creative but rather disorganized community – but incapable of expressing clear threads and trajectories that would be widely recognized as a critical tradition of sorts.

We thus decided to open this reflection on the past of Italian critical geography by focusing on two highly interrelated 'roots' of critical thought in Italian geography: the work of Lucio Gambi and the experiments of the abovementioned *Geografia Democratica*, both characterized by a complicated legacy for the decades that followed. Starting from there, we tentatively trace some of the threads that, by and large, originated directly or indirectly from those two original projects, and emerged in the 1990s and 2000s with projects driven by a critical stance and strongly influenced by topics and approaches developed internationally and 'imported' into the Italian context. We then reflect on what is perhaps the most meaningful manifestation of the fact that critical theory has for a long time failed to 'penetrate' the key sites where official Italian academic geography was certified: that is, the substantial lack of engagement with gender issues on the part of the broader establishment of Italian geography. In selectively focusing on these moments and threads, we are fully aware of having marginalized other important moments and sites of critical geography which would have deserved a much more substantial engagement. We think of the far-reaching project on the geographies of complexity led by Angelo Turco (1986, 1988) around which a number of alternative geographies of Africa have been produced in L'Aquila (where he was based), and also in Padua and Turin; but also, of his theory on territory and territorialization, developed in those texts, which will later become an integral part of Italian geographical research. We also think of the school of social and development geography created in Naples by Pasquale Coppola and his cohort of disciples, or even, at a more individual level, of the influence of Claude Raffestin's (1981) work on power and geography, and also of some of the experiments initiated by Clara Copeta (1986a, 1986b) or Gabriele Zanetto (1989) in the 1980s. All these sites and events (and possibly many others) represent examples of Italian critical work in geography that would be important to analyse in depth but that, within the limited space of this chapter, it is impossible to discuss in detail (see, for example, dell'Agnese, 2008; Fall and Minca, 2012; Minca, 2005, 2012b).

We conclude with a few considerations about an episode of radical geographical practice, namely the involvement of a group of geographers in spatial practices of resistance against the neoliberal reform of the university system in 2010. We discuss their attempts to create alternative sites – compared to the more institutional ones – where geography can return to reflect, both in terms of theory and activism, on major social and political issues.

Lucio Gambi's critical geographies and other 'heretical' surroundings

'Ce n'est que le début': Geography and the student movement

Milan, 1968. That city and that year may be considered in many ways as the birthplace and the birthdate of critical geography in Italy. Lucio Gambi had already published relevant work providing a critical perspective on the discipline a few years before (Gambi, 1962, 1964). However, at the height of the *soixante-huitard* movement, something much more significant happened: the same Lucio Gambi – who at the time was teaching at the University of Milan and was deeply engaged with the student movement – published *Geografia e Contestazione* (*Geography and Contestation*) (Gambi, 1968), a 'manifesto' of sorts for a new, critical approach for the discipline.

The book, or better the booklet (52 pages), was 'revolutionary' in many aspects, including its unconventional writing style. The first essay, authored by Gambi himself, tellingly opens with a provocative sentence: 'Ce n'est que le début, continuons le combat' [It's only the beginning, let's carry on the fight] – an explicit reference to well-known graffiti that appeared in the same period on the walls of the Sorbonne in Paris. The rest of the book consists of a collection of 'the voices of the students', collected by Gambi in a survey entitled 'Relazione su di un questionario rivolto agli studenti universitari di Milano intorno alle funzioni della geografia' [Report about a survey on the role of geography conducted with the students at the University of Milan] and deliberately printed without much editing in terms of style or structure. In many ways, *Geografia e Contestazione* was an exploratory essay of participatory 'critical geography', an explicit challenge to the conventional and rather conservative approach to the discipline practiced in Italy until then. Students – mostly freshmen taking their very first geography courses – were asked to answer two simple questions: 'What do you think of geography as a school subject and as a topic in popular discourse?' and 'Which should be the role of geography in contemporary society?'. The rather rough assemblage of their reactions makes for interesting reading, since geography is accused of being simply 'a collection of notions', quite 'chaotic and mnemonic' in nature and lacking 'a rigorous methodological approach' – but also, remarkably, to be a field of knowledge instrumental to power. One student even defined geography as 'an encyclopaedical almanac *ad usum delphini* [for the use of the heir]', that is, a collection of information from different sources to provide members of the future ruling class with strategic knowledge of the territories they will govern one day. In those pages, the two most controversial aspects of traditional geography were clearly exposed: first, its 'methodological weakness'; second, its 'relationship with institutional power'.

Lucio Gambi, between geography and history

In the aftermath of the student protests of those years, many changes occurred in Italian academia, including the fact that the former 'University of the Barons' was forced to engage with the demands of the students and to make room (to some extent at least) for new generations of scholars. However, despite the new climate, a path towards a more critical approach to geography was not marked out. Notwithstanding its growing popularity, Gambi's work to unveil the ideological nature of the discipline and radically rethink its methodologies was resisted by most Italian geographers of his generation. Indeed, as Gambi wrote in the preface of his best-known work, *Una geografia per la storia*: 'This [new] way of interpreting geography, while rejected by a great number of old and bombastic geographers, has represented a source of ideas and questions among the younger practitioners of the discipline' (Gambi, 1973, vii). As a result of these difficulties, Gambi soon adopted what many have defined as an academic stance of 'splendid isolation'. He distanced himself from the 'depressed' geography of his time and became attuned to what other disciplines, like history and sociology, were doing in those days in Italy; disciplines related to geography, but more critical and open to the new challenges.

Outside the circles of academic geography, Gambi's innovative work soon gained currency and succeeded in attracting a broader readership, something unusual for an Italian geographer. At the beginning of the 1970s, after having published most of his previous work with a small local press (F.lli Lega, based in Faenza, near Bologna), he began writing for what is considered the most prestigious Italian publisher, Giulio Einaudi Editore. This very fact granted his books a remarkable reception among academics and intellectuals nationally. In 1972, he was asked to author the opening essay of the first volume of the prestigious *Storia d'Italia* (1972), a monumental 30-tome-opus intended to comprehensively cover the history, the landscapes, and the different culture(s) of the country. For Einaudi he also edited a collection of his own methodological essays, *Una geografia per la storia* (1973, *A Geography for History*) and the *Atlante* – that is, the sixth volume of *Storia d'Italia* (AAVV, 1976) – together with authors from other, cognate disciplines, but also with a team of talented younger geographers, including Franco Farinelli, Teresa Isenburg, Massimo Quaini and Paola Sereno. Notwithstanding the title, the *Atlante* was far from being a simple collection of maps. On the contrary, it was a fascinating exercise in understanding cartography as 'a way of seeing', and in this sense it stood as an important move towards one of the future developments of critical geography in Italy that related to the critique of cartographic reason led by Farinelli and developed in the following decades by many others. Even if mostly overlooked by academic geographers of his time, Gambi's work left a 'fertile track' for the decades to come, especially by showing to a younger generation of researchers that 'a different kind of geography was possible'.

Geografia Democratica

In 1975 Gambi left the University of Milan and moved to Bologna. In the same period, some of his most enthusiastic students launched a discussion forum, dubbed 'Geografia Democratica' (Democratic Geography) (see Cavallo, 2007; dell'Agnese, 2005, 2008; Farinelli, 2006; Governa, 2007). The group comprised young scholars from different areas of Italy, who periodically gathered not only to debate Gambi's ideas but also to rework the most stimulating suggestions coming from authors like Yves Lacoste and Jean-Bernard Racine or, from across the Atlantic, William Bunge and David Harvey. They also strove to jolt Italian geography out of its academic 'conventions', trying to connect it to the many political and social problems facing Italy in those years. For Turin-based geographer Giuseppe Dematteis (1981), one of the initiators of this critical movement, the aim of Geografia Democratica was to 'disseminate anxieties in the corporation of academic geographers, in order to set the discipline's potentialities free, since they were repressed by a predominant a-critical paradigm, and also in order to reconnect geographical research with the problems, sometimes dramatic, of social reality' (see also Fall and Minca, 2012).

In its relatively short span of life (1974–1980), Geografia Democratica produced some important work, especially in terms of providing a critical perspective on theory and methods in geography. In 1978, under the direction of Massimo Quaini, the group launched an Italian version of Yves Lacoste's radical journal *Herodote*, named *Hérodote/Italia*, possibly the first organized attempt at promoting 'alternative geographical knowledge' in Italian academia. In 1980, for example, the journal published a special issue on the relationship between 'geographical discourse' and the work of philosopher Michel Foucault, presenting in this way some of the key questions famously posed by Foucault to *Hérodote* France in 1976, to the potentially large audience of Italian geographers. Regrettably, Foucault's questions received very little attention and that brave attempt at an international dialogue went almost unnoticed, both on the part of the above-mentioned group and on the part of the geographical community at large. Perhaps one possible explanation for this failed engagement is that most members of Geografia Democratica, in those days, were too involved in Marxist debates to be interested in Foucault's ideas about power (the intervention was opened by this sentence: 'Between Foucault and Hérodote is Marx') (Geografia Democratica, 1980). Most Italian geographers, however, simply did not read the journal. So, as Massimo Quaini (2007) would later state, Foucault's reflections were bound to remain, for a long time, a 'treasure trove largely unexplored'.

However, a positive reaction came from Giuseppe Dematteis, who suggested that a full engagement with Foucault's notion of power offered great potential to geography and would have been of great use for geographers: 'This idea paves the way to stimulating research into the local specificities of power, interstices and deviances... The idea of power as a diffuse and local alternative... opens a new perspective on the respective roles of established knowledge and theory, on the one hand, and the related intellectuals, on the other' (Dematteis, 1980, 12). Dematteis, on that occasion, also reflected on the role of geography as a way of geo-graphing the world: 'The ideological function of "textbook geography" is more complex than the one acknowledged by Lacoste. Not only does it make geographical knowledge look like an innocent form of knowledge and teaches people that which exists is *natural* and *cannot* be changed, but also, and more subtly, ... that it is natural *because it is normal*' (Dematteis, 1980, 10). Dematteis would further develop some of these ideas in his influential *Le Metafore della Terra* (1985, *The Metaphors of the Earth*), to which we will return later.

In 1979, Geografia Democratica organized a conference entitled *Inchiesta sul terreno in geografia* [Fieldwork as a tool for geographical research], aimed at promoting alternative methodological tools as part of a new research agenda for the discipline (Canigiani et al., 1981). Partially influenced by the radical geographies populating a journal like *Antipode* and by the 'geographical expeditions' envisaged by William Bunge, the group worked with the specific intent of remaking geography as a social science capable of overcoming what was perceived as a problematic distance between researchers and the objects of their research. Again, Dematteis's (1981) contribution to that discussion was particularly relevant, since it focussed on a new concept of 'territory', a concept inspired by the political debates developed within the '68 Movement, and capable of challenging the traditional descriptive approach adopted by the discipline until then. From his perspective, 'territory' had to be understood as a 'system' interconnecting the spatial variations of the working conditions and of labor exploitation, including the important effects of the growing presence of international corporations and the peculiar twist given, in the Italian context, by a pervasive 'grey economy'. Beyond the new centrality assigned to fieldwork, and the relevance assigned to the territory-as-a-network-of-power-relations, the conference shed new light on the importance of alternative sources of geographic knowledge, for example, literature. Along this line, and quite innovatively, Massimo Quaini analysed the literary representation of borders, as provided by the work of Italian writer Mario Rigoni Stern (see Gruppo Hérodote, 1981).

The momentum of Geografia Democratica found its first and foremost institutional recognition at a conference organized in Varese by the newly established AGeI (Association of Italian Geographers) (see Corna Pellegrini and Brusa, 1980). The aim of the conference was to assess the progress of the discipline over the previous two decades, and soon became the opportunity, also thanks to the inclusive stance adopted by the then-President of the Association, Giacomo Corna Pellegrini, to officially open the doors to the self-defined 'heretical streams' in the discipline, from Gambi's critical geographies to the Marxist-inspired work of Geografia Democratica (Dematteis, 2007). However, and somehow paradoxically, in that very same year Geografia Democratica began a sharp and sudden decline. Despite the Varese conference, in fact, both *Hérodote/Italia*'s engagements with class struggle and radical geographies, and the experimental work of *L'inchiesta sul terreno in geografia* experienced fierce resistance (and neglect) on the part of the broader academic community of Italian geographers. As a result of this resistance, the journal, renamed *Erodoto* in 1982, did not have enough subscriptions by University libraries to survive and was forced to shut down in 1984 (Antonsich, 1997). The discipline as an institutional body was probably not ready for such a challenge and its main gatekeepers – the 'old and bombastic geographers', to use Gambi's words – soon found ways to marginalize and de-potentiate that (critical) momentum. Such lack of engagement with critical thought in many ways influenced Italian geography in a decisive way for the decades to come, resulting also in a general isolation and marginalization of the discipline from mainstream international debates, despite the excellent work (albeit only partially critical, as we shall see) of several individual geographers, some of them affiliated with the complicated legacy of Geografia Democratica, and, more in general, of Lucio Gambi.

Challenging the paradigm? Massimo Quaini,
Giuseppe Dematteis, Franco Farinelli

Apart from the above-described collective effort, which turned into a substantial failure in terms of its impact on the discipline, some protagonists of
that 'critical' momentum managed to pursue important individual projects
clearly driven by a critical stance. One good example is Massimo Quaini,
who, after publishing important work on the relationship between *Marxism
and Geography* (1974), focussed his attention on alternative forms of geographical knowledge in another book, *Dopo la Geografia* (*After Geography*,
1978). There, Quaini (1978, 10) stresses the importance of understanding
'the history of inequality between popular culture and hegemonic culture,
distinguishing between the knowledge of the winner and the knowledge of
the loser'. From this perspective, he offers a powerful reinterpretation of the
development of 'official geography' as a 'modern science', a science capable
of forcibly translating the 'qualitative' space of everyday life into a rigid
mathematical framework, while, at the same time, disempowering the geographical quotidian knowledge of the subaltern classes.

Another significant contribution in rethinking the role of the discipline
as a 'critical geo-graphy' was Giuseppe Dematteis's abovementioned *Le
metafore della Terra* (1985). Here, Dematteis continues to elaborate his ideas
about the *normalizing* role of geography, and to discuss several key topics
in the history of the discipline, while adopting a critical stance towards the
quantitative turn, the emergence of regional science, and, more in general,
all forms of organicism and all spatial theories based on 'objectivist' perspectives (see Fall and Minca, 2012). Dematteis criticizes particularly the
relations between space and power that the work of many geographers
appeared to support while remaining seemingly oblivious of its far-reaching
political and epistemological implications. He is also deeply critical of geographers' newly (then) discovered love for calculators (Dematteis, 1985, 12),
describing it as a mere attempt to regain legitimacy and overcome their
sense of inferiority compared to putatively more 'rigorous' social sciences.

Gambi's move to Bologna had an important impact on the work of geographers working in that University. Among them was Franco Farinelli,
who developed what is perhaps the most poignant analysis of the workings of cartographic reason ever published (Farinelli, 1985, 1992, 2003,
2009). Farinelli, in the last decades, has literally re-written (and somehow
subverted) the history of the discipline, deconstructing the determinis-
tic approaches that dominated a great part of the 1900s, while recovering
the (largely forgotten) ontological dimension of the *Erdkunde* project, and
forcefully condemning the 'cartographic prison' to which Western thought
has been constrained since Anaximander's early geographical experiment
(see dell'Agnese, 2008; Minca, 2007a). Farinelli's theory of cartographic
reason is indeed an extraordinary contribution to the understanding of the
'deep' nature of geography and the genealogies of power that have always

accompanied it. Farinelli's work is, overall, highly admired in Italy, and has also been influential outside of Italy (see, for example, Olsson, 2007; Pickles, 2004): his reflection on the relationship between Western thought and the reduction of reality into the two-dimensional plane of the map remains today a path-breaking contribution to critical thought in Italian geography.

Thanks to the intellectual engagement of these authors, and some others that we do not have the space here to analyse in detail, Gambi's original lesson was not lost and the intellectual (and political) ambitions of Geografia Democratica were not entirely abandoned. On the contrary, in the following years Massimo Quaini's 'alternative geographies', Giuseppe Dematteis' ideas on territory and geo-graphical metaphors and Franco Farinelli's work on the power of cartographic reasoning, while not dialoguing with each other as one would have expected, did indeed influence in a significant way several streams of Italian geographical thought.

While some suggestions coming from the 'heretical geographies' of those lively years have somehow only implicitly penetrated 'the officially recognized paradigm' of Italian geography (at least in the view of Dematteis, 2007, 276), some 'heretical geographers' have reached important positions in the discipline: many former members of Geografia Democratica, in Milan, Bologna, Turin, Pavia, Genoa, Naples and Florence have in fact become full professors, while Franco Farinelli was later elected President of the Association of Italian Geographers. This should not suggest, however, that they brought into the hallowed halls of Italian academia the political fervour of their years on the academic 'barricades'. On the contrary, while their work and their academic 'local schools' still reflect some of the highest standards expressed in Italian geographical research, their political engagement, both inside and outside of academia, has been largely watered down, if not entirely disappeared. Whether this imbroglio of personal intellectual and career trajectories, intellectual ambitions and political commitment has been the result of a strategic appeasement towards 'the establishment' on the part of many protagonists of that season of critical thought, or, instead, the consequence of their penetration and related 'normalization' in the core of the discipline, is a question that remains unanswered. But it is a question that deserves further scrutiny, especially in a moment when new academic trends and new intellectual ambitions have begun emerging in the discipline, also as a reaction of the growing impact of neoliberal policies on the national University system and of the ongoing internationalization of all academic work.

Moving forward

The early years of the twenty-first century have been a period of soul-searching for critical thought in Italian geography. The period has been marked both by the re-elaboration of (also critical) ideas produced in the previous decades and, perhaps more importantly, by the reception of new ideas and theoretical perspectives derived from international debates and their consequent incorporation into the work of a new generation of geographers. Two significant examples among others are initiatives that took place respectively in Palermo and Venice. The Palermo group of cultural geographers led by Enzo Guarrasi has developed a project entitled *Atlante Virtuale* (Virtual Atlas) (de Spuches, 2002), that famously included an extraordinary itinerant seminar organized in Palermo in 2000 during which a series of (virtually) interconnected 'events' and 'moments' explicitly toyed with ideas of 'performance' and geography. 'Bringing together speakers/performers, discourses, and material context (whether the city theatre, the botanical garden, the insane asylum or archaeological site), the event playfully proposed an experimental 'nomadic critical geography' (see Minca, 2005, 931). Unfortunately, again, this grand 'gesture' of critical geography had limited influence on disciplinary debates despite its great potential, to the point that one may start asking questions about the very existence of national debates as such in the last decade or so.

Another perhaps similar example was the 1999 conference *Postmodern Geographical Praxis*, which brought to Venice some of Italy's most prominent geographers to meet top international critical thinkers – such as Denis Cosgrove, Michael Dear, Cindi Katz, Don Mitchell, Gunnar Olsson, Neil Smith and Edward Soja – to discuss the 'postmodern' in geography. The well-attended meeting was intended also as an attempt at mapping out Italian critical thought by engaging with geographers who played a key role in the preceding decades, like Giuseppe Dematteis, Franco Farinelli, Enzo Guarrasi, who engaged in debates with their foreign counterparts, several of whom – Cosgrove and Olsson above all – were certainly not strangers to Italian geography (for a discussion, see Minca, 2000, 2001a; Samers and Sidaway, 2000). This event was accompanied by the publication of two books on the same topic, *Postmodern Geography* (Minca, 2001a) and *Introduzione alla Geografia Postmoderna* (Minca, 2001b), in which this dialogue between Italian and international critical thought continued. The books and 'the postmodern in geography' sparked further debate, since they were discussed in a workshop organized by the *Società Geografica Italiana* in 2002, and in a special issue of the Society's journal, with interventions penned by Giuseppe Dematteis (2003), Enzo Guarrasi (2003), Massimo Quaini (2003) and Adalberto Vallega (2003).

In the years to follow, new books were published by Italian authors, such as *Spazio e Politica* (Minca and Bialasiewicz, 2004) and *Geografia Politica Critica* (dell'Agnese, 2005), where critical perspectives of the English-speaking post-structuralist debate were confronted with the critical (Italian) traditions discussed above and with new contributions coming from Italian political philosophy, most recently preoccupied with questions of space and power. *Spazio e Politica*, for example, was the first book in Italian geography to engage directly with biopolitics and the work of philosopher Giorgio Agamben (see, also, Minca, 2006, 2007b), while reflecting on the importance of a critical approach to geography and politics along the lines delineated by Dematteis and Farinelli. *Geografia Politica Critica*, on the other hand, was the first book in Italian geography to deal systematically with the ideas expressed by Geografia Democratica, and to compare them with the work of Yves Lacoste and his school, with Raffestin's theory of territoriality and, more generally, with the field of Critical Geopolitics. Both books have been (and still are) adopted in many Italian universities, hopefully stimulating newer generations of Italian geographers to a fuller engagement with the developments in the discipline, while at the same time elaborating a specifically 'Italian' perspective on it.

Later, more critical 'glimpses' emerged reflecting on cartography and political power (Boria, 2007), teaching in geography (Squarcina, 2009) and alternative tourism (Borghi and Celata, 2009) were published. Meanwhile, a renewed interest in the legacy of Geografia Democratica ignited a discussion that has found hospitality in several journal articles (Cavallo, 2007; Farinelli, 2006), and in a conference organized in Turin in 2005, where a roundtable, chaired by Massimo Quaini, was dedicated to the assessment of this complicated but important piece of disciplinary history (see Dansero et al., 2007). In that same context, another session aimed at rethinking the situation of gender studies in Italy, and especially its (lack of) role in geography; something we now turn to in order to show, with a telling example, how the history of Italian critical geography is also a history of missed opportunities and of agendas that remained implicit for possibly too long.

A geography of difference?

If there is a putative date for the birth of Italian critical geography, there is also a place (even if virtual) that can be considered as the emblematic site of the difficult journey of Italian gender geography. Or, perhaps trying to be less dramatic, at least the demonstration that Italian geography has for long shown an extraordinary resistance not only to the incorporation of feminist theory into its theoretical body, but also to the largely accepted conventions of a non-gender-biased disciplinary language. This 'place' is the website of the 2012 Festival of Geography, a festival devoted, as stated by its logo, to 'Man, Environment, Resources' (orig. *Uomo, Ambiente, Risorse*), a title that would be considered either ironic (for its nineteenth-century echoes) or simply politically unacceptable in most contemporary international fora. The issue of gendered language is (in 2021) still present in the title of conferences, and publications, and of a PhD program (at the University of Milan) as well.

One would be tempted to say that this is simply the result of the absence of gender geography – or, better, of any gender awareness in Italian geography. But this is untrue, because thanks to the engagement of a few forerunners, like Maria Luisa Gentileschi (1983), Paola Bonora, Anna Segre and Gabriella Arena, a 'geography of women' took its first important steps already in the 1980s, while in the last decade or so, an explicit 'geography of gender' has emerged and entered a somewhat complicated dialogue with mainstream Italian geography. However, the efforts of this small but very active group of researchers (mostly women), the associated constitution of a research group specifically focused on gender and the publications of a series of related works have not been enough to even scratch the surface of a discipline that has remained largely conservative and problematically gender-biased, at least in its institutional formations. As we shall see, the feminist revolution of the last thirty years or so has had little impact on the core academic business of Italian geography, on its methodologies and theoretical perspectives, or even on its jargon.

Feminism and geography in Italy

Italian feminism has a relatively long history, with its origins lying in the 1970s. The key moments of political visibility of the feminist turn in Italian politics can be broadly traced back through the political battles for the introduction of divorce (1973), for the radical reform of family law (1975), for more equality in employment conditions (1977) and for the right to abortion (1978). The important and in many cases successful struggles of the 1970s were accompanied by an equally productive intellectual atmosphere. Some of the most relevant and innovative 'theories of sexual difference' that began circulating were the result of the work of several groups of critical scholars. Particularly influential was the activity of the *Libreria delle Donne* (based in Milan) and of the *Comunità Diotima* (based in Verona), the latter founded by a group of scholars including Luisa Muraro, Adriana Cavarero and Chiara Zamboni. However, despite the vitality of these movements, until the 1990s feminist scholarship gained very little institutional legitimacy in Italy (Dell'Abate Çelebi, 2009, 29). In part, this limited impact may be attributed to feminist scholars themselves and to their separatist strategies. However, a general lack of resources, spaces and actual power on the part of feminist academics certainly did not facilitate the introduction of women and gender studies in Italian academia.

In this unfavourable context – unlike history and sociology – geography was no exception. Not only did geography remain almost totally absent from key debates, but also when it slowly and selectively opened its doors to some feminist suggestions, it did it more as an answer to challenges coming from the international academic scene than as a reaction to national debates. As a reflection of this, apart from the few pioneering works mentioned above, the first volume on gender studies in geography, published in 1990, was in fact a translation of the path-breaking collection of essays *Geography and Gender*, published by the British Women and Geography Study Group in 1984. Editor Gabriella Arena made this volume available to Italian readers with the title *Geografia al Femminile* (*Women Geography*) (Arena, 1990), since the term 'gender' was at the time entirely unfamiliar to Italian geographers. In the translation, gender was tellingly replaced by terms like 'sex' and 'sexual difference'.

Gender geographies

Since then, gender geography has timidly come to the shores of Italian geography. In 1993, the first 'formal' recognition of an Italian geography of gender was given by a panel coordinated by Gisella Cortesi, as part of a workshop on Population Geography organized by Maria Luisa Gentileschi in Cagliari, in collaboration with the Institute of British Geographers. In the same year, a research project entitled 'Il ruolo della componente femmi- nile nell'organizzazione del territorio: casi di studio in Italia' [The Role of Women in Territorial Organization: Italian Case Studies] was approved and funded by the Italian Ministry of Research and Education (MURST). One output of the project was the volume *Donne e Geografia* (*Women and Geography*), edited by Cortesi and Gentileschi (1996).

This 'gendered' approach, integrated within the broader field of popu- lation geography, was explicitly committed to social change and justice: 'A geography [...] striving to produce proposals for change as well as new knowledge [...] to single out the objectives and aims for greater social jus- tice' (Cortesi and Gentileschi, 1996, 16). Not surprisingly, the main topics analysed by Italian gender geography in its initial phase were the conditions of the labor market (Cortesi and Marengo, 1991; Gentileschi, 1983), ques- tions of migration (Cortesi et al., 1999; Marengo, 1997) and the geography of history and travel (Rossi, 1995). We should also mention here Anna Segre's leading role, between 1991 and 1996, in a research project entitled: 'Women in processes of development. Interdisciplinary research on micro and macroeconomics, demographic, psycho-sociological and anthropological aspects of development'. Several chapters on the theme of 'Gender, devel- opment and territory' are included in the book by Dansero et al. (2007), published to commemorate Anna Segre and recognize her contribution to the introduction of gender perspectives in the discipline nationally. The fol- lowing decade was opened by an important milestone, the organization of a session entitled 'For a Geographical Perspective on Gender' (chaired by Gentileschi) at the XXVIII Italian Geographical Congress (Rome, 2000) (Gentileschi, 2003). That event was followed by other important moments of institutional 'presence'. In 2003, the International Geographical Union sponsored an international seminar on 'Gendered cities: identities, activi- ties, networks' ['La città delle donne'], hosted by the Italian Geographical Society in Rome (Cortesi et al., 2006). The research group 'Geography and Gender' was established in 2005 (initially convened by Gisella Cortesi, and later by Marcella Schmidt di Friedberg and Giulia de Spuches) affiliated to the Association of Italian Geographers (AGeI). In 2007, this group published a special issue of *Geotema* (the official journal of the AGeI) on 'Luoghi e identità di genere' (Identity, Places and Gender, 2009, ed. by Gisella Cortesi) and 'Sguardi di genere' (Gender Views, 2017, ed. by Marcella Schmidt di Friedberg, Marina Marengo and Valeria Pecorelli). In 2006, Gisella Cortesi published 'Donne, società, territorio: il quadro generale' (Women, society, territory: a general picture), the first review essay on the state of the art of gender geography in Italy, published in a book on social geography edited by Daniela Lombardi (Lombardi, 2006).

New perspectives in a traditional context

Notwithstanding these promising openings, gender geography in Italy remained, for long, a 'women studying women' affair. It was 2007 before Italian geography was introduced to the first explicit reflections on the role of masculinity in relation to national identity (dell'Agnese and Ruspini, 2007). The first systematic book on gender and geography (*Geografie di Genere*) was published two years later (Borghi and Rondinone, 2009). Also in 2009, the international Conference 'Spaces of Difference' discussed questions of spatial heteronormativity and the relationship between queer theory and geography for the first time (see Borghi and Schmidt di Friedberg, 2011). In the same context, Borghi (2011) introduced the concept of heteronormativity in Italian geography, as well as other perspectives on sexual geography, including LGBTQ performance theory and the post-porno movement. A few years later, Valeria Pecorelli, with sociologist Elisabetta Ruspini and psychologist Mario Inghilleri, critically addressed the subject of fatherhood (Ruspini, Inghilleri and Pecorelli, 2017).

All these initiatives must be praised and recognized for their role in introducing questions of gender in the Italian context. At the same time, their impact on the discipline at large has been relatively limited. One of the reasons for this is that the theoretical framework provided by the philosophy of sexual difference has still to find its place in Italian geography. Furthermore, Italian gender geography continues to suffer from significant institutional limitations, from lack of pressure groups and from the limited space offered to this perspective in teaching and research within universities. Whilst internationally it has been largely recognized that 'human geography is incomplete without considering gender' (Longhurst, 2002, 549), in Italy the theme has yet to find full recognition. The persistence of 'quite scarce numbers of gender geographers in Italy' (Rondinone, 2003, 69) leads to the conclusion that more work should be done 'in relation to the resistance of a scientific community that has been very traditional from this point of view and which, while now finally considering these studies without arrogance or irony, at the same time does not consider them *also* worthy of comparison in terms of the research that they produce' (Rossi, 2011, 40).

However, some Italian geographers have been actively involved in imaginative spatial practices of resistance and contestation. A good example is Palermo-based geographer Giulia de Spuches, who was one of the organizers of the 2012 Gay Pride parade in that same city. Furthermore, in 2015, the

3rd European Geographies of Sexualities conference was hosted in Rome, marking a further milestone for Italian geography. In this context, a group of inspired early career scholars contributed discussions of LGBT issues from a variety of angles. Finally, at the XXXII Italian Geographical Congress held in Rome in 2017, gender geography has appeared as an established institutional presence, a recognition confirmed by a successful and well-attended session entitled 'Narrating the body/narrated body: Itineraries of gender geography between revolutions and reforms'. As Marcella Schmidt di Friedberg and Valeria Pecorelli put it, 'the two Rome conferences and the second special issue on gender published by *Geotema* have provided the Italian geography community with new insightful perspectives in the realm of gender geography' (Schmidt di Friedberg and Pecorelli, 2019, 1144). Regrettably, despite these developments in the academic context, gender discrimination remains dramatically present in the country, which in 2020 was ranked 76th out of 153 countries in the Global Gender Gap (GGG) index of the World Economic Forum.

Coda: Critical geographies of/on the roof

We would like to conclude this tentative review of critical work in Italian geography by recalling a significant episode of political activism that, in 2010, was associated with the work of Italian critical geographers. Year 2010 was a year marked by protest and unrest in Italian universities related to the implementation of the so-called 'Riforma Gelmini', a reform perceived by a large part of Italian academic as a poorly camouflaged effort to cut funds and jobs in Italian universities in the name of efficiency and 'meritocratic values'. The reform sparked different reactions, which included the participation and the actions of some geographers. In particular, some early career geographers were actively involved in realizing 'Rete 29 aprile' (Network, April 29), a network of researchers that sought to resist the Riforma Gelmini – and the ongoing neoliberalization of the national university system – by occupying a very visible spot in Rome, the roof of the Faculty of Architecture at the University of Rome 'La Sapienza', for 35 days. This initiative received significant media attention.

Arguably, this occupation was also inspired by geographer Angelo Turco's conceptualizations of the process of territorialization. The idea was to convert an empty and anonymous space (the roof of a university building) into a social and meaningful political space through actions of 'critical territorialization'. Turco's book *Verso una Geografia della Complessità* (Turco, 1988) was considered a 'manual for action' by some activists, with the book's three key stages of territorialization (denomination, reification and structuration) turned into geographical political practice. Following this initiative in Rome, about 50 other roofs of university buildings were taken across the country, sometimes for a day, at other times for longer periods. Inspired by these actions, several groups of students occupied other meaningful sites, which were visited by celebrities and politicians to express solidarity with the activists. The Riforma Gelmini was approved despite these forms of resistance. However, resistance to this neoliberal university reform represented an important opportunity for some critical geographers to turn theory into action and at the same time inspired the emergence of new potential spaces for critical thought (Maida, 2011; Tabusi, 2012). These radical geographies of/on the roof represent perhaps a specifically 'Italian' case of critical geographical practice, which possibly shows how the somewhat fragmented critical traditions and experiments we have outlined in this chapter have left an important mark on the work of many Italian geographers. Whether these 'legacies' represent a good starting point for connecting to emerging international critical geographical approaches is perhaps the most significant question that will confront Italian critical geographers in the years to come.

References

AAVV. 1976. *Atlante, Storia d'Italia*. Turin: Einaudi.

Antonsich, M. 1997. La geopolitica italiana nella rivista "Geopolitica", "Hérodote/ Italia (Eurodoto)" e "Limes". *Bollettino della Società Geografica Italiana* 3, 411–8.

Arena, G. 1990. *Geografia al Femminile*. Milan: Unicopli.

Barbera, F., A. Corsi, E. Dansero, P. Giaccaria, C. Peano and M. Puttilli. 2014. Cosa c'è di alternativo negli Alternative Food Networks? Un'agenda di ricerca per un approccio interdisciplinare. *Scienze del territorio* 2, 35–44.

Bonora, P. 2015. *Fermiamo il Consumo di Suolo*. Bologna: Il Mulino

Borghi, R. 2011. *Chi ha paura dell'eteronormatività spaziale? La geografia della sessualità in Italia*. Storia dei lesbismi e studi lgbtq in Italia Conference, Department of Geographical and Historical Studies, University of Florence, Florence, Italy, 25 February 2011. http://vimeo.com/serverdonne (27 June 2012).

Borghi, R. and F. Celata. 2009. *Turismo Critico. Immaginari Geografici, Performance e Paradossi sulle Rotte del Turismo Alternativo*. Milan: Unicopli.

Borghi, R. and A. Rondinone. 2009. *Geografia di Genere*. Milan: Unicopli.

Borghi, R. 2020. *Decolonialità e Privilegio. Pratiche Femministe e Critica al Sistemamondo*. Milan: Meltemi.

Borghi, R. and M. Schmidt di Friedberg. 2011. Lo spazio della differenza. *Bollettino della Società Geografica Italiana* 8(4), 159–64.

Boria, E. 2007. *Cartografia e Potere. Segni e Rappresentazioni negli Atlanti Italiani del Novecento*. Turin: Utet.

Brambilla, C. 2015. Exploring the critical potential of the borderscapes concept. *Geopolitics* 20, 14–34.

Canigiani, F., M. Carazzi and E. Grottanelli (eds.). 1981. *L'inchiesta sul Terreno in Geografia*. Torino: Giappichelli Editore.

Casti, E. 2013. *Cartografia Critica. Dal Topos alla Chora*. Milan: Guerini

Cavallo, F.L. 2007. Quelle insegne un po' scomode e parecchio ingombranti. Appunti su Geografia democratica. *Rivista Geografica Italiana* 114(1), 1–25.

Celata, F. and R. Coletti. 2019. Borderscapes of external Europeanization in the Mediterranean neighbourhood. *European Urban and Regional Studies* 26(1), 9–21.

Cerreti, C., I. Dumont and M. Tabusi (eds.). 2012. *Geografia Sociale e Democrazia*. Rome: Aracne.

Colombino, A. and P. Giaccaria. 2016. Dead liveness/living deadness: Thresholds of non-human life and death in biocapitalism. *Environment and Planning D: Society and Space* 34(6), 1044–62.

Copeta, C. 1986a. *Esistere e Abitare*. Milan: Franco Angeli.

Copeta, C. 1986b. *L'Uomo e la Terra*. Milan: Unicopli.

Corna Pellegrini, G. and C. Brusa (eds.). 1980. *La Ricerca Geografica in Italia*. Varese: Ask Edizioni.

Cortesi, G. 2006. Donne, società, territorio: il quadro generale. In, D. Lombardi (ed.), *Percorsi di Geografia Sociale*. Bologna: Patron, pp. 315–31.

Cortesi, G. (ed.) 2009. Luoghi e identità di genere. *Geotema* 33. Bologna: Patron.

Cortesi, G., F. Cristaldi and J. Droogleever Fortuijn. 2006. *La Città delle Donne. Un Approccio di Genere alla Geografia Urbana*. Bologna: Pàtron.

Cortesi, G. and M.L. Gentileschi. 1996. *Donne e Geografia*. Milan: Franco Angeli.

Cortesi, G., C. Ghilardi and M. Marengo. 1999. Esperienze migratorie a confronto: donne italiane all'estero e donne straniere in Italia. In, C. Brusa (ed.), *Atti del Convegno Immigrazione e Multicultura nell'Italia di Oggi*. Milan: Franco Angeli, pp. 156–71.

Cortesi, G. and M. Marengo. 1991. La differenziazione spaziale dell'attività femminile in Italia. *Rivista Geografica Italiana* 3, 381–407.

Dansero, E., G. di Meglio, E. Donini and F. Governa. 2007. *Geografia, Società, Politica*. Milan: Franco Angeli.

Dell'Abate Çelebi, B. 2009. Italian feminist thought at the periphery of the empire. *LITERA, Journal of Western Literature* 22(1), 17–36.

dell'Agnese, E. 2005. *Geografia Politica Critica*. Milan: Guerini.

dell'Agnese, E. 2008. Geo-graphing, writing words. In, K. Cox, M. Low and J. Robinson (eds.), *The Handbook of Political Geography*. London: Sage, pp. 439–453.

dell'Agnese, E. and E. Ruspini. 2007. *Mascolinità all'Italiana. Costruzioni, Narrazioni, Discorsi*. Turin: Utet.

Dematteis, G. 1980. Fra Foucault e Hérodote c'è di mezzo Marx. *Hérodote/Italia* 2/3, 9–13.

Dematteis, G. 1981. Il "terreno" come lotta di classe: la "scoperta" del territorio nel 1968-'69. In, Geografia Democratica (ed.), *L'Inchiesta sul Terreno in Geografia*. Turin: Giappichelli, pp. 135–144, pp. 397-400.

Dematteis, G. 1985. *Le Metafore della Terra*. Milan: Feltrinelli.

Dematteis, G. 2003. La metafora geografica è postmoderna? *Bollettino della Società Geografica Italiana* 127, 947–954.

Dematteis, G. 2007. Inseguire i fantasmi o stare dentro al mondo? In, E. Dansero et al. (eds.) *Geografia, Società, Politica. La Ricerca in Geografia come Impegno Sociale*. Milan: Franco Angeli, pp. 275–7.

de Spuches, G. 2002. *Atlante Virtuale*. Palermo: University of Palermo.

Di Feliciantonio, C. 2017. Spaces of the expelled as spaces of the urban commons? Analysing the re-emergence of squatting initiatives in Rome. *International Journal of Urban and Regional Research* 41(5), 708–25.

Fall, J. and C. Minca. 2012. Not a geography of what doesn't exist, but a counter-geography of what does: Rereading Giuseppe Dematteis' *Le Metafore della Terra*. *Progress in Human Geography* 37(4), 542–63.

Farinelli, F. 1985. De la crise à la critique de l'imagination géographique. In, *Actes du Colloque, L'Imagination Géographique*. Université Genève-Lausanne.

Farinelli, F. 1992. *I Segni del Mondo*. Florence: La Nuova Italia.

Farinelli, F. 2003. *Geografia*. Turin: Einaudi.

Farinelli, F. 2006. A proposito di geografia democratica. *Rivista Geografia Italiana* 112, 163–5.

Farinelli, F. 2009. *La Crisi della Ragione Cartografica*. Turin: Einaudi.

Ferretti, F. 2011. The correspondence between Élisée Reclus and Pëtr Kropotkin as a source for the history of geography. *Journal of Historical Geography* 37(2), 216–22.

Ferretti, F. 2013. "They have the right to throw us out": Élisée Reclus' *New Universal Geography*. *Antipode* 45(5), 1337–55.

Ferretti, F. 2017. Evolution and revolution: Anarchist geographies, modernity and poststructuralism. *Environment and Planning D: Society and Space* 35(5), 893–912

Gambi, L. 1962. *Geografia Regione Depressa*. Faenza: Fratelli Lega.

Gambi, L. 1964. *Questioni di Geografia*. Naples: E.S.I.

Gambi, L. 1968. *Geografia e Contestazione*. Faenza: Fratelli Lega.

Gambi, L. 1972. I valori storici dei quadri ambientali. In, C. Vivanti and R. Romano (eds.), *Storia d'Italia, Vol. 1, I Caratteri Originali*. Turin: Einaudi, pp. 5–60.

Gambi, L. 1973. *Una Geografia per la Storia*. Turin: Einaudi.

Gentileschi, M.L. 1983. Special focus on the role of women in population redistribution. Guest editorial. *Population Geography* 1, 1–3.

Gentileschi, M.L. 2003. Per una prospettiva geografica di genere: le donne nella città e gli spazi della cultura, del lavoro e del tempo libero. In, M.L. Gentileschi (ed.), *Vecchi Territori, Nuovi Mondi: La Geografia nelle Emergenze del 2000*, Atti XXVIII Congresso Geografico Italiano, vol. III, pp. 3393–507

Geografia Democratica. 1981. *L'Inchiesta sul Terreno in Geografia*. Turin: Giappichelli.

Giubilaro, C. 2017. (Un)framing Lampedusa: Regimes of visibility and the politics of affect in Italian media representations. In, L. Odasso, and G. Proglio (eds.), *Border Lampedusa. Subjectivity, Visibility and Memory in Stories of Sea and Land*. London: Palgrave Macmillan, pp. 103–17.

Grasso, M. 2020. Oily politics: A critical assessment of the oil and gas industry's contribution to climate change, *Energy Research & Social Science* 50, 106–16.

Governa, F. 2007. Ricordando geografia democratica: ripensare il passato per immaginare il futuro. In, E. Dansero, G. di Meglio, E. Donini, and F. Governa (eds.), *Geografia, Societa, Politica*. Milan: Franco Angeli, pp. 237–40.

Gruppo Hérodote. 1981. Fonti e metodi alternativi dell'inchiesta geografica. In, Geografia Democratica (ed.), *L'Inchiesta Sul Terreno in Geografia*. Turin: Giappichelli, pp. 287–323.

Guarrasi, V. 2003. Paesaggio di teorie. *Bollettino della Società Geografica Italiana* 127, 955–66.

Isenburg, T. 2000. *Legale/illegale: una Geografia*, Milano: Edizioni Punto Rosso.

Lombardi, D. 2006. *Percorsi di Geografia Sociale*. Bologna: Patron.

Longhurst, R. 2002. Geography and gender: a "critical" time? *Progress in Human Geography* 26(4), 544–52.

Maida, B. 2011. *Senti che Bel Rumore. Un Anno di Lotta per l'Università Pubblica*. Turin: Accademia.

Marengo, M. 1997. La donna nei luoghi di immigrazione. In, C. Brusa (ed.), *Immigrazione e Multicultura dell'Italia di oggi. Il Territorio, i Problemi, la Didttica*. Milan: Angeli, pp.163–81.

Minca, C. 2000. Venetian geographical praxis. *Environment and Planning D: Society and Space* 18, 285–9.

Minca C. 2001a. *Introduzione alla Geografia Postmoderna*. Padua: CEDAM.

Minca C. (ed.). 2001b. *Postmodern Geography: Theory and Praxis*. Oxford: Blackwell.

Minca, C. 2003. Critical peripheries. *Environment and Planning D: Society and Space* 21, 160–7.

Minca, C. 2005. Italian cultural geography or, the history of a prolific absence. *Social and Cultural Geography* 6(6), 927–49.

Minca, C. 2006. Giorgio Agamben and the new biopolitical nomos. *Geografiska Annaler, Series B: Human Geography* 88, 387–403.

Minca, C. 2007a. Humboldt's compromise, or the forgotten geographies of landscape. *Progress in Human Geography* 31, 179–93.

Minca, C. 2007b. Agamben's geographies of modernity. *Political Geography* 26, 78–97.

Minca, C. 2012a. Carlo Galli, Carl Schmitt and contemporary Italian political thought. *Political Geography* 31(4), 250–3.

Minca, C. 2012b. Claude Raffestin's Italian travels. *Environment and Planning D: Society and Space* 30, 142–58.

Minca, C. and L. Bialasiewicz, 2004. *Spazio e Politica: Riflessioni di Geografia Critica*. Padua: CEDAM.

Olsson, G. 2007. *Abysmal: A Critique of Cartographic Reason*. Chicago: University of Chicago Press.

Pickles, J. 2004. *A History of Spaces: Cartographic Reason, Mapping and the Geo-Coded World*. London: Routledge.

Quaini, M. 1974. *Marxismo e Geografia*. Florence: La Nuova Italia.

Quaini, M. 1978. *Dopo la Geografia*. Milan: Espresso strumenti.

Quaini, M. 1982. *Marxism and Geography*. Totowa, New Jersey: Barnes and Noble Books.

Quaini, M. 2003. Postmodernismo or rivisitazione critica della modernità? Ovvero è mai esistita una geografia veramente moderna? *Bollettino della Società Geografica Italiana* 127, 973–88.

Quaini, M. 2007. Riflessioni post-marxiste sul fantasma di Geografia Democratica. In, E. Dansero et al. (eds.), *Geografia, Società, Politica. La Ricerca in Geografia come Impegno Sociale*. Milan: Franco Angeli, pp. 241–54.

Raffestin, C. 1981. *Per una Geografia del Potere*. Milan: Unicopli.

Rondinone, A. 2003. Le donne mancanti: lo squilibrio demografico di genere in India. *Rivista Geografica Italiana* 110(1), 69–96.

Rossi, L. 1995. Per la storia del viaggio al femminile. Una prima riflessione sulle viaggiatrici in Oriente e Africa. *Notiziario del Centro Italiano per gli studi Storico-Geografici* 3(1), 15–26.

Rossi, L. 2011. *L'altra Mappa. Esploratrici Viaggiatrici Geografe*. Reggio Emilia: Diabasis.

Rossi, U. 2017. *Cities in Global Capitalism*. Cambridge, UK and Malden, MA: Polity Press.

Samers, M. and Sidaway J. 2000. Exclusions, inclusions and occlusions in "Anglo-American geography": reflections on Minca's "Venetian Geographical Praxis". *Environment and Planning D: Society and Space* 18(6), 663–6.

Ruspini, E, M. Inghilleri and V. Pecorelli (eds.). 2017. *Diventare Padri nel Terzo Millennio*. Milan: Angeli.

Schmidt di Friedberg, M., M. Marengo and V. Pecorelli (eds.). 2017. Sguardi di Genere. *Geotema* 53.

Schmidt di Friedberg, M. and V. Pecorelli. 2019. Gender and geography in Italy. *Gender, Place & Culture* 26, 1137–48.

Squarcina, E. 2009. *Didattica Critica della Geografia*. Milan: Unicopli.

Tabusi, M. 2012. Da una mailing list alla "territorializzazione" di un tetto. Ricercatori e studenti in lotta per una università pubblica, libera e aperta. In, C. Cerreti, I. Dumont and M. Tabusi (eds.), *Geografia Sociale e Democrazia. La Sfida della Comunicazione*. Rome: Aracne, pp. 183–96.

Torre S., M. Benegiamo and A. Dal Gobbo. 2020. Il pensiero decoloniale: dalle radici del dibattito ad una proposta di metodo. *ACME* 19, 448–68.

Turco, A. 1986. *Geografie della Complessità in Africa: Interpretando il Senegal*. Milan: Unicopli.

Turco, A. 1988. *Verso una Teoria Geografica della Complessità*. Milan: Unicopli.

Vallega, A. 2003. Postmoderno, postmodernismo, postmodernità. Teoria e prassi in geografia. *Bollettino della Società Geografica Italiana* 8, 909–46.

Vanolo, A. 2015. The image of the creative city, eight years later: Turin, urban branding and the economic crisis taboo. *Cities* 46, 1–7.

Zanetto, G. 1989 (ed.). *Les Langages des Représentations Géographiques*. Venice: University of Venice.

Zinzani, A. 2020. L'Ecologia Politica come campo di riconcettualizzazione socio-ambientale: governance, conflitto e produzione di spazi politici, *Geography Notebooks* 3, 1–7

12 Moments of renewal
Critical conversions of Nordic *samhällsgeografi*

Ari Lehtinen and Kirsten Simonsen

Introduction: A 'Nordic' critical geography

In 1979 the Swedish leftist journal *Häften för kritiska studier*, involving young geographers from the three Scandinavian countries, published a special issue on 'Geography and society: capitalism and the analysis of space'. A group of geographers from University of Uppsala, Sweden, decided to arrange a small two-day seminar around the topics of the issue. This event can be claimed to mark the beginning of a 'Nordic' critical geography. The seminar surprisingly attracted more than 100 participants from Denmark, Finland, Iceland, Norway and Sweden, and fruitful discussions rendered the participants convinced that it was an occasion worthy of repetition. This evolved into the annual Nordic meetings (or symposia) of critical geographers.

Now, a designation like 'Nordic' should be used with caution. In accordance with Benedict Anderson's (1991) identification of nations as imagined communities, doubt must necessarily also be cast upon calling anything inherently Nordic. National as well as Nordic identities are phenomena of discourse, constructed at distinctive junctures in time and space for specific purposes. The invention of 'the Nordic' can, in fact, be traced back to nationalist-Romantic movements of the nineteenth century. Alongside the spread of a nationalist discourse within the individual Nordic countries, a transnational ideology extolling a Scandinavian or Nordic spirit of community arose. Interest in a uniquely Nordic past was part of national as well as Nordic imaginations.

However, acknowledging 'the Nordic' as an imagined community does not mean that this construction does not work in a material sense, or that particular connections and collaborations do not exist. First, even though the Nordic countries are not as different from other European countries as ideology would sometimes have us believe, certain distinctions do stand out. It is difficult to talk about a 'Nordic model' per se, but throughout the twentieth century the Nordic countries, perhaps excluding Finland, have probably undergone a smoother modernization than most countries in Europe, and the welfare state has stood its ground. The background to that might be

DOI: 10.4324/9781315600635-12

traced in the tradition of the labour movement as well as other popular movements and their relative success in being accepted as negotiating partners in the development of the welfare regimes. Moreover, Nordic cooperation has a long tradition, also when it comes to cultural collaboration rooted in the civil society. Alongside more formalized organizations, cooperation has developed between movements – e.g., the folk high school movement, labour movements and feminist movements – and between professions, scientists, painters and writers who have maintained close ties through Nordic conferences and inter-Nordic journals. It is in this connection we shall see the development of a 'Nordic' critical geography: it is a network constituted of common academic (and social) practices and it has had its ups and downs during a period of more than 40 years.

While similarities within the Nordic context constituted some common conditions for critical endeavours, different national academic milieus entered the conversation with rather different backgrounds and theoretical approaches (see Asheim, 1987; Folke, 1985; Öhman, 1994). Danish geographers in opposition to mainstream thinking had already from the early 1970s developed a radical geography based in Marxist theory. In Sweden, the critique followed a dominant empirical line, but differed by focusing on exposed groups in processes of industrial change. Both Norway and Finland were marked by a regional perspective, but in different ways. In Norway, an affiliation to a socially hegemonic 'local community' discourse infused a centre-periphery model, while in Finland critical geographers questioned the empiricist mainstream of the disciplinary practices, both in academic milieus and in relation to the development of regional policy instruments. These different approaches at times brought about heated discussions, but the discussions also levelled some of the differences and created transnational 'language games'. In this chapter we will attempt to tell this story, first by way of a concentrated historical outline of events that marked the rise of critical geography in and between Nordic countries, and, secondly, by picking out three themes which we due to their current status think deserve a more developed treatment.[1]

Connections and critical renewal

The 1968 student revolts, with their social critique and demands for educational renewal, turned into pressures to re-forge the curricula of the Nordic universities, including geography institutions. Consequently, Nordic branches of critical geography were founded in the 1970s, under the titles (in different languages) of *samfundsgeografi, samfunnsgeografi, samhällsgeografi* and *yhteiskuntamaantiede*. The renewing parts of geography were identified as a societal geography and it became intimately linked to the development of other critical social sciences. This was a radical move in the Nordic milieus where geographers were mostly accustomed to historical and regional descriptions maybe supplemented by classificatory systematizations.

Danish students and younger scholars started re-forging geography by developing a genuine Marxist radical geography. Since most radical geography places were only in its germ, the development in Denmark went off relatively autonomously. An early inspiration was the East German geographer Gerhard Schmidt-Renner's book *Elementare Theorie der ökonomischen Geographie* (1966), which put the relationship between mode of production and the territorial structure as the core of geographical investigation. However, the group quickly both became active within and inspired by the discussions going on within the journal of *Antipode* (e.g. Folke, 1972; Buch-Hansen and Nielsen, 1976). In particular David Harvey's work and his shift of paradigm became a source of inspiration. Locally, the critique was initially launched against the hierarchical organization and the curricula of the universities (dominated by descriptive approaches and physical geography), the lack of self-determination of the people for instance in Greenland and Vietnam, and urban planning not taking into account the needs and desires of people.

An organization connecting students and young teachers (Fagligt Forum) was launched in 1971 in the Copenhagen department with the purpose of co-developing disciplinary critique and alternative approaches within the subject as well as constituting a connection between academic critique and social activism. As part of this effort, a regular publication of a journal called *Kulturgeografiske Hæfter* was started in 1973. Another initiative, a textbook *Om Geografi* (Buch-Hansen et al., 1975) presenting the core contents of capitalism critique and global uneven development, turned popular in the secondary schools for several years and even got translated into German. The major topics in this initial period concerned uneven development, mostly around the issues of (uneven) regional development and imperialism, both based in a Marxist theoretical approach, but also other topics such as housing problems and urban planning and the relationship between society and nature were addressed.

These discussions also became influential when a new university was opened in Roskilde in 1972. Having its starting point in cross-disciplinary research and teaching and student project group work as its dominating pedagogical method, Roskilde University was known as Denmark's 'red university' and employed many young scholars that had been a part of the student movement. In this context, a geography department was established in 1974. Following the discussions of the period, the topics of uneven regional development and studies of imperialism came to dominate research and teaching in the new department. In addition, an ecologically oriented milieu was founded addressing resource exploitation and bio-geographical issues. On the whole, radical geography of this period gained considerable impact at the Danish universities and in teaching in secondary schools. Steen Folke (1985) asks himself why of all places this could happen in a small complacent country like Denmark, and he summarizes a combination of explanations such as a liberal tradition in the education system, the strength of the student revolt, a weak mainstream geography dominated by physical geography and an outdated 'man-land' paradigm and the political role of different social movements.

In Norway, on the other hand, critical geography was less prominent. It started from a mainstream geography not so different from the Danish one, and its first attempts took form of government commissioned studies of regional inequality and living standards (1972–1976), which emphasized the social relevance of geographical research. This demand of relevance helped to strengthen the links of Norwegian geography with other social sciences, a connection that became most fruitful in research projects dealing with regional planning and economic restructuring (Asheim, 1985). Bjørn Asheim himself followed a radical path, even though much of the Norwegian research followed populist left-wing ideas proposed by the social anthropologist Ottar Brox. Brox' book *Hva skjer i Nord-Norge* (1966) (What happens in Northern Norway), dealing with problems and potentialities of the peripheral regions, founded a strong paradigm in Norwegian social science and planning.

Within Swedish geography, which due to an early connection to regional planning gained an applied social science status already in the 1920s, a turn towards critical social relevance also took place in the 1970s. This happened in the early 1970s in the form of studies of industrial restructuring and closures and their consequences for the labour power and the local communities. It was mostly in the University of Uppsala that these critical studies were conducted (e.g. Berger, 1973; Gonäs, 1974). It also took the form of action research in cooperation with local labour unions performed simultaneously with critical analyses of industrial companies (Axelsson et al., 1980; Gonäs et al., 1979). Similarly to Norway, then, research in industrial restructuring and regional planning became the core of critical geography (Buttimer and Mels, 2006, 83–102).

The Finnish launch into critical geography mostly emerged from the relatively peripherally sited University of Joensuu. The work was initially not only grounded in German influences (Vartiainen, 1984) but also co-inspired by the intensifying Nordic cooperation. Vartiainen's (1984) detailed 590-pages-long treatise on the historical constitution of geography as a critical social science fuelled the Finnish renewal, and it also had an impact on the Nordic and Anglophone re-thinking of geography (e.g. Vartiainen, 1986, 1987). One specific characteristic was that it, more than the critical works from the other Nordic countries, tried to fuse a socio-economic thinking with a more humanistic one. Another innovative contribution was a collective action-orientated research conducted in a number of villages in the Finnish countryside (Rouhinen, 1981).

The annual Nordic meetings of critical geographers started in 1979 in Uppsala, Sweden, and it continued with yearly seminars until 1999 as a forum for exchange of ideas, theoretical and empirical co-inspiration and network building. An important outgrowth of seminars was the journal *Nordisk Samhällsgeografisk Tidskrift*, which was for the first many years (1984–1999) edited in Uppsala and later on in Roskilde. Initially the languages of both the meetings and the journal were primarily Scandinavian (Danish, Norwegian and Swedish – and various spoken mixtures), but gradually, in the 1990s English gained ground. Papers from the meetings have during the years composed a substantial part of the content of the journal. Another offspring of the Nordic symposia was a series of cooperative PhD courses between the Nordic countries starting in 1993.

The early formulations of Nordic critical geography were mostly Marxist and concentrated on critiques of capital accumulation. A returning theme within the Nordic symposia during the first ten years was the regional restructuring of manufacturing in the Nordic countries and its relationship to changes in technology. Issues such as uneven development, spatial divisions of labour, theories of technology, discussions of the concept of the region, consequences in local living conditions and local-global relationships were covered. These themes together created an economic geography emphasizing social and critical dimensions, which became the driving force of the initial development of the Nordic critical geography (e.g. Friis and Maskell, 1981). Already in the late 1970s and early 1980s, however, analyses of particular local political restructurings, social movements and action groups as well as rural and urban conflicts were included in the critical research agenda (e.g. Simonsen et al., 1982). Intense debates on the conceptualizations of the relationship between society and nature also surfaced early in Denmark (Brandt et al., 1976; Nielsen, 1976; Olwig, 1976). They did, however, primarily appear as an appendage to the economic geography aiming at clarifying the role of natural factors in economic and regional development. Nature was mostly understood as combinations of local environmental assets and conditions passively influencing the regional industrialization. Finally, an important event was the emergence of feminist geography (see Box 12.1).

As it evolved, the Marxist orientation became enriched by cultural approaches, in particular by studies focusing on the contested landscape representations (Olwig, 1984), changes in forms of life and everyday practises (Bærenholdt, 1991; Simonsen, 1991; Sørensen and Vogelius, 1991) and representational, social and societal natures (Lehtinen, 1991; Seppänen, 1987). This reorientation did not take the form of an opposition between social theory and cultural studies, as it often was the case in Anglophone geography. In general, cultural issues such as difference and identity have in the Nordic context been theorized and explored through the lens of critical social theory, and the social, the material and the cultural have generally not been separated (Häkli, 1996; Paasi, 1996; Simonsen, 1999). Parallel to this development, but impossible to put in any of its 'boxes', runs the

Box 12.1

Feminist geography

Seen in relation to feminist studies in other social sciences in the Nordic countries, feminist geography was a latecomer. At the Nordic seminar in Norway 1982, however, a group of female participants, who found it difficult to draw attention to feminist issues, decided to set up separate meetings. Such meetings were successfully undertaken the next five years. It was the starting signal to a gender research, which before that had only been touched upon through the lens of time-geography. The early contributions to a feminist geography focussed mostly on labour markets and regional development. One important question concerned the connection and/or mutual integrations of regional and gender division of labour. Concretely, this was often implemented in studies of industrial restructuring and its consequences for women's employment (Forsberg, 1989; Gonäs, 1989; Valestrand, 1982). This major theme was later supplemented by others such as gender division of urban space (Simonsen, 1990), fear and surveillance in public space (Flemmen, 1999; Koskela, 1999; Listerborn, 2002) and gender and rurality (Berg and Forsberg, 2003; Sireni, 2008).

Theoretically, Nordic feminist geography, on the one hand, has developed along lines similar to other settings: from structural theories of patriarchy, over emphasis on gender relations, to poststructuralist ideas of subject positions and femininity and masculinity. On the other hand, however, it has taken inspiration from concepts and approaches specifically developed in the Nordic context. They have, for example, focussed on the concept of *rationality of responsibility or care* applied both in relation to family life and labour market, a spatialization of the concepts of *gender system* and *gender contract* originally proposed by the Swedish historian Yvonne Hirdman, and an integration of the Norwegian philosopher Toril Moi's critique of the sex/ gender division and her understanding of *the body as a situation*.

geographical/philosophical work of Swedish geographer Gunnar Olsson (see Box 12.2).

The 1990s heralded a professionalization and a mainstreaming of Nordic critical geographies, which, in part, were due to success in gaining key chairs in geography departments. But this also led to gradual weakening of the critical potential. As *samhällsgeografi* became fully established as part of academic institutions, it also grew closer to the trends of mainstream human geography. Social relevance was increasingly seen as synonymous with skills of cooperating with the administration of the welfare state and gaining external funding for the applied research projects. Particularly economic geography followed this path. As mentioned above, an emphasis on uneven development and social consequences of economic development was a driving force in the launch of Nordic critical

---

Box 12.2

Gunnar Olsson

Gunnar Olsson's contribution can be highlighted through the three books: *Eggs in Bird/Birds in Egg* (1980), *Lines of Power/Limits of Language* (1991) and *Abysmal: A Critique of Cartographic Reason* (2007). Olsson was originally based in the spatial analysis paradigm of the 1960s calibrating distance-decay models of spatial interaction, but parallel and simultaneously with David Harvey he broke out of that cage and followed a very different path. It is impossible to represent Olsson's rich thinking in a textbox, but an attempt might be to see it as led by two related trajectories. The first one is an exploration of the deep structure (or the grammar) of human thought and action; the second an attempt to escape from two allegedly constraining languages – those of social science and practical action. Between them he is continuously searching new languages that can transcend the limitations of the other two, 'other' ways of writing and entering into language. In pursuing that goal, he in some sense draws on his past as a spatial scientist by using a conceptual geometry including points, lines, plans, squares, boundaries etc., to investigate the 'cartography of thought', in the end turning the logic of 'cartographic reason' on itself.

---

geography. Now, however, the focus turned toward regional economic growth, innovation and industrial success. This shift was accompanied by a theoretical move from viewing economic development through the lens of broader social theories towards a more reduced political economy that externalized social values and human interaction. Moreover, and similar to the development of Anglophone critical geographies (e.g. the 'What's left' debate in *Antipode* 1991-1992, see Hadjimichalis 1991; Smith 1991; Sayer 1992), contests over academic merits silenced critical voices worried about the dominating social and environmental trends in society. In the Nordic context, this development had an additional dimension. The increasing pressure for 'international' (or, rather, Anglophone) publishing unintentionally reinforced *Anglocentrism* – an implicit assumption of the superiority of knowledge produced in Anglo-American contexts – hampering the emphasis on 'local' knowledge. After some years of fighting for its survival, this development closed down *Nordisk Samhällsgeografisk Tidskrift* in 2007. Even if Nordic critical geographers have confronted the power-knowledge system of the so-called 'international' writing spaces (see Gregson et al., 2003; Paasi, 2005; Setten, 2008), none of us escapes the power processes working through the media of language, institutional arrangements and social practices of inclusion/exclusion and through the political economy of international publishing.

Notwithstanding setbacks and difficulties, however, we claim that the 2000s witnessed a rise of Nordic critical geographies partly nurtured by the institution of the International Critical Geography Group (ICCG) and by the Nordic Geographers Meetings (NGM), which, even if they are broader in scope than the lapsed Nordic critical symposia, contain clear stands of critical geography. Amongst others, fresh approaches are co-developed in relation to urban questions, socio-environmental problems and postcolonial approaches. This progression can be interpreted as three major paths of renewal within *samhällsgeografi*, and they will below be presented under separate subtitles. This grouping of sections does injustice to practical linkages across the paths, of course, and marginalizes some trends in Nordic critical geography. Still, we find that these three paths can illustrate important moments of critical renewal.

Current strands of critical geography

The right to the city

As described above, the urban question was an early part of Nordic critical geography, in particular in the form of concrete critiques of urban planning and an interest in urban social movements. Later, this was replaced by broader concerns for urban politics and urban everyday life. However, critical urban studies also became the object of the previously mentioned professionalization and mainstreaming, leading to a dominance of applied research. But critical perspectives have been revived in new forms over the latest decades.

We have chosen to call this revival 'the right to the city', thus paraphrasing Henri Lefebvre's book *Le droit à la ville* (1968) that was translated into Swedish in 1982. Lefebvre was introduced into Nordic urban studies in the 1980s both through his urban writings and through *La production de l'espace* (1974) (see Simonsen, 1988, 1991). Later employments take more specific grips of Lefebvre. Kirsten Simonsen (2005) explores his contribution to the spatiality and the temporality of the body and the tension between the critical potential of the lived body, on the one hand, and the history and the abstraction of the body on the other one. Jussi Semi (2010) leans on Lefebvre while exploring the intergenerational differences in urban experiences of place. He uses Lefebvre's conceptual triad of spatiality to explore specific urban assemblages according to intergenerational dynamics. On this basis he develops a multigenerational view on urban change that grows into a critique of dominating routines in urban planning. Jan Lilliendahl Larsen (2007) uses Lefebvre in an original reconstruction of the field of urban politics. Taking off from the vitality and potentials embedded in urban wastelands, he seeks to reconstruct a *political urbanity* overcoming the dualism between 'politics-as-usual' and antagonistic contradictions. Empirically, it involves participatory or action research in the development of a wasteland in the harbour of Copenhagen. Theoretically, Lefebvre's critiques of everyday life and social space are worked up to conceptualize such a reconstruction of 'political urbanity'. These employments illustrate the broadness of the idea of the right to the city. It starts form urban residents' participation in the urban society and involves new ways of life, new social relations and possibilities for political struggles. Therefore, Lefebvre's rights, at once ethical and political projects, constitute a good starting point for critical urban studies.

One important part of critical urban studies within the Nordic countries concerns the complex relationship between welfare politics and social inequality and segregation (e.g. Andersson, 1999; Andersen and Clark, 2003; Hansen, 2003; Lilja, 2010; Wessel, 2000). The contributions explore welfare regimes, the significance of egalitarian values of the welfare state and processes of segregation and social polarization within Nordic cities, and they discuss the paradoxes arising when the welfare politics not only fail to counteract segregation/polarization processes, but also directly or indirectly are contributing to them.

Furthermore, connected to an increasing international orientation and exchange, a group of works (initiated at the University of Lund) has developed, which can be described as a kind of new urban political economy. An illustrative example might be a set of articles collected under the suggestive title of *Space Wars and the New Urban Imperialism* (Lund Hansen, 2006). The metaphor of 'space wars' (borrowed from Zygmunt Bauman) is chosen to emphasize conflicts around the control over space in the study of urban transformation. The main topics of the collection are globalization of the commercial property market, changes in urban governance and changes in social geography, explored in Copenhagen, Lisbon and New York. In this sense, urban space wars are explored in connection to the social architecture of capitalism – a 'vagabond' capitalism (cf. Katz, 2001), where capital moves through all imaginable scales. The combination in this collection of empirical work from different parts of the world and theoretical emphasis on the global-local nexus is characteristic for the approach. Two keywords might summarize its contributions.

The first issue is *gentrification*. Eric Clark introduced the term into the Nordic housing discussion in 1987, where he employed 'rent gap theory' on Swedish cities. Since then, gentrification has become part of the repertoire of urban studies in the other Nordic countries as well (e.g. Jauhiainen, 1997; Larsen and Lund Hansen, 2008; Wessel, 1988). In the later employments, however, the ideas of gentrification have undergone several extensions and changes. Larsen and Lund Hansen explore the connection between public supported urban renewal and processes of gentrification. They show how the result of urban renewal, even in cases where the official point of departure is improvement of the conditions of the current inhabitants, ends up by a dislocation of disadvantaged inhabitants. Hedin et al. (2012) describe an extension of the concept well beyond inner city working class residential space. It has become a global urban strategy, they argue, which may more adequately be understood as a generic form of 'accumulation by dispossession' (cf. Harvey, 2006). In accordance with the original rent-gap thesis, this refers to any kind of 'underutilized' land onto which the flow of capital facilitates 'highest and best' land uses.

The second (and related) keyword is *neoliberalism*. Much of the research on gentrification, segregation and urban polarization are connected to issues of neoliberalization of the housing sector, for example, when Clark and Johnson (2009) write about the circumvention of the 'circumscribed neoliberalism' of the Swedish housing system. While the term in these housing studies refers to a rather classic thematic of commodification, Baeten (2012) develops a more sophisticated conception of neoliberalism in his work on 'neoliberal planning'. In continuation of an analysis of a planning project in Malmö, Sweden, he identifies neoliberalism, not as a contrast to planning, but as an attitude applying the market metaphor to cities and installing cost-benefit logic into the planning process. Neoliberal planning in this optics is not a hegemonic project but rather a pragmatic strategy capable of co-existing with other urban regimes. It eats into the vulnerable sides of the welfare state (for example, by regulating the poor), but leaves significant parts of it intact. Generally, Baeten argues, it undermines ideas of the city as *right* and replaces it with those of 'freedom', opportunity or prospect.

Finally, other approaches supplements and complicates the 'urban political economy' one in relation to planning and the right to the city. Feminist and postcolonial planning critiques raise the question of the voices heard in planning debates and planning processes (e.g. Forsberg, 2006; Listerborn, 2007). In addition, poststructuralist approaches explore struggles over representation and the potentials of the 'eventalization' of urban space (e.g. Pløger, 2010a, 2010b).

Generally, the Nordic geography studies concerning the right to the city are developed in close connection to discussions going on internationally, in particular discussions combining activism and research performed in the international conferences in critical geography (ICCG).

Society-nature

As we have seen, the society-nature debate was in the 1970s already included in the Nordic critical geography. However, in these early writings nature was mostly reduced to a passive background for the development of territorial structures. The 1980s witnessed an opening of the structural emphasis in the form of studies that covered both the material changes in the realm of society-nature and the contested representations of social and ecological natures.

An inspiring combination of material and representational natures was in the mid-1980s sketched by Maaria Seppänen (1987) who in her studies of the historical changes of the pre-colonial production modes in Andean societies advocated a broad outline of the linkages between humans and nature. She, as inspired by Perttu Vartiainen's (1984, 351–356) analysis of the historical conceptualization of nature, regarded social nature as the human relationship to nature that is continually re-modified in social practises. Moreover, social nature cannot, according to her, be divorced from representational nature that she defined as 'assemblages of religious, symbolic and mythical constructions of nature' (Seppänen, 1987, 15). Similar type of conceptual grouping was later applied in a specific (post)colonial setting, detailing the contested articulations of nature – now explicitly: social and societal natures – attached to the development of the Sámi homeland in the Nordic North under the pressures of expanding forest industry (Lehtinen, 1991, 64–73).

The transformation of Nordic critical geography of society and nature thus did not follow the drama of radical leaps. The particular Nordic mode of renewal, favouring moderate broadening instead of dramatic turns, caused the rise of some impatient voices among the younger generation of critical geographers (Birkeland, 1998; Nynäs, 1990), but it also guaranteed a useful footing for developing hybrid approaches and conceptions that could help to grasp the interconnections between material and representational natures. Hence, instead of radical epistemic leaps we can identify several waves of conceptual diversification in Nordic society-nature research. This extension brought along, for example, critical analyses of nature connections (*natursammenhænge* in Danish) as part of changing modes of life (Bæhrenholdt, 1989), historical inter-folding of social and nature contracts (Bladh, 1995, 370), urban politics of nature (Häkli, 1996), hybrid geographies of woodlands and riverscapes (Kortelainen, 1999) and nature and body politics (Olwig, 2002). The material grounding, including its terrestrial (Hansen, 1990; Häkli, 1996; Larsen, 2008; Vartiainen and Vesajoki, 1991) and industrial dimensions (Eskelinen and Kautonen, 1997; Kortelainen, 1996; Lehtinen, 1991; Sæther, 1999), was explicitly kept along and further cultivated. In the Nordic setting, accordingly, the debate on the changing role of the material processes of nature in the discourses on the environment grew into a central continuity factor.

In Nordic critical geography, the intimate coupling of material and representational aspects of socio-environmental change has resulted in approaches where changing particular articulations of society-nature are seen as thoroughly embedded in pressures of mobile trans-localization launched both by the new economy of intensifying financial speculation and international efforts of co-regulation initiated by (inter)governmental and non-governmental organizations. The hybrid conceptualization of society and nature has therefore been located at the interfaces of local and trans-local developments. This emphasis is exemplified below by identifying some central trends in Nordic critical geography focusing on the changes in the politicization of nature as part of the globalizing forest industry. The traces of renewal are identified here only at the level of critical re-conceptualizations.

The *client state* researchers have, for example, analysed the profiles of countries or provinces whose political and financial spaces are largely dominated by the interests of the enterprises dominating their export operations (see Lehtinen, 2006, 32–38; Sandberg, 1992). Second, signs of forest-industrial de-territorialization, or *de-linking,* have been critically examined by concentrating on firm-specific reasons for selectively moving operations beyond traditional home areas and in this way shed light on the motives and constrains for partial de-embedding from local path-dependencies (Donner-Amnell, 2001; Moen, 1998; Moen and Lilja, 2001; Sæther, 1998, 1999, 176–223, 2007; see also Oinas, 1997, 1999). Third, the changes in the regulative settings of globalizing forest companies have been approached by identifying traces of *de-centred or discursive regulation.* According to this approach, the forest industry is seen becoming less regulated by formal legislation and state authorities but increasingly through the market and international socio-environmental pressure-building (Berglund, 1997; Donner-Amnell et al., 2004; Lehtinen, 2008; Rytteri, 2002; 2006). Fourth, latent forest dependencies have been highlighted while examining those aspects of forest-based development that need to be explained by *absent factors,* that is: factors that are influential due to their non-presence. Old-growth forests, for example, have gained their particular value due to avoided loggings, and similar-type of 'non-linkages' can be widely documented in forest-based development in general, as elsewhere at the interfaces of society and nature (Kortelainen, 2010; Lehtinen, 2010). Fifth, *forest regimes* have been researched by identifying forest-industrial formations rooting, on the one hand, from local conventions of governing the complex inter-articulations between various forest actors, both human and non-human and, on the other hand, embedded on the vested presence of international principles and agreements of co-regulation (Lehtinen et al., 2004, 11–13). Sixth, studies of biodiversity prospecting, or *bioprospecting,* have focussed on corporate performances that favour investing in forest extractivism, biobusiness and ecotourism, in the name of making particular environmental riches, both material and immaterial, economically profitable commodities with options for growing international market value (Nygren, 1998; also Pitkänen, 2008; Rytteri and Puhakka, 2009; Vepsäläinen and Pitkänen, 2010). And finally, symptoms of *decoupling, or double logic,* have been criticized in forest-industry and environmental management, while identifying actor-specific production of green or otherwise positive images strikingly deviating from the related socio-environmental performances (Donner-Amnell et al., 2004, 268; Heikkilä, 2008, 9–10).

This type of 'Nordic' re-conceptualization attached to critical forest research has broadly been non-recognized within the broader circles of (critical) geography. The debate has co-inspired only the work of those geographers and neighbouring scholars who have studied different aspects of forest-based development in the Nordic countries, Russia and Canada. Some contacts to forest researchers of the global North-South relations have in addition been established.

Postcolonialism and minorities

A paradoxical characteristic of Nordic critical geography is its limited engagement with the conditions of indigenous people within own borders. It seems as if the problems identified in other parts of the world have turned difficult to recognize within the borders of the welfare states. A predecessor can be found in a 'Greenland-report' produced by a group of activist geographers in Copenhagen 1972, but the successors are sparse. One example is, as mentioned above, the study that analysed confrontations between representational natures as they surface through industrial pressure on indigenous Sámi communities. Later, Juha Ridanpää, as part of a study of the imaginary geographies of the Nordic North, examines how literary fiction, simultaneously working as a device constructing socio-spatial inequality and as a tool that deconstruct Otherness, manage to decolonize Northern cultures. Postcolonialism for him, then, becomes both a continuity of processes initiated by European imperial aggression and a form of social criticism. Literary irony has the potential to turn into an emancipatory way of criticizing an unevenly developed world (Ridanpää, 1998, 2007, 2010; also Lehtinen, 2006). Another interesting research addresses Greenland as one of the few countries in the world that has experienced an implementation of a Social Democratic programme by its colonial power (Bærenholdt, 2007). Bærenholdt explores the 'Danification' aspect of the post-war Social Democratic project for the modernization of Greenland and the subsequent Greenlandic nationalism emphasizing the indigenous way of life as one based on a subsistence lifestyle. He describes the ambivalent and complex relationship between Denmark and Greenland created in these processes with the term *colonial modernity* (cf. Gregory, 2004) – a relationship performing exclusion by way of inclusion. It involves a form of cultural recognition that is conditional on tacit acceptance of an asymmetric, if not paternalistic relationship.

Postcolonial thinking has gained a stronger foothold in works connected to the increasing non-Western immigration into the Nordic welfare states. This (relatively sparse) inflow has enforced a theoretical-empirical recognition of the degree to which the solidarity of the welfare state has been based on a cultural concurrence of equality and likeness and a presupposition of a relatively homogeneous population (Gullestad, 2002; Koefoed and Simonsen, 2007; Molina, 1997). The first contributions were three dissertations from the University of Uppsala emphasizing the discursive construction of immigrants (Mattsson, 2001; Molina, 1997; Tesfahuney, 1998). In particular Irene Molina has since then had a central position in the debate. Her major contribution evolves around the *racialization* of Swedish towns and cities; that is, processes and practises on the housing market producing a differentiated residential spatiality according to imagined racial differences and ascribed racial affiliations (e.g. Andersson and Molina, 2003). Interestingly, she relates these processes to practises of normalization and to a social hygienist discourse in Swedish housing policy. Later, ideas of intersectionality between race, gender and class, are included – for example, in explorations of the construction of 'the immigrant woman' and the potential for a 'subversive, antiracist feminism' (Molina, 2004). Another addition comes from the Danish quarter. Koefoed and Simonsen (2010) employ postcolonial theory when studying the habitability of different spatial formations for ethnic minorities. They do so, however, in a modified way by merging Said's more representational approach with a phenomenological inspired practice theory (Simonsen, 2007). This move opens to analyses of experiences and emotions arising in the multiplicity of (more or less mediated) bodily encounters, for instance, analytically grasped through intermediate concepts such as *experienced otherness* and *practical orientalism* (also Haldrup et al., 2006; Simonsen, 2008). Included in these analyses are the possibilities of identification afforded by spatial formations from the scale of the body, to place, the city, the nation and the global scale. Hidle (2004) has also analysed relationships to place performed by ethnic minorities.

Other uses of postcolonial theory concern the power and the contested politics of place naming (Berg and Vuolteenaho, 2009) and the performance of tourism (Buciek, 2003; Hottola, 1999; Pedersen and Larsen, 2009; Tesfahuney and Schough, 2010). The latter issue is primarily drawing on the 'exoticism' side of Orientalism. In particular Petri Hottola provides a detailed theoretical and empirical analysis of conflicts and 'culture confusion' created in the embodied encounters between mutual gendered stereotypes in the context of Western backpacker tourism to South Asia – for instance, when the body of (Western, tourist) woman becomes a factor of confusion in intercultural encounters in India and Sri Lanka.

This can lead us to another group of works, not so much based in postcolonial theory, but on questions of democratic rights and minorities within postcolonial settings. South Asia, especially Sri Lanka, is also a major setting of research for Kristian Stokke. His research has dealt with ethnic conflicts, warfare and peace promotion and it has especially emphasized the changes in civil society, socio-economic rights and challenges of democratization. Stokke, who has also participated in research in sub-Saharan Africa, especially South Africa, has thematically much concentrated on the emerging forms of local civic activities under the pressures of changing state governance and neoliberal privatization. This orientation has, for example, resulted in pointed critique of the current indifference to theories of democracy in Western political geography in general (see Stokke, 2009; Stokke and Sæther, 2010). Correspondingly, Cathrine Brun has centred on ethnic conflicts and post-tsunami recovery in Sri Lanka and on this ground participated in the critical debate on forced migration and risky positions of refugees, in particular in relation to internal displacement. Her key research questions, formulated as part of humanitarian aid projects, have entangled around the drama of migrants and refugees' violently changing conditions of home-making and identity-building. She has consequently contributed to the renewal of the debates on displacement, camps, integration and re-territorialization (Brun, 2003, 2008). Moreover, Brun's praxis orientation meets Darley Jose Kjosavik's action research profile, which is empirically connected to Kerala, South India. Kjosavik has studied the practises of exclusion grounded on local categorizations of class, caste and ethnicity while especially underlining the fate of indigenous groups in her study area. She has accordingly contributed to the critical geographical debate on intersectionality, identity and land issues, as well as changes both in the sphere of institutions and at the level of political economy (Kjosavik, 2009a; 2009b).

These writings show how postcolonial thinking has gained a position within Nordic critical geography in line with the increasing importance of migration and the problems meeting minorities in various settings.

Concluding remarks

By way of conclusion, we would argue that Nordic critical geography is characterized by a certain in-betweenness. This is to say, situated 'local' knowledges that are continuously intertwined with both continental European and Anglophone sources of inspiration. This position partly stems from the necessary openness and linguistic proficiency of small language communities, such as the Nordic ones, but also from European connections of the student movements of the 1970s. Thus, the Nordic critical geography of the 1970s, while awakened by Danish geographers also leant on both German scholarship and radical Anglophone geography – e.g. as discussed in *Antipode*. This development secured a co-enrichment of several approaches, stretching from Marxism to critical humanism, cultural approaches and varying interarticulations of society-nature. The early co-enrichment also secured a multilingual atmosphere which, however, gradually and increasingly has become Anglophonized.

Nordic critical geography has also been, literally, *samhällsgeografi* – a societal geography. It has emphasized disciplinary critique and renewal as well as social critique and involvement. Even 'the cultural turn' has, for the most part, simultaneously been social. The moments of disciplinary renewal have been thorough, challenging earlier geographies of historical and regional descriptions or versions of spatial analysis. Ambitions of social engagement strongly characterized the early years of critical geography, both in the form of connections to social activism and in the form of action research as research methodology. The early 'activism' has later turned into more moderate forms, such as examinations of dialogical relations between researchers and the administration, enterprises, urban and rural civic campaigns, and environmental movements and organizations. Much of this approach more mainstream forms of applied research. But, as we have suggested, critical potentials are being revived by a new generation of critical researchers.

Where does that leave us today? We have told a story about a group of enthusiastic young geographers developing the field during the 1970s and 1980s. After the euphoria of these first decades, an ambiguous mainstreaming occurred. On the one hand, the critical geographers came to occupy central positions within academic geography in the different countries, in this way opening for the integration of the critical reflexivity into the broad understanding of the subject. On the other hand, however, this mainstreaming did not happen without some adaptation and watering down of the critical potential where critique developed in the direction of application. Today, this partial setback exists side by side with new critical approaches and discussions nurtured by international networks of critical geography. As regards content, Nordic critical geographers will still claim different forms of inbetweennes – between internationalization and situated knowledges, in particular related to the local welfare regimes; between Anglophone and continental inspiration; and between representation and materiality, not drawing any of them to their extremes.

Note

1 The 'Nordic' is conventionally seen as comprising Denmark, Finland, Iceland, Norway and Sweden as well as the Faroe Islands, Greenland and the Åland Islands. Sometimes, for example, in relation to the Nordic Geographers Meeting (NGM), also the Baltic states of Estonia, Latvia and Lithuania are included. Due to our linguistic competences and experiences, and to delimit our scope, this chapter focuses on critical geographies emerging from geographers in Denmark, Finland, Norway and Sweden.

References

Andersen, H.T. and E. Clark. 2003. Does welfare matter? Ghettoisation in the welfare state. In, J. Öhman and K. Simonsen (eds.), *Voices from the North: New Trends in Nordic human Geography*. Aldershot: Ashgate, pp. 91–102.

Anderson, B. 1991. *Imagined Communities: Reflections on the Origin and Spread of Nationalism*. London: Verso.

Andersson, R. 1999. "Divided cities" as a policy-based notion in Sweden. *Housing Studies* 14, 601–24.

Andersson, R. and I. Molina. 2003. Racialization and migration in urban segregation processes. In, J. Öhman and K. Simonsen (eds.), *Voices from the North: New Trends in Nordic Human Geography*. Aldershot: Ashgate, pp. 261–282.

Asheim, B.T. 1985. The history of geographical thought in Scandinavia. *Meddelser fra Geografisk Institutt, Ny kulturgeografisk serie 15*. Oslo: Universitet i Oslo.

Asheim, B.T. 1987. A critical evaluation of postwar development in human geography in Scandinavia. *Progress in Human Geography* 11, 333–54.

Axelsson, B., S. Berger and J. Hogdal. 1980. *Vikmanshyttan: Lära för framtida bruk*. Stockholm: Prisma.

Baeten, G. 2012. Normalising neoliberal planning: the case of the Hyllie urban development project in Malmö, Sweden. In, T. Tasan-Kok and G. Baeten (eds.),

Contradictions of Neoliberal Planning: Cities, Policies and Politics. Berlin: Springer, pp. 21–42.

Berg, L. and J. Vuolteenaho (eds.). 2009. *Critical Toponymies: The Contested Politics of Place Naming.* Farnham: Ashgate.

Berg, N.G. and G. Forsberg. 2003. Rural geography and feminist geography: discourses on rurality and gender in Britain and Scandinavia. In, J. Öhman and K. Simonsen (eds.), *Voices from the North: New Trends in Nordic Human Geography.* Aldershot: Ashgate, pp. 173–90.

Berger, S. 1973. Företagsnedläggning – konsekvenser för individ och samhälle, *Geografiska regionstudier nr. 9.* Uppsala: Uppsala Universitet.

Berglund, E. 1997. Lost in the woods? Competing knowledges in Finland's forest debates. *Discussion papers, College of Natural Resources.* Berkeley: University of California.

Birkeland, I. 1998. Nature and the 'cultural turn' in human geography. *Norsk Geografisk Tidsskrift* 52, 229–40.

Bladh, G. 1995. Finnskogens landskap och människor under fyra sekler. En studie as samhälle och natur i förändring. *Forskningsrapport* 95:11. Karlstad: Högskolan i Karlstad.

Brandt, J., R. Guttesen, E. Hove, A. Jørgensen, R.O. Rasmussen and P. Sonne. 1976. Dialektisk materialisme og geografi. *Kulturgeografiske Hæfter* 9, 79–104.

Brox, O. 1966. *Hva skjer i Nord-Norge.* Oslo: Pax Forlag.

Brun, C. 2003. *Finding a Place: Local Integration and Protracted Displacement in Sri Lanka.* Colombo: Social Scientist's Association.

Brun, C. 2008. Birds of freedom: Young people, LTTE and representations of gender, nationalism and governance in northern Sri Lanka. *Critical Asian Studies* 40, 399–422.

Buch-Hansen, M., H. Folke, S. Folke, J. Gottlieb, F. Hansen, A.-M. Hellmers, P. Krøijer and B. Nielsen. 1975. *Om geografi: En introduktionsbog.* København: Hans Reitzel.

Buch-Hansen, M. and B. Nielsen. 1977. Marxist geography and the concept of territorial structure. *Antipode* 9, 35–44.

Buciek, K. 2003. Post-colonial reason – selected perspectives. *Nordisk Samhällsgeografisk Tidskrift* 37, 45–65.

Buttimer, A. and T. Mels. 2006. *By Northern Lights: On the Making of Geography in Sweden.* Aldershot: Ashgate.

Bærenholdt, J.O. 1989. Livsformer og natursammenhænge i kritisk geografi. *Nordisk Samhällsgeografisk Tidskrift* 9, 15–25.

Bærenholdt, J.O. 1991. Bygdeliv: Livsformer og bosætningsmønstre i Nordatlanten. *Om fiskeriafhængige bygder i Island og Færøerne. Publikationer fra Institut for geografi, samfundsanalyse og datalogi. Forskningsrapport no. 78.* Roskilde: Roskilde University.

Bærenholdt, J.O. 2007. *Coping with Distances: Producing Nordic Atlantic Societies.* Oxford/New York: Berghahn Books.

Clark, E. 1987. *The Rent Gap and Urban Change: Case Studies in Malmö 1860–1985.* Lund: Lund University.

Clark, E. and K. Johnson. 2009. Circumventing circumscribed neoliberalism: The 'system switch' in Swedish housing. In, S. Glynn (ed.), *Where the Other Half Lives: Lower Income Housing in a Neoliberal World.* London/New York: Pluto Press, pp. 173–94.

Donner-Amnell, J. 2001. To be or not to be Nordic? How internationalisation has affected the character of the Nordic forest industry and forest utilisation in the Nordic countries. *Nordisk Samhällsgeografisk Tidskrift* 33, 87–124.

Donner-Amnell, J., A. Lehtinen and B. Sæther. 2004. Comparing the forest regimes in the conifer North. In, A. Lehtinen, J. Donner-Amnell and B. Sæther (eds.), *Politics of Forests: Northern Forest-Industrial Regimes in the Age of Globalization.* Aldershot: Ashgate, pp. 255–84.

Eskelinen, H. and M. Kautonen. 1997. In the shadow of the dominant cluster – the case of furniture industry in Finland. In, H. Eskelinen (ed.), *Regional Specialisation and Local Environment – Learning and Competitiveness.* NordREFO 1997 3, 89–123.

Flemmen, A.B. 1999. *Mellomromserfaringer: En analyse av kvinners frykt for seksualisert vold.* Dissertation. Tromsø: University of Tromsø.

Folke, S. 1972. Why a radical geography must be Marxist. *Antipode* 4, 13–8.

Folke, S. 1985. The Development of radical geography in Scandinavia. *Antipode* 17, 13–19.

Forsberg, G. 1989. Industriomvandling och könsstruktur. Fallstudier på fyra lokala arbetsmarknader. *Geografiska regionstudier nr.* 20. Uppsala: Uppsala Universitet.

Forsberg, G. 2006. Genusanalys – en utmaning för framsynt storstadsplanering. In, E. Gunnarsson, A. Neergaard and A. Nilsson (eds.), *Kors och tvärs: Intersektionalitet och makt i storstadens arbetsliv.* Stockholm. Normal Förlag, pp. 93–116.

Friis, P. and P. Maskell. 1981. *Teknologi- og regionaludvikling – en nordisk antologi.* Roskilde: Roskilde Universitetsforlag/GeoRuc.

Gonäs, L. 1974. Företagsnedläggning och arbetsmarknadspolitik. En studie av sysselsättningskriserna vid Oskarshamns varv. *Geografiska regionstudier nr.* 10. Uppsala: Uppsala Universitet.

Gonäs, L. 1979. *Emmaboda-rapporten – En studie om inflytande i strukturomvandling.* Stockholm: Arbetslivscentrum.

Gonäs, L. 1989. *En fråga om kön.* Stockholm: Arbetslivscentrum.

Gullestad, M. 2002. *Det norske sett med nye øyne. Kritisk analyse av norsk innvandringsdebatt.* Oslo: Universitetsforlaget.

Gregory, D. 2004. *The Colonial Present: Afghanistan, Palestine, Iraq.* Oxford: Blackwell.

Gregson, N., K. Simonsen and D. Vaiou. 2003. On writing (across) Europe: writing spaces, writing practices and representations of Europe. *European Urban and Regional Studies* 10, 1–17.

Hadjimichalis, C. 1991. "What's left to do" A view from southern Europe. *Antipode* 23 (4), 403–05.

Haldrup, M., L. Koefoed and K. Simonsen. 2006. Practical Orientalism – bodies, everyday life and the construction of Otherness. *Geografiska Annaler* 88B, 173–85.

Hansen, F. 1990. Nature in geography. *Nordisk Samhällsgeografisk Tidskrift* 11, 87–97.

Hansen, F. 2003. Welfare states and social polarization? In, J. Öhman and K. Simonsen (eds.), *Voices from the North: New Trends in Nordic Human Geography.* Aldershot: Ashgate, pp. 69–90.

Harvey, D. 2006. *Spaces of Global Capitalism: Towards a Theory of Uneven Geographical Development.* London/New York: Verso.

Hedin, K., E. Clark, E. Lundholm and G. Malmberg. 2012. Neoliberalization of housing in Sweden: gentrification, filtering and social polarization. *Annals of the Association of American Geographers* 102, 443–63.

Heikkilä, K. 2008. *Teaching Through Toponymy: Using Indigenous Place-names in Outdoor Science Camps.* Saarbrücken: VDM Verlag.

Hidle, K. 2004. *Migrasjon og stedsmyte. Sted, migrasjonserfaringer og romlige forståelser i Kristiansand.* PhD-thesis. Bergen: University of Bergen.

Hottola, P. 1999. The intercultural body: western woman, cultural confusion and control of space in the South Asian travel scene. *Department of Geography Publications* no 7. Joensuu: University of Joensuu.

Häften för kritiska studier. 1979. Geografi och samhälle: Kapitalismen och analysen av rummet (special issue). *Häften för kritiska studier* 12, nos. 2–3.

Häkli, J. 1996. Culture and politics of nature in the city. *Capitalism, Nature, Socialism* 7, 125–38.

Jauhiainen, J. 1997. Urban development and gentrification in Finland. *Scandinavian Housing and Planning Research* 14, 71–81.

Katz, C. 2001. Vagabond capitalism and the necessity of social reproduction, *Antipode* 33, 708–27.

Kjosavik, D.J. 2009a. Articulating identities in the struggle for land: The case of the indigenous peoples (Adivasis) of Kerala, India. In, P.-Y. Le Meur and J.-P. Jacob (eds.), *At the Frontier of Land Issues: Citizenship, Belonging and Land.* Paris: Karthala, pp. 113–32.

Kjosavik, D.J. 2009b. Standpoints and intersections: towards an indigenist standpoint epistemology. In, D.J. Rycloft and S. Dasgupta (eds.), *Becoming Adivasi: Indigenous Pasts and the Politics of Belonging.* New Delhi: Oxford University Press, pp. 189–203.

Koefoed, L. and K. Simonsen. 2007. The price of goodness: Everyday nationalist narratives in Denmark. *Antipode* 39, 310–30.

Koefoed, L. and K. Simonsen. 2010. *"Den fremmede", byen og nationen.* Frederiksberg: Roskilde Universitetsforlag.

Kortelainen, J. 1996. Tehdasyhdyskunta talouden ja ympäristötietoisuuden murrosvaiheissa. Joensuu: *Joensuun yliopiston yhteiskuntatieteellisiä julkaisuja* 24.

Kortelainen, J. 1999. The river as an actor-network and river systems. *Geoforum* 30, 235–47.

Kortelainen. J. 2010. Old-growth forests as objects in complex spatialities. *Area* 42, 494–501.

Koskela, H. 1999. *Fear, Control and Space: Geographies of Gender, Fear of Violence and Video Surveillance.* Publications of the Department of Geography at the University of Helsinki, A 137. Helsinki: Department of Geography, University of Helsinki.

Larsen, H.G. 2008. Scaling the Baltic Sea environment. *Geoforum* 39, 2000–2008.

Larsen, H.G. and A. Lund Hansen. 2008. Gentrification – gentle or traumatic? Urban renewal policies and socio-economic transformations in Copenhagen. *Urban Studies* 45, 2429–2448.

Larsen, J.L. 2007. *Politisk urbanitet. Projekter, planer, protester og supertanker på Krøyers Plads.* PhD-thesis. Roskilde: Department of Environment, Society and Spatial Change.

Lefebvre, H. 1968. *Le droit à la ville.* Paris: Éditions anthropos.

Lefebvre, H. 1974. *La production de l'espace.* Paris: Éditions anthropos.

Lehtinen, A. 1991. Northern natures: a study of the forest question emerging within the timber-line conflict in Finland. *Fennia* 169, 57–169.

Lehtinen, A. 2006. *Postcolonialism, Multitude, and the Politics of Nature: On the Changing Geographies of the European North.* Lanham: University Press of America.

Lehtinen, A. 2008. Lessons from Fray Bentos: forest industry, overseas investments and discursive regulation. *Fennia* 186, 69–82.

Lehtinen, A. 2010. Mobility, displacement and multilingualism – remarks on the Sámi imagination of place and space. *Finnish Journal of Ethnicity and Migration* 5, 4–9.

Lehtinen, A., J. Donner-Amnell and B. Sæther (eds.). 2004. *Politics of Forests: Northern Forest-Industrial Regimes in the Age of Globalization*. Aldershot: Ashgate.

Lilja, E. 2010. Urban space and segregation: A socio-spatial perspective. In, S. Gaddoni (ed.), *Spazi pubblici e parchi urbani nella città contemporanea*. Bologna: Pàtron Editore.

Listerborn, C. 2002. *Trygg stad: diskurser om kvinnors rädsla i forskning, policyutveckling och lokal praktik*. Dissertation. Gothenburg: Chalmers University of Technology.

Listerborn, C. 2007. Who speaks? And who listens? The relationship between planners and women's participation in local planning in a multi-cultural urban environment, *GeoJournal* 70, 61–74.

Lund Hansen, A. 2006. *Space Wars and the New Urban Imperialism*. Meddelanden från Lunds universitets geografiska institution. Avhandlingar CLXVIII. Lund: Lund University.

Mattsson, K. 2001. (O)likhetens geografier – marknaden, forskningen och de Andra. *Geografiska regionstudier* no 45. Uppsala: Uppsala University.

Moen, E. 1998. *The Decline of the Pulp and Paper Industry in Norway, 1950-1980*. Oslo: Scandinavian University Press.

Moen, E. and K. Lilja. 2001. Constructing global corporations: contrasting national legacies in the Nordic forest industry. In, G. Morgan, P.H. Kristensen and R. Whitley (eds.), *Organizing Internationally: Restructuring Firms and Markets in the Global Economy*. Oxford: Oxford University Press, pp. 152–179.

Molina, I. 1997. Stadens rasifiering. Etnisk boendesegregation i folkhemmet. *Geografiska regionstudier no.* 32, Uppsala: Uppsala University.

Molina, I. 2004. Intersubjektivitet och intersektionalitet för en subversiv antirasistisk feminism. *Sociologisk Forskning* 3, 19–24.

Nielsen, B. 1976. Naturens betydning for territorialstrukturens udvikling. *Fagligt Forum - Kulturgeografiske Hæfter* 9, 68–78.

Nygren, A. 1998. Environment as discourse: Searching for sustainable development in Costa Rica. *Environmental Values* 7, 201–22.

Nynäs, H. 1990. Hur är det möjligt att ha monopol på kritik? *Nordisk Samhällsgeografisk Tidskrift* 12, 72–3.

Oinas P. 1997. On the socio-spatial embeddedness of business firms. *Erdkunde* 51, 23–32.

Oinas, P. 1999. The embedded firm? Prelude for a revived geography of enterprise. *Acta Universitatis Oeconomicae Helsingiensis* A-143.

Olsson, G. 1980. *Birds in Egg: Eggs in Bird*. London: Pion.

Olsson, G. 1991. *Lines of Power/Limits of Language*. Minneapolis, MN: University of Minnesota Press.

Olsson, G. 2007. *Abysmal: A Critique of Cartographic Reason*. Chicago: The University of Chicago Press.

Olwig, K. 1976. Menneske/natur problematikken i geografi. *Kulturgeografiske hæfter* 9, 5–15.

Olwig, K. 1984, *Natures ideological landscape*. London: George Allen & Unwin.

Olwig, K. 2002. *Landscape, Nature, and the Body Politic*. Madison: The University of Wisconsin Press.

Öhman, J. (ed.). 1994. *Traditioner i Nordisk kulturgeografi*. Uppsala: Nordisk Samhällsgeografisk Tidskrift.

Paasi, A. 1996. *Territories, Boundaries and Consciousness: The Changing Geographies of the Finnish-Russian Border*. Chichester: Wiley.

Paasi, A. 2005. Globalisation, academic capitalism, and the uneven geographies of internal journal publishing space, *Environment and Planning A* 37, 769–89.

Pedersen, M.H. and J. Larsen. 2009. *Tourism, Performance and the Everyday: Consuming the Orient*. London: Routledge.

Pitkänen, K. 2008. Second-home landscape: The meaning(s) of landscape for second-home tourism in Finnish Lakeland. *Tourism Geographies* 10, 169–92.

Pløger, J. 2010a. Presence experiences – the eventalisation of urban space. *Environment and Planning D: Society and Space* 28, 848–65.

Pløger, J. 2010b. Contested urbanism: struggles about representation. *Space and Polity* 14, 143–65.

Ridanpää, J. 1998. Postcolonialism in a polar region? Relativity concerning a postcolonialist interpretation of literature from Northern Finland. *Nordia* 27, 67–77.

Ridanpää, J. 2007. Laughing at northernness: Postcolonialism and metafictive irony in the imaginative geography. *Social & Cultural Geography* 8, 907–28.

Ridanpää J. 2010. A masculinist northern wilderness and the emancipator potential of literary irony. *Gender, Place & Culture* 17, 319–35.

Rouhinen, S. 1981. A new social movement in search of new foundations for the development of the countryside: the Finnish action-oriented Village Study 76 and 1300 village committees. *Acta Sociologica* 24, 265–78.

Rytteri, T. 2002. *Metsäteollisuusyrityksen luonto: tutkimus Enso-Gutzeitin ympäristö- ja yhteiskuntavastuun muotoutumisesta*. Joensuu: *Joensuun yliopiston yhteiskuntatieteellisiä julkaisuja 10*.

Rytteri, T. 2006. *Metsän haltija. Metsähallituksen yhteiskunnallinen vastuu vuosina 1859-2005*. Helsinki: Suomen tiedeseura.

Rytteri, T. and R. Puhakka. 2009. Formation of Finland's national Parks as a political issue. *Ethics, Place & Environment* 12, 91–106.

Sandberg, L.A. (ed.). 1992. *Trouble in the Woods. Forest Policy and Social Conflict in Nova Scotia and New Brunswick*. Fredericton: Acadiensis Press.

Sayer, A. 1992. What's left to do?: A reply to Hadjimichalis and Smith. *Antipode* 24 (3), 214–17.

Schmidt-Renner, G. 1966. *Elementare Theorie der ökonomischen Geographie*. Gotha: Haach.

Semi J. 2010. *Sisäiset sijainnit. Tutkimus sukupolvien paikkakokemuksista*. Dissertations in Social Sciences and Business Studies 2. Joensuu: University of Eastern Finland.

Seppänen, M. 1987. Mikä luonnoista luonnollisin. Ihmisen "kolme luontoa" *Keski-Andeilla keramiikan keksimisestä conquistadoreihin*. Helsinki: *Kehitysmaantieteen yhdistyksen toimitteita 16*.

Setten, G. 2008. Encyclopaedic vision: Speculating on the dictionary of human geography. *Geoforum* 39, 1097–104.

Simonsen, K. 1988. Henri Lefebvre og 'det urbane'. In, J. Tønboe (ed.), *Storbyens Sociologi*. Aalborg: Aalborg Universitetsforlag, pp. 215–34

Simonsen, K. 1990. Urban division of space: A gender category! *Scandinavian Housing and Planning Research* 7, 143–53.

Simonsen, K. 1991. Towards an understanding of the contextuality of mode of life, *Environment and Planning D: Society and Space* 9, 417–31.

Simonsen, K. 1999. Difference in Human Geography – Travelling through Anglo-Saxon and Scandinavian Discourses, *European Planning Studies* 7, 9–25.

Simonsen, K. 2005. Bodies, sensations, space and time: the contribution from Henri Lefebvre, *Geografiska Annaler* 87B, 1–15.

Simonsen, K. 2007. Practice, spatiality and embodied emotions: an outline of a geography and practice. *Human Affairs* 17, 168–82.

Simonsen, K. 2008. Practice, narrative and the 'multicultural city' – a Copenhagen case. *European Urban and Regional Studies* 15, 145–59.

Simonsen, K., H.T. Jensen and F. Hansen (eds.). 1982. *Lokalsamfund og sociale bevægelser.* Roskilde: Roskilde Universitetsforlag/GeoRuc.

Sireni, M. 2008. Agrarian femininity in a state of flux: Multiple roles of Finnish farm women. In, I.A. Morell and B.B. Bock (eds.), *Gender Regimes, Citizen Participation and Rural Restructuring.* Oxford: Elsevier, pp. 33–55.

Smith, N. 1991. What's left? A lot's left. *Antipode* 23 (4), 406–18.

Stokke, K. 2009. Human geography and contextual politics of substantive democratisation. *Progress in Human Geography* 33, 739–42.

Stokke, K. and E. Sæther. 2010. Political geography in Norway: current state and future prospects. *Norsk Geografisk Tidsskrift* 64, 211–5.

Sæther, B. 1998. Environmental improvements in the Norwegian pulp and paper industry – from place and government to space and market. *Norsk Geografisk Tidskrift* 52, 181–94.

Sæther, B. 1999. *Regulering og innovasjon: miljøarbeid i norsk treforedlingsindustri 1974-1998.* Dr. Polit. Dissertation. Oslo: Department of Sociology and Human Geography, Oslo University.

Sæther, B. 2007. From researching regions at a distance to participatory network building: integrating action research and economic geography. *Systemic Practice and Action Research* 20, 115–26.

Sørensen, O.B. and P. Vogelius. 1991. *Arbejdets transformation og livsformers forandring.* PhD dissertation. Copenhagen: Department of Geography, University of Copenhagen.

Tesfahuney, M. 1998. Imag(in)ing the Other(s). Migration, Racism and the Discursive Construction of Migrants. *Geografiska regionstudier no 34.* Uppsala: Uppsala University.

Tesfahuney, M. and K. Schough. 2010. *Det globala reseprivilegiet,* Lund: Sekel Bokförlag.

Valestrand, H. 1982. "Då konene miste jobben". Om situasjonene for en gruppe "stedbunden, marginal" arbeidskraft etter en nedleggelse. In, K. Simonsen, H.T. Jensen and F. Hansen (eds.), *Lokalsamfund og sociale bevægelser.* Roskilde: Roskilde Universitetsforlag/GeoRuc, pp. 59–71.

Vartiainen, P. 1984. *Maantieteen konstituoitumisesta ihmistieteenä.* Joensuu: *Joensuun yliopiston yhteiskuntatieteellisiä julkaisuja* 3.

Vartiainen, P. 1986. Om det geografiska i samhällsteorin. *Nordisk Samhällsgeografisk Tidskrift* 3, 3–16.

Vartiainen, P. 1987. The strategy of territorial integration in regional development: defining territoriality. *Geoforum* 18, 117–26.

Vartiainen, P. and H. Vesajoki. 1991. Ekologisen maantieteen haasteista. In, P. Hakamies, V. Jääskeläinen and I. Savijärvi (eds.), *Saimaalta Kolille. Karjalan tutkimuslaitos 1971-1991*. Joensuu: Karelian Institute, Joensuu University, pp. 112–39.

Vepsäläinen, M. and K. Pitkänen. 2010. Second-home countryside: Representation of the rural in Finnish popular discourses. *Journal of Rural Studies* 26, 194–204.

Wessel, T. 1988. Gentrification som forskningsfelt – utviklingen av et samfunnsteoretisk mikrokosmos. *Nordisk Samhällsgeografisk Tidskrift* 8, 39–55.

Wessel, T. 2000. Social polarisation and socio-economic segregation in a welfare state: the case of Oslo. *Urban Studies* 37, 1947–1967.

13 Critical approaches and the practice of geography in Spain

Abel Albet and Maria-Dolors García-Ramon

Introduction

One could be tempted to think that the political and social history of Spain over the past 40–45 years created a climate and context highly beneficial for the birth of critical thinking in the social sciences, including in human geography. However, we do not believe that this is the case. Today, Spanish geography still retains some of the features of the Vidalian tradition of regional geography (descriptive, uncritical, apolitical), although having developed in the last two decades a strong technical and applied component derived, above all, from the emergence of GIS. Therefore, even if we can find some interest in critical thinking over the last three decades, such perspectives have always been – and remain – very marginal to the mainstream of Spanish geography. The reasons for this lack of critical geography – or, at least, of a steady trend of critical thinking – are complex. In order to analyse this issue, and to contribute with our own long experience within Spanish academia, we will in this chapter look at the development of critical perspectives within the general evolution of Spanish geography since the death of Franco in 1975, and we will pay particular attention to studies of gender as a way of doing critical geography.

Critical geography in the evolution of post-Franco geography

With just a handful of exceptions, until the early 1970s Spanish geography remained anchored in Vidalian geography and was otherwise isolated from new influences from abroad. Generally speaking, it was essentially a descriptive and acritical discipline from which social engagement and public relevance had been uprooted. Geography was mainly used in primary and secondary schools to consolidate a national Spanish identity along with the principles of Franco's fascist dictatorship. In fact, academic geography was limited to the development of certain issues mainly linked to history or to geomorphology, and it lacked any critical methodology or approach. However, in the early 1970s, a few young geographers were influenced by the radical trends from both Latin America and the Anglo-American world,

DOI: 10.4324/9781315600635-13

where they had studied or worked. These new ideas, combined with the anti-Franco political upheavals and concerns with serious urban development problems, aroused a social conscience in small circles within Spanish geography. This was reflected in several publications, which are the earliest manifestations of critical geography in Spain.

The pioneer was the journal *Documents d'Anàlisi Urbana* (*Documents on Urban Analysis*), which in 1974 started being published under the impetus of Enric Lluch at the Autonomous University of Barcelona, established in 1968 based on open concepts of academic and managerial organization. Jordi Borja and Laura Zumín were the editors of the three issues that were published on the themes of urban movements, capitalist urban planning and urban agents. These issues included articles by both Borja and other European and Latin American theorists (such as Enrique Browne, Jorge Hardoy, Jean Lojkine, Christian Topalov, François Ascher, Edmond Preteceille), whom Borja had met during his stays in Paris, Santiago de Chile and Buenos Aires (Borja, 1974; Tulla et al., 2020). Some years later – in 1977 and 1978 – a new series of the same journal, *Documents d'Anàlisi Metodològica en Geografia* (*Documents on Methodological Analysis in Geography*), published two issues following the same critical vein. The first was edited by Maria Dolors García-Ramon and was devoted to "La geografía radical anglosajona" (Radical Anglo-American Geography) featuring articles by David Harvey, James Blaut, William Bunge and Richard Peet translated into Spanish (García-Ramon, 1977). This issue had a major impact, as it was the first time that both English-language radical geography and the journal *Antipode* were aired in Spain. The second issue was devoted to a Marxist analysis of the peasantry with the majority of articles written by Spanish geographers and historians (Frutos, 1980; García-Ramon, 1985).

It should be borne in mind that these were times of immense political upheaval: Franco's regime was forcefully besieged by activism of trade unions, political parties (which were illegal at the time) and citizen mobilizations. Urban and neighbourhood movements, mainly led by the communist parties, became extremely important in cities like Madrid and Barcelona where hundreds of thousands of immigrants from southern Spain during the two previous decades had been forced into the suburbs. Franco died in 1975 and the Communist Party was legalized in the spring of 1977. Shortly afterward, the first legislative elections were held, yielding the 1978 Constitution that is still in force. The first democratic local elections were called in 1979 and the first elections to the regional parliaments ("autonomous communities") were held between 1980 and 1981. This was significant for Spanish geography, as these new institutions of local and regional power offered a new kind of jobs (outside the primary and secondary schools) for recent geography graduates.

Between mid-1970s and mid-1980s, Spanish universities experienced years of heightened political activism. Most geography professors were involved (specially in Barcelona and some others in Madrid) but not many students, as the degree in geography only existed at the two universities of Barcelona from the early 1980s. (In the rest of Spain, the BA in geography did not exist until the early 1990s.) During this troubled period at Spanish universities, an important milestone for geography was the 1977 Spanish translation of Yves Lacoste's *La géographie ça sert d'abord à faire la guerre* (*The Purpose of Geography is, Above All, to Make War*). Lacoste's critical vision had significant impact in some departments and among students in general. In part, this was because he came from French geography, which was viewed as closer to Spanish geography than Anglo-American geography. (In fact, Anglo-American geography was until recently seen as a "foreign" tradition.) Closely related to this was the 1977 publication of the book *Geografías, ideologías, estrategias espaciales* (*Geographies, Ideologies, Spatial Strategies*), edited by Nicolás Ortega from the Autonomous University of Madrid, which contained a series of translations of articles from *Hérodote*, the radical geography journal founded by Lacoste in 1976, which was fairly well-known in Spanish geography (Ortega, 1977; Reche and Rodríguez, 1978; Rodríguez, 1979).

Another important event was the 1976 launching of the journal *Geocrítica. Cuadernos Críticos de Geografía Humana* (*Geocriticism. Critical Journal on Human Geography*), edited by Horacio Capel at the University of Barcelona, which became quite popular among students despite misgivings on the part of most of the faculty. The editorial in the first issue stated the need for a criticism *of* the field of geography *from* the field of geography. This goal was addressed in the first issues, which were mostly devoted to translated articles on geographical thought and epistemology as a way to challenge the prevailing conception of geography as an inductive, historicist, empirical and uncritical field rooted in the Franco era. Especially in its early years, *Geocrítica* played a crucial role in arousing a reflective conscience and a conceptual concern among the Spanish geography community. Likewise, the influence of *Geocrítica* was significant at the earliest student and young geographers' conferences, which began to be organized around that time and examined the issue of social and political commitment in geography. However, as Bosque-Maurel (1986) points out, *Geocrítica*'s aim to use geography as a critical weapon against the social and political reality did not fare as well. *Scripta Nova*, the online heir to *Geocrítica*, is the most internationally referenced geographical journal published in Spain. In its statement of purpose, the journal clearly aims for geography as a social science. In the context of the Spanish discipline, which is generally only linked to history, to humanities or eventually as part of urban and regional planning (Capel, 1998), this can be seen as a critical repositioning. In spite of this, only few articles in *Scripta Nova* contain some critical component, all of them by South American researchers.

From 1973 to 1985 Horacio Capel (University of Barcelona) edited a book series called "Realidad Geográfica" (Geographical Reality), published by Los Libros de la Frontera, which was devoted to the study of space as a product of social structure and as an object of appropriation by the different social groups. A number of texts on urban problems in socialist countries (by Soviet, Chinese, French and Italian geographers) as well as on urban planning and political praxis in capitalist countries were published. The series also included texts by Spanish radical geographers and other social scientists that became much used in geography and in the emerging criticism of environmental and planning problems in the last years of Francoism and the subsequent political transition (Capel, 1974; Costa, 1980; Gavira and Grilló, 1975; Sánchez, 1981; Vicente-Mosquete, 1983).

In the 1980s, Enric Lluch (Autonomous University of Barcelona) was the editorial director of Oikos-Tau that became a leading publisher of innovative geography in Spanish. Oikos-Tau also translated several key texts by contemporary radical geographers such as Pierre George, Derek Gregory, Yves Lacoste, Massimo Quaini and Milton Santos. It is worth mentioning that since the 1990s, Spanish translations of texts by David Harvey, Richard Peet and Edward Soja were undertaken on the initiative of architects (i.e. Barcelona's renowned critical school of architecture and urban planning) and economists rather than geographers, and mainly through Akal publishing company. Enric Lluch was also the main editor of a critical and innovative encyclopaedia that, following radical concepts and methodologies, placed geography within the social sciences, breaking its traditional descriptiveness and narrow relationship to the field of history (Albet, 2016; Lluch, 1981–1984).

At the beginning of the twentieth century, the strong Spanish anarchist movement spread the geography of Elisée Reclus and Piotr Kropotkin into many workers' associations. Some of those ideas were implemented in the 1936 Spanish Revolution. In the second half of the 1970s, there was a short revival of this anarchist tradition in Spanish geography. At the 1977 congress of Spanish geographers in Granada, Maria Dolors García-Ramon and Nicolás Ortega delivered papers on the contribution of anarchist ideas on the organization of space. At that time, a Spanish translation of *Antipode*'s 1978 special issue on anarchism (edited by Myrna Breitbart) was proposed, but for a number of reasons the resulting book was not published until a decade later with the title of *Anarquismo y geografía* (*Anarchism and Geography*) (Breitbart, 1988). It should be mentioned that Breitbart's doctoral thesis (Clark University) examined the anarchist collectivizations in Republican Catalonia, and that María-Teresa Vicente-Mosquete (University of Salamanca) submitted a thesis on Reclus' *L'homme et la Terre* that earned him a doctorate in Paris with Yves Lacoste as supervisor. The thesis was subsequently published in Barcelona (Vicente-Mosquete, 1983). In that same period, a selection of texts by Reclus was published by a group of teachers and students of the Autonomous University of Barcelona (Colectivo de Geógrafos, 1980). Only in the 2000s, a new and ephemeral anarchist publication (only five issues) appeared online and on paper: *Anarco-Territoris: Revista Anarquista de Pensament Territorial* (*Anarco-Territories: Anarchist*

Journal on Territorial Thought). It included articles on the classical topics of anarchist geography (particularly Kropotkin) as well as new ones from the conceptual and theoretical perspective of academic geography that goes beyond the agitprop approaches common to many such texts (Arnau et al., 2007; Blanco, 2010; Oliveras, 2010).

A landmark event in Spanish radical geography was a 1983 colloquium in Madrid called "Geografía y marxismo" (Geography and Marxism) – a daring title within the context of Spanish geography at the time. The colloquium was organized by Aurora García Ballesteros (Complutense University of Madrid) and included papers by Spanish, French and British geographers. The majority of the contributions were status reports as opposed to empirical studies along Marxist lines (Albet, 1988; García-Ballesteros, 1986). But around 1980 there also appeared several Marxist-inspired doctoral theses, which not only are notable because they are practically the only ones of their kind, but they also engaged in empirical studies. The leftist militancy of the authors – in a highly politicized academic climate – was a decisive factor in their decision to instigate these studies. First, we would like to highlight the thesis by Carles Carreras, submitted at the University of Barcelona in 1978, on real estate agents in Barcelona's working-class neighbourhoods. The study was highly influenced by the work of Manuel Castells (Spanish by birth but exiled in France; see Castells, 1972), as well as by Marxist urban planners (both French and Italian; see Campos-Venuti, 1967). Likewise, in 1981, two theses on rural geography from a Marxist vantage point were submitted at the Autonomous University of Barcelona. These were Antoni Tulla's study of family farms in the Pyrenees (Tulla, 1993), which drew on Anglo-American Marxism and Helena Estalella's investigation of the evolution in landownership in Northern Catalonia (Estalella, 1984). In 1983, a thesis by Joan-Eugeni Sánchez (University of Barcelona) – a sociologist by training with a strong Althusserian influence – on the formation of Catalan space can also be classified into this category (Sánchez, 1991). These critical contributions also reached geographical teaching at primary and secondary school (Ascon, 1990; Benejam, 1992).

The majority of key studies in the late 1970s and early 1980s are much closely related to critical tendencies in European social sciences than to the emergence of Anglo-American critical geography, despite the fact that the work of David Harvey was relatively well known, for example (Gómez-Mendoza, 1988). An exception was the 1986 thesis on landownership in Santa Cruz de Tenerife by Luz Marina García (University of La Laguna), who had spent some time at Johns Hopkins University with David Harvey and was clearly inspired by his work (García-Herrera, 1988). Also, in relation to Anglo-American Marxist geography, some Spanish geographers participated in international debate on radical perspectives of nature, for example, David Saurí who obtained his PhD from Clark University (Saurí, 1988). Later, two doctoral theses with notable critical or Marxist component appeared, one by Pere López (University of Barcelona) focusing on a historical and critical appraisal of the development of Barcelona (López, 1993), the other by Víctor Martín (University of La Laguna) on the tourist development of Tenerife (Martín, 1999).

In 1982, the Socialist Party won the legislative elections, and for the first time since the imposition of the Franco regime, the so-called "left" returned to power in Spain. In 1986, Spain joined the European Community, which triggered an influx of foreign capital and European funding. In a brief time-span, Spain witnessed a surge in both public investment and per capita income. It was as if all of this led society to forget former essential political claims (condemnation of the Francoist dictatorship, a real welfare state, social justice, etc.) and a clear process of depoliticization of Spanish society got underway – and carried geography with it.

The 1990s witnessed a steep rise in the technical approaches that afforded more or less secure job opportunities, a process that ran parallel to a rising neglect of the critical dimension and commitments to social change (Segrelles, 1998, 2001). The cultural-humanistic approach had in the mid-1980s gained some popularity, yet it soon declined and left the field to the technical perspectives, especially GIS. The development of new curricula in the early 1990s marked an important milestone in Spanish geography. First, the bachelor degree in geography was established, permanently leaving behind the joint degree in geography and history. Second, the content of geography studies changed substantially, as instrumental subjects were added (notably remote sensing and GIS) and regional planning became the main focus. Many students and younger faculty members warmly embraced this transition. A possible interpretation of this enthusiasm is that for many Spanish geographers, the modernization of the discipline came through specialization. But this entailed an abandonment of more historical, social, cultural and critical approaches, and there were few efforts to combine a concern with professionalism with the preservation of a critical dimension in the training of these specialized professionals.

The culmination of this process of professionalization might be the 2004 curriculum, which aims to adapt the degree in geography to the guidelines of the European Higher Education Area and the Bologna Declaration. The new degree is in most universities called "Geography and Regional Planning", and its basic underlying idea is employability, a fashionable concept in the European Union. The procedure used to draw up this new curriculum is itself significant. It began by making a list of possible jobs and their professional orientations, and based on this list the educational profiles (and subjects and courses in the degree programme) were developed. The result is a heavy emphasis on methodological and instrumental subjects, which leaves a huge void in relation to theoretical and critical subjects. Spanish geography came to be seen in a highly restrictive sense; that is, as nothing more than land-use planning. The university seems to have become a vocational school where students' critical spirit no longer has to be nurtured. Therefore, after the 1990s, only few scattered critical voices were heard in Spain. Postmodern, postcolonial and cultural approaches, which came to the fore in that decade, contributed new vantage points in geographical research (Artigues, 2002; García-Ramon et al., 2003; Puente, 2009). Some authors have also stated that environmental studies might entail a critical contribution to geographical thinking (Alió and Bru, 1991–1992; Casellas, 2010; Gómez-Mendoza, 2002; March, 2013). However, these critical contributions to the discipline became the exception rather than the rule. Maybe gender geography is the only organized source of direct criticism of the dominant paradigms, both with respect to theory (García-Ramon, 2005) and methodology (Baylina, 2004).

Geography and gender as a way of doing critical geography

A good part of the research that has been carried out in the field of gender geography can be considered a contribution to critical thinking in Spanish geography. Firstly, it should be borne in mind that many geographers that have worked on gender issues in Spain come from what has been called radical or Marxist geography. Thus, it is not surprising that the reading and interpretation of Spanish gender geography, which clearly arrived from the English-speaking world at the end of the 1980s, often had a more marked social and critical leaning than where it originated. Secondly, we should remember that the theoretical framework of gender geography (in Spain and in other countries) frequently has challenged the conceptual bases of geography and, quite often, has contributed to destabilize the "establishment" of the discipline. Spanish gender geography can thus be considered the strongest and the most consolidated way in which critical geography has developed in Spain, both in terms of people involved, academic production (research and publications) and social engagement.

The starting point of the interests in gender was, on one hand, some informal contacts with the Women and Geography Study Group (WGSG) of the Institute of British Geographers (IBG) as well as the impact of the 1984 book *Geography and Gender*. Also, since 1988, the Gender and Geography Commission of the International Geographical Union (IGU) – where several Spanish geographers have been very active – gave impetus to this approach, together with many other contacts coming from outside the Anglo-American gender geography (García-Ramon, 1989). This interest for other "peripheral" gender geographies has been one of the characteristics of gender geographers in Spain, who have multiplied the contacts with Greece, Italy, Portugal, France, Hungary, Argentina, Mexico and Brazil, to mention only the countries with stronger links. Here, we will not provide an exhaustive overview but only examine some main contributions of Spanish gender geography.

Processes of rural and regional restructuring

Initially, two main lines of research examining women's work in rural environments were developed: (1) women's contribution to work on family-run farms and (2) the role of women in the economic diversification and restructuring of rural areas. Within the former, a series of studies demonstrated that women's contributions to work are important when both productive and reproductive labour is taken into account (García-Ramon and Cruz, 1996). This has also contributed to theoretical debates on the division between productive and reproductive labour, making clear that any attempt to explain women's work has to take into consideration the patriarchal control of the work process and the ownership of the means of production (Nieto, 2006).

The second line of studies focusses on non-agricultural work by women in the rural areas (rural tourism, putting out industrial work, etc.). This work is characterized by being informal, labour intensive and badly remunerated, but it cannot simply be seen as marginal or as remnants of work forms bound for extinction. In many cases these work practices open new possibilities and can be seen as forms of resistance against the competitive pressures of the market. In this sense, women's work can be crucial to local rural development (Casellas and Pallarès, 2005; Pallarès et al., 2003). As researchers committed to critical thinking and equal opportunities, we must not marginalize their role; rather, we must make these women more visible and condemn the specific circumstances in which they must do their jobs and insist on the need for a substantial change in gender roles and relations (Sabaté, 2002).

Time, work and urban spaces

Inspired by debates coming from the Italian left, studies of women's daily routines in the city were undertaken as early as the 1990s (Prats and García-Ramon, 2004). Given the challenge of the massive influx of women into the labour force and the rigidity of lifestyles and timetables in cities, especially in Southern Europe, an attempt was made to consider how women managed their time and to study the schedules of businesses and services and their appropriateness in light of the demand. The purpose was to provide recommendations aimed at fostering greater freedom and equality in the use of time. Most studies have focussed on the city of Barcelona, as the local administration – at that time quite leftist – fostered initiatives along this line, following the model of left-wing Italian municipalities.

The role of women in defending the urban environment has also been approached in studies of female citizens' movements and women-led movements (Bru, 1996). This work has involved comparative regional research on mobilizations protesting environmental risks (related to the presence of industrial waste) in municipalities in different regions. It has revealed that women's views of the environment and their way of doing "local politics" were quite different from usually accepted rigid, androcentric views.

Another recent avenue of research involves analysing a series of urban planning actions undertaken in Catalan cities of varying sizes (Díaz-Cortés et al., 2008; García-Ramon et al., 2004, 2014). Women have been absent from urban planning, not only as users of public spaces, but also as urban planners. The research examined a series of actions aimed at rehabilitating urban spaces, which had included among their objectives the fight against socio-spatial exclusion. The potential of public spaces as an integrating force has been shown, as has the fact that the design of public spaces is a crucial element in fostering the presence of women – and minority groups – in efforts to overcome traditional forms of social and spatial isolation; in short, to encourage processes of emancipation (Díaz-Cortés, 2012; Estévez, 2012; Rodó-de-Zárate, 2018). This research also questions the hegemony of the knowledge of professional planners that assume "absolute" values in physical design and do not take into account cultural diversity and, above all, relations of power within communities. One recent and innovative line of this research on uses of public spaces has focussed on children, youth and gender, seeing places as educational sites and learning in everyday life (Baylina, et al., 2006).

Openings for a critical revival

There has since round about the turn of the century been promising signs of a wider revival of Spanish critical geography. This started with an international seminar on "Geografías disidentes" (Dissident Geographies) that was held in Girona in 2001. The seminar included approaches capable of generating alternatives to institutionalized knowledge. This included critical, postmodern, postcolonial and gender approaches, and the seminar triggered theoretical and methodological debates that had not been seen in Spanish geography for quite some time (García-Ramon and Nogué, 2002). Likewise, the 18th congress of the Association of Spanish Geographers, held in Bellaterra in 2003 under the clearly political slogan of "Geografías para una sociedad global: diversidad, identidad y exclusión social" (Geographies for a Global Society: Diversity, Identity and Social Exclusion), included several sessions with a critical orientation (Albet et al., 2004). Doreen Massey and Neil Smith were invited speakers. This was an isolated event, however, and geography congresses in Spain usually do not include critical panels or papers.

Another promising event was an international seminar held in 2006 at the Autonomous University of Barcelona to discuss, from a critical perspective, the hegemony of Anglo-American geography (gender geography was taken as an example). Ironically, when Anglo-American geography is focusing on ideas about exclusion, marginality, periphery, situated knowledge, differences and the politics of identity and place, it has not systematically turned the gaze on ways in which institutionalized discursive and material practices of Anglo-American geography marginalize geographic knowledge and practices from other geographical traditions. It was also made clear that Anglo-American hegemony lies in the power of language, and that this hegemony is not only produced in the centre but also reproduced in the places that it tends to dominate. For example, to get an academic promotion it is necessary to publish in "international" journals, which implies Anglophone ones. Two publications on the seminar in two non-hegemonic journals (Belgian and Spanish) gave the opportunity for international researchers to have a general picture of the critical work being done outside the Anglophone world, in this case on work in gender geography (García-Ramon and Monk, 2007).

Abalar: A Xeografía galega en construcción (*Abalar: Galician Geography under Construction*) was a critical journal published in Galician between 2003 and 2006 in which students play a central role. The journal did not strive to be academic; rather, it emerged in an atmosphere laden with protest against the new law on universities in late 2001 and the outrage at the mismanagement of the devastation triggered by the sinking of the *Prestige* off the Galician coasts and the resulting petrol pollution on the coastline (Abalar, 2003). The publication in 2006 of the book *Las otras geografías* (*Other Geographies*) edited by Joan Nogué (University of Girona) and Joan Romero (University of Valencia) is another good example of emerging critical thinking in Spanish geography. It is a textbook on human geography that is attractive, unconventional and aims critically to give voice to those who normally are voiceless or invisible (Nogué and Romero, 2006; Romero, 2004).

Since 2010, Icaria, one of the most renowned critical publishers in Spain, publishes the book series "Espacios Críticos" (Critical Spaces) run by Abel Albet (Autonomous University of Barcelona) and Núria Benach (University of Barcelona). This series introduces the ideas of non-Spanish authors (mainly from geography) that approach social reality from critical spatial perspectives. The aim is to spread critical spatial thought in Spanish speaking countries and to foster debate on the role of space in interpretations of society. The first volumes of the series deal with Edward Soja, Doreen Massey, Richard Peet, Franco Farinelli, Neil Smith, Francesco Indovina, Jean-Pierre Garnier, Neil Brenner, David Harvey, Claude Raffestin, William Bunge, Yi-Fu Tuan, Horacio Capel and Maria-Dolors García-Ramon. Each volume includes an anthology of translated texts, an unpublished text by the author written specifically for the volume and a bibliographical study together with a critical essay on his/her contribution. This book collection, which includes works by geographers and others writing with a spatial perspective, is perhaps the most relevant contribution to critical thought in Spanish geography in the present decade (see Albet, 2019; Albet and Benach, 2012;

Benach, 2012, 2017; Benach and Albet, 2010, 2019; Benach and Carlos, 2016; García-Herrera and Sabaté-Bel, 2015; Lladó, 2013; Nel·lo, 2012; Nogué, 2018; Sevilla-Buitrago, 2016; Tello, 2016). Together with the book series, "Espacios Críticos" has a wide research and activist agenda on critical spatial issues (see http://espaiscritics.org).

There are in such ways encouraging signs of a wider reassertion of critical thought in Spanish geography. But several factors have made it difficult for critical geographers to actively participate in public administration – and influence society more widely. The first students with geography as their only subject obtained their degrees in the beginning of the 1980s; that is, at time when local councils became democratically elected and that regional governments were created with competences in fields closely related to geography (urban and regional planning, environment, tourism, forestry, etc.). At the beginning, there were some ideological reluctances to participate in projects of the public administrations, which was seen as a kind of collaboration with power or right-wing governments. That changed very soon, as geographers saw it as an opportunity transform society and territory, even at a small scale and to a limited extent. It was in the 1990s, and more specially after 2000, that some geographers reached high-level jobs, which enabling them to make decisive contributions beyond the mere implementation of technical processes (Nel·lo, 2011). But generally speaking, the influence of critical geographers on Spanish spatial policies has been limited, either because of their specific position in public administration and/or because of the low political profile of geography in society. At the local level the role of critical geographers has been more active, but still without much visibility (Zoido, 2002). In a similar way, a significant number of students, academics and professional geographers are active in neighbourhood associations, NGOs and various territorial conflicts (Nel·lo, 2003), but in spite of the fact that conflicts often are intensely geographical (i.e. territorial or environmental), only in very few cases have geographers gained significant influence. Architects, environmental researchers, biologists and political scientists, among others, have demonstrated a stronger power of mobilization (Agüera, 2008; Zusman, 2004).

Concluding thoughts

In the 1970s several Spanish geographers saw the need to confront and solve problems created by both an unregulated capitalism and the political dictatorship in rural areas and especially in cities. This stimulated theoretical thinking on urban development and on agrarian reform, and on the role of academics in the transformation of the society. At the same time, a few young Spanish geographers established strong links with prominent international critical academics, and some of their published works were decisive to introduce new epistemologies, concepts and methodologies of radical geography. However, they were not only very few but also concentrated in two or three universities. The mainstream Spanish geography remained comfortably within the conservative and descriptive Vidalian tradition (Mendizàbal and Albet, 2005).

In the 1990s, after joining the European Union and in the midst of a boom in academic and professional geography, some expected that increasing contacts with foreign universities (i.e. through the Erasmus Program) could lead to a more open attitude towards new forms of theoretical thinking in geography and enhancing new and critical approaches (García-Ramon et al., 1992; Gómez-Mendoza, 2001). That did not happen. The new curricula, which basically designed geography as training for jobs related to geographical techniques and planning, has maintained description as a fundamental feature of Spanish geography. This emphasis on quantitative methodologies and techniques has favoured a passive field, which does not question itself and has further subordinated geography to initiatives of public administrations and to private and business interests (Segrelles, 2002). Although with some significant exceptions, international contacts have generally not been a source of theoretical innovation, and the contribution of Spanish geographers to international research have mostly been in the form of case studies rather than conceptual reflections.

In some cases, "critical" geography has been understood merely as the introduction of new issues and methodologies – even humanistic geography has been considered as one of the present forms of critical geography (Benito, 2004). In the context of Spanish geography, so conservative and hostile to innovation, the introduction of new approaches could surely be taken as a critique of established methodologies and the rigidity of academic power (Santos, 2002). But such innovations are very often empty of ideological and even theoretical content, and they are far from the radicalism of earlier times in Spanish geography (Albet and Zusman, 2009; Ortega, 2007).

The "competitive" spirit and the search for "excellence" in academia have seduced and co-opted a significant part of formerly critical geographers. Expectations of professional stability and pressures to publish in mainstream and "indexed" journals have had a sedative effect on the critical role intellectuals and academics should perform. Although several groups have produced doctoral theses and researches with critical inspiration and with innovative methodologies and approaches, this is more an exception than a rule. Moreover, such research has a rather uneasy position in the present academic context.

The short-sightedness of established Spanish geography, and particularly the technical and applied narrowness of the new curricula, makes it difficult for critical voices to be heard. There is not a demand for critical perspectives among most students, and the academic context does not present many options to understand geography beyond techniques and descriptions. Of course, there are some new critical voices in Spanish geography, but it is revealing that in many cases they come from researchers with undergraduate degrees in parallel disciplines (philosophy, environmental sciences, political sciences, sociology) or from young geographers trained outside of Spain. In both cases, a probable explanation is that they have enjoyed a more open academic environments and less rigidly professionalized curricula (Clua and Zusman, 2002; Puente, 2011).

Among students, and more broadly in society, geography is not perceived as a discipline with critical potentials. Rather, the field is conceived as a descriptive and encyclopaedic science, mainly based on GIS and remote sensing. Spanish society neither expects nor demands from geography a critical appraisal of what happens in the global world or in the street nearby, despite its obvious spatiality: severity of social and economic crisis, geopolitical tensions, migrations, etc. (Albertos and Sánchez, 2014; Cairo, 2009; Ferrer and Gabrielli, 2018; Fraguas, 2016; Lois, 2010; Nel·lo and Durà, 2021; Nogué and Vicente, 2001; Méndez, 2004, 2011). Geography and geographers have a very limited presence in activisms, debates and research related to powerful discourses as the commons, *indignados*, the neoliberal political and economic crisis, new urban social movements, etc. (Boira, 2015; Casellas, 2016; Díaz-Cortés and Sequera, 2015; Murray, 2020; Nel·lo, 2015) and it is only somehow present in some major urban concerns such as housing, planning or gentrification (Jover and Almisas, 2015; Ortiz and Gómez, 2017; Tutor, 2019; Vives and Rullán, 2014) and through certain methodological proposals as citizen participation or critical cartography (Canosa and García, 2017; Font-Casaseca, 2020; Lladó, 2016). Nevertheless, "geography", as a perspective, it is not always appealed.

The loss of critical spirit, and the reign of contentment, fear, disappointment or simply lack of interest, is, admittedly, common features of our times. But in the case of Spanish geography, it is more problematic given its low profile in society, which leaves to other social sciences (mainly anthropology, sociology, architecture, urban planning, environmental sciences) the critical role that geography does not play, due to its long conservative path and its marginal role in Spanish society (Vez, 2005). Nevertheless, critical potentials are still present in Spanish geography. These potentials are ready to be used, not only as an antidote against the increasing hyper-specialization, which cannot explain nor transform present day complex spaces, but also as a way to build alternative proposals of civic consciousness, social and spatial justice, environmental balance, cultural identity and solidarity (Nogué, 2007).

References

Abalar. 2003. Manifesto. *Abalar* 0, 2–3.
Agüera, M. 2008. L'activisme femení en conflictes ambientals. Reflexions en clau feminista i apunts per a la gestió del medi. *Documents d'Anàlisi Geogràfica* 51, 13–37.

Albertos, J.M. and J.L. Sánchez (eds.). 2014. Geografía de la crisis económica en España. Valencia: Publicacions de la Universitat de València.

Albet, A. 1988. Valoració dels lligams entre Geografia Radical i Geografia Humanística. *Documents d'Anàlisi Geogràfica* 13, 5–18.

Albet, A. 2016. Enric Lluch i Martín (1928-2012). Geographers. Biobibliographical Studies 35, 95–119.

Albet, A. (ed.). 2019. *Maria Dolors García-Ramon. Geografía y género, disidencia e innovación*. Barcelona: Icaria.

Albet, A. and N. Benach (eds.). 2012. *Doreen Massey. Un sentido global del lugar*. Barcelona: Icaria.

Albet, A., N. Benach, L.-M. García-Herrera and X.-M. Santos-Solla. 2004. Del postmodernismo a las nuevas geografías culturales. *Treballs de la Societat Catalana de Geografia* 57, 141–58.

Albet, A. and P. Zusman. 2009. Spanish language geography. In, R. Kitchin and N. Thrift (eds.), *International Encyclopedia of Human Geography*. New York: Elsevier, vol. 10, pp. 296–301.

Alió, M.-À. and J. Bru. 1991–1992. L'esquerda ecologica: residus industrials i geografia humana. *Documents d'Anàlisi Geogràfica* 19-20, 11–31.

Anarco-Territoris: *Revista Anarquista de Pensament Territorial*. <http://www.berguedallibertari.org/anarco-territoris>.

Arnau, X., L. Calvo, Á. Girón and F. Nadal (eds.). 2007. *Ciència i compromís social. Élisée Reclus (1830-1905) i la geografia de la llibertat*. Barcelona: Residència d'Investigadors CSIC-Generalitat de Catalunya.

Artigues, A.-A. 2002. Capitalisme global, crítica postmodernista i pensament geographic. *Estudis d'Història Econòmica* 19, 3–35.

Ascon, R. 1990. La introducció dels conceptes de la geografia crítica a l'ensenyament secundari. *Documents d'Anàlisi Geogràfica* 16, 79–91.

Baylina, M. 2004. Metodología para el estudio de las mujeres y la sociedad rural. *Estudios Geográficos* 65, 5–28.

Baylina, M., A. Ortiz and M. Prats. 2006. Children in playgrounds in Mediterranean Cities. *Children's Geographies* 4, 173–83.

Benach, N. (ed.). 2012. *Richard Peet. Geografía contra el neoliberalismo*. Barcelona: Icaria.

Benach, N. (ed.). 2017. *William Bunge. Las expediciones geográficas urbanas*. Barcelona: Icaria.

Benach, N. and A. Albet (eds.). 2010. *Edward W. Soja. La perspectiva postmoderna de un geógrafo radical*. Barcelona: Icaria.

Benach, N. and A. Albet (eds.). 2019. *David Harvey. La lógica geográfica del capitalismo*. Barcelona: Icaria.

Benach, N. and A.-F.-A. Carlos (eds.). 2016. *Horacio Capel. La ciudad en tiempos de crisis*. Barcelona: Icaria.

Benejam, P. 1992. La didàctica de la geografia des de la perspectiva constructivista. *Documents d'Anàlisi Geogràfica* 21, 35–52.

Benito, P. 2004. Planteamientos críticos y alternativos en geografía. *Finisterra* 39, 47–62.

Blanco, M. 2010. Actualidad y vigencia del pensamiento geográfico de Élisée Reclus. *Treballs de la Societat Catalana de Geografia* 70, 225–36.

Boira, J.V. 2015. Deconstruyendo el mapa conservador de la geografía en el siglo XXI. *Boletín de la AGE* 67, 233–55.

Borja, J. 1974. Introducción. *Documents d'Anàlisi Urbana* 1, 11.

Bosque-Maurel, J. 1986. Presencia y significado de la revista *Geo-crítica* de la Universidad de Barcelona. In, A. García-Ballesteros (ed.), *Geografía y marxismo.* Madrid: Editorial de la Universidad Complutense, pp. 197–221.

Breitbart, M. (ed.). 1988. *Anarquismo y geografía.* Vilassar de Mar: Oikos-Tau.

Bru, J. 1996. Spanish women against industrial waste: A gender perspective on environmental grassroots movements. In, D. Rocheleau, B. Thomas Slayter and E. Wangari (eds.), *Feminist Political Ecology.* London: Routledge, pp. 105–24.

Cairo, H. 2009. Geopolítica crítica. In, R. Reyes (ed.), Diccionario crítico de ciencias sociales. Madrid: Plaza y Valdés & Universidad Complutense de Madrid, pp. 34–6.

Campos-Venuti, G. 1967. *Amministrare l'urbanistica.* Roma: Einaudi.

Canosa, E. and Á. García. 2017. Cartografías críticas de la ciudad. Treballs de la Societat Catalana de Geografia 84, 145–60.

Capel, H. 1974. *Capitalismo y morfología urbana en España.* Barcelona: Los Libros de la Frontera.

Capel, H. 1998. Presentación. *Scripta Nova. Revista de Geografía y Ciencias Sociales.* http://www.ub.edu/geocrit/sn-pres.htm (11 December 2020).

Casellas, A. 2010. La geografía crítica y el discurso de la sostenibilidad: perspectivas y acciones. *Documents d'Anàlisi Geogràfica* 56, 573–81.

Casellas, A. 2016. Desarrollo urbano, coaliciones de poder y participación ciudadana en Barcelona: una narrativa desde la geografía crítica. *Boletín de la AGE* 70, 57–75.

Casellas, A. and M. Pallarès. 2005. Capital social como estructura de análisis. Validaciones en perspectivas de género y territorio. *Cuadernos de Geografía. Universitat de València* 78, 177–90.

Castells, M. 1972. *La question urbaine.* Paris: François Maspéro (First Spanish translation: *La cuestión urbana.* Madrid: Siglo XXI, 1974).

Clua, A. and P. Zusman. 2002. Más que palabras: otros mundos. Por una geografía cultural crítica. *Boletín de la AGE* 34, 105–17.

Colectivo de Geógrafos. 1980. *La geografía al servicio de la vida (Antología). Eliseo Reclús.* Barcelona: Editorial 7 1/2.

Costa, P. 1980. *Nuclearizar España.* Barcelona: Los Libros de la Frontera.

Díaz-Cortés, F. 2012. Mujeres, barrio e investigación: ejercicio de autoreflexión desde una trayectoria investigadora y activista en Geografía (2002-2011). Revista Latinoamericana de Geografia e Gênero, Ponta Grossa 3(2), 30–48.

Díaz-Cortés, F., A. Albet and M.-D. García-Ramon. 2008. Old and new migrant women in Ca n'Anglada: Public spaces, identity and everyday life in the Metropolitan region of Barcelona. In, J.N. DeSena (ed.), *Gender in an Urban World.* Bingley: Jai-Emerald Press, pp. 263–84.

Díaz-Cortés, F. and J. Sequera. 2015. Introducción a 'Geografías del 15-M: crisis, austeridad y movilización social en España'. *ACME: An International Journal for Critical Geographies*, 14(1), 1–9.

Estalella, H. 1984. *La propietat de la terra a les comarques gironines.* Girona: Col·legi Universitari de Girona.

Estévez, B. 2012. La idea de espacio público en geografía humana. Hacia una conceptualización (crítica) contemporánea. *Documents d'Anàlisi Geogràfica* 58(1), 137–63.

Ferrer-Gallardo, X. and L. Gabrielli (eds.). 2018. Estados de excepción en la excepción del estado. Ceuta y Melilla. Barcelona: Icaria.

Font-Casaseca, N. 2020. Prácticas cartográficas para una geografía feminista: los mapas como herramientas críticas. Documents d'Anàlisi Geogràfica 66(3), 565–89.

Fraguas, R. 2016. Manual de geopolítica crítica. València: Tirant Humanidades.

Frutos, M.-L. 1980. Una penetración en España de la geografía radical. *Norba* 1, 99–122.

García-Ballesteros, A. (ed.). 1986. *Geografía y marxismo.* Madrid: Editorial de la Universidad Complutense.

García-Herrera, L.-M. 1988. El acceso al suelo de la clase trabajadora canaria: las parcelaciones marginales. *Ciudad y territorio: Revista de Ciencia Urbana* 75, 107–18.

García-Herrera, L.-M. and F. Sabaté-Bel. 2015. *Neil Smith. Gentrificación urbana y desarrollo desigual.* Barcelona: Icaria.

García-Ramon, M.-D. 1977. La geografía radical anglosajona. *Documents d'Anàlisi Metodològica en Geografia* 1, 59–69.

García-Ramon, M.-D. 1985. *Teoría y método en la geografía humana anglosajona.* Barcelona: Ariel.

García-Ramon, M.-D. 1989. Para no excluir a la mitad del género humano: un desafío pendiente en geografía humana. *Boletín de la AGE* 9, 27–48.

García-Ramon, M.-D. 2005. Enfoques críticos y práctica de la geografía en España. Balance de tres décadas (1974-2004). *Documents d'Anàlisi Geogràfica* 45, 139–48.

García-Ramon, M.-D., A. Albet and P. Zusman. 2003. Recent developments in social and cultural *geography in Spain. Social and Cultural Geography* 4, 419–31.

García-Ramon, M.-D. and J. Cruz. 1996. Regional welfare policies and women's agricultural labour in Southern Spain. In, M.-D. García-Ramon and J. Monk (eds.), *Women of the European Union. The Politics of Work and Daily Life.* London: Routledge, pp. 247–62.

García-Ramon, M.-D. and J. Monk (eds.). 2007. Feminist Geographies around the World (special issue). *Belgeo* 3, 247–398.

García-Ramon, M.-D. and J. Nogué (eds.). 2002. Geografies dissidents (special issue). *Documents d'Anàlisi Geogràfica* 40, 17–9.

García-Ramon, M.-D., J. Nogué and A. Albet. 1992. *La práctica de la geografía en España (1940-1990).* Vilassar de Mar: Oikos-Tau.

García-Ramon, M.-D., A. Ortiz and M. Prats. 2004. Urban planning, gender and the use of public space in a peripherical neighbourhood of Barcelona. *Cities,* 21, 215–23.

García-Ramon, M.-D., A. Ortiz and M. Prats (eds.). 2014. *Espacios públicos, género y diversidad: Geografías para unas ciudades inclusivas.* Barcelona: Icària.

Gavira, M. and E. Grilló. 1975. *Zaragoza contra Aragón.* Barcelona: Los Libros de la Frontera.

Gómez-Mendoza, J. 1988. Las expediciones geográficas radicales a los paisajes ocultos de la América urbana. In, J. Gómez-Mendoza and N. Ortega (eds.), *Viajeros y paisajes.* Madrid: Alianza Universidad, pp. 151–64.

Gómez-Mendoza, J. 2001. La Geografía española: final y principio de capítulo. In, *Actas del XVII Congreso de Geógrafos Españoles.* Oviedo: Asociación de Geógrafos Españoles, pp. 19–27.

Gómez-Mendoza, J. 2002. Disidencia y Geografía en España. *Documents d'Anàlisi Geogràfica* 40, 131–152.

Jover, J. and S. Almisas. 2015. Recuperando espacios y resignificando el concepto patrimonio desde los movimientos sociales. El caso del CSOA La Higuera (Cádiz, Andalucía). Documents d'Anàlisi Geogràfica 61(1), 91–112.

Lladó, B. (ed.). 2013. *Franco Farinelli. Del mapa al laberinto.* Barcelona: Icaria.

Lladó, B. 2016. Del qüestionari obrer a la cartografia militant. Quadern de les idees, les arts, i les lletres 205, 8–11.

Lluch, E. (ed.). 1981-1984. *Geografia de la Sociedad Humana*. Barcelona: Planeta.

Lois, M. 2010. Estructuración y espacio: la perspectiva de lugar. Geopolítica(s) 1(2), 207–31.

López, P. 1993. *Un verano con mil julios y otras estaciones: Barcelona, de la Reforma Interior a la Revolución de Julio de 1909*. Madrid: Siglo XXI.

March, H. 2013. Neoliberalismo y medio ambiente: una aproximación desde la geografía crítica. *Documents d'Anàlisi Geogràfica* 59(1), 137–53.

Martín, V. 1999. Los grandes propietarios de la tierra ante el desarrollo urbano-turístico en el Sur de Tenerife. *Ería* 49, 185–202.

Méndez, R. 2004. Geografía económica: la lógica espacial del capitalismo global. Barcelona: Ariel.

Méndez, R. 2011. El nuevo mapa geopolítico del mundo. València: Tirant lo Blanch.

Mendizàbal, E. and A. Albet. 2005. Una aproximació a la geografia dels Països Catalans (1985-2005). *Afers. Fulls de recerca i pensament* 50, 153–75.

Murray, I. 2020. De las geografías del capital a las geografías poscapitalistas. In, J. Farinós (ed.), Desafíos y oportunidades de un mundo en transición. Una interpretación desde la Geografía. València: Publicacions de la Universitat de València, pp. 285–306.

Nel·lo, O. 2003. *Aquí no! Els conflictes territorials a Catalunya*. Barcelona: Empúries.

Nel·lo, O. 2011. La ordenación de las dinámicas metropolitanas. El Plan Territorial Metropolitano de Barcelona. *Scripta Nova. Revista Electrónica de Geografía y Ciencias Sociales* 362. http://www.ub.edu/geocrit/sn/sn-362.htm (11 December 2020).

Nel·lo, O. (ed.). 2012. *Francesco Indovina. Del análisis del territorio al gobierno de la ciudad*. Barcelona: Icaria

Nel·lo, O. 2015. *La ciudad en movimiento. Crisis social y respuesta ciudadana*. Madrid: Díaz & Pons.

Nel·lo, O. and A. Durà. 2020. Geographical presences and absences. The role of Spanish academic geography in geopolitical debates. In, R. Lois (ed.), Geographies of Mediterranean Europe. New York: Springer, pp. 357–92.

Nieto, C. 2006. Las mujeres y el cooperativismo en los procesos de desarrollo local: algunos ejemplos de la provincia de Málaga. *Documents d'Anàlisi Geogràfica* 47, 31–52.

Nogué, J. 2007. Las otras geografías. *La Vanguardia (suplement Culturas)*, 273 (12 September), 2–3.

Nogué, J. (ed.). 2018. *Yi-Fu Tuan. El arte de la geografía*. Barcelona: Icaria.

Nogué, J. and J. Romero. (eds.). 2006. *Las otras geografías*. Valencia: Tirant lo Blanch.

Nogué, J. and J. Vicente. 2001. Geopolítica, identidad y globalización. Barcelona: Ariel.

Oliveras, X. 2010. L'arrelament al territori: una perspectiva anarquista. In, *Anarquisme i pobles*. Bellaterra: Federació d'Estudiants Llibertàries (UAB) & Edicions Anomia, pp. 9–22.

Ortega, J. 2007. La Geografía para el siglo XXI. In, J. Romero (ed.), *Geografía humana. Procesos, riesgos e incertidumbres en un mundo globalizado*, second edition. Barcelona: Ariel, pp. 27–55.

Ortega, N. (ed.). 1977. *Geografías, ideologías, estrategias espaciales*. Madrid: Dédalo.

Ortiz, S. and J.D. Gómez. 2017. La producción de un espacio cooperativo. *Boletín de la AGE* 73, 77–98.

Pallarès, M., M. Pallarès and A.-F. Tulla. 2003. *Capital social i treball de les dones als Pirineus. El cas de l'Alt Urgell.* Barcelona: Institut Català de la Dona.

Prats, M. and M.-D. García-Ramon. 2004. Emploi du temps et vie quotidienne des femmes adultes à Barcelone. *Espace, Populations, Societés* 1, 71–9.

Puente, P. 2009. Viajes por los paisajes urbanos posmodernos. O de cómo ubicarse en medio del caos. *Boletín de la AGE* 51, 275–304.

Puente, P. 2011. La reconstrucción de los enfoques críticos contemporáneos y el rol del espacio. Una visión desde la Geografía. *Documents d'Anàlisi Geogràfica* 57, 223–54.

Reche, A. and J. Rodríguez. 1978. La geografía radical: una nueva alternativa, un proyecto de trabajo. *Paralelo* 37(2), 47–56.

Rodó-de-Zárate, M. 2018. Hogares, cuerpos y emociones para una concepción feminista del derecho a la ciudad. In, M.G. Navas and M. Makhlouf (eds.), Apropiaciones de la ciudad. Género y producción urbana: la reivindicación del derecho a la ciudad como práctica espacial. Barcelona: Pol·len, pp. 45–74.

Rodríguez, J. 1979. *Radical Geography*: una nueva corriente de la geografía anglosajona. *Estudios Geográficos* XL, 213–22.

Romero, J. (ed.). 2004. *Geografía humana. Procesos, riesgos e incertidumbres en un mundo globalizado.* Barcelona: Ariel.

Sabaté, A. 2002. Rural development is getting female: Old and new alternatives for women in rural areas in Spain. *Antipode* 34, 1004–6.

Sánchez, J.-E. 1981. *La geografía y el espacio social del poder.* Barcelona: Los Libros de la Frontera.

Sánchez, J.-E. 1991. *Espacio, economía y sociedad.* Madrid: Siglo XXI.

Santos, X. 2002. Espacios disidentes en los procesos de ordenación territorial. *Documents d'Anàlisi Geogràfica* 40, 69–104.

Saurí, D. 1988. Cambio y continuidad en la geografía de los riesgos naturales: la aportación de la geografía radical. *Estudios geográficos* 49, 257–70.

Segrelles, J.-A. 1998. ¿Tiene sentido actualmente una geografía marxista en la universidad española? *Papeles de la Fundación de Investigaciones Marxistas* 10, 161–80.

Segrelles, J.-A. 2001. Hacia una enseñanza comprometida y social de la Geografía en la universidad. *Terra Livre* 17, 63–78.

Segrelles, J.-A. 2002. Luces y sombras de la geografía aplicada. *Documents d'Anàlisi Geogràfica* 40, 153–72.

Sevilla-Buitrago, Á. (ed.). 2016. *Neil Brenner. Teoría urbana crítica y políticas de escala.* Barcelona: Icaria.

Tello, R. (ed.). 2016. *Jean-Pierre Garnier. Un sociólogo urbano a contracorriente.* Barcelona: Icaria.

Tulla, A.-F. 1993. *Procés de transformació agrària en àrees de muntanya.* Barcelona: Institut Cartogràfic de Catalunya.

Tulla, A.-F., M.D. García-Ramon and H. Estalella. 2020. La geografia a la Universitat Autònoma de Barcelona: un projecte d'Enric Lluch (II). *Documents d'Anàlisi Geogràfica* 66(1), 3–23.

Tutor, A. 2019. Nuevas legitimidades en la arena urbana. *Crítica Urbana* 4, 31–34.

Vez, M. del [pseudonym of Abel Albet, Núria Benach, Anna Clua]. 2005. 'Conocimientos situados': reflexión sobre las geografías de la Geografía. (Crónica de un viaje al 100 Congreso de la Association of American Geographers). *Documents d'Anàlisi Geogràfica* 45, 131–8.

Vicente-Mosquete, M.-T. 1983. *Eliseo Reclus: La geografía de un anarquista.* Barcelona: Los Libros de la Frontera.

Vives, S. and O. Rullán. 2014. La apropiación de las rentas del suelo en la ciudad neo-liberal espanyola. *Boletín de la AGE* 65, 387–408.

Zoido, F. 2002. Geografía y territorio. El papel del geógrafo a escala local. In, M. Blázquez, M. Cors, J.M. González and M. Seguí (eds.), *Geografía y Territorio. El papel del geógrafo en la escala local*. Palma de Mallorca: Universitat de les Illes Balears, pp. 13–5.

Zusman, P. 2004. Activism as a collective cultural praxis: Challenging the Barcelona Urban Model. In, D. Fuller and R. Kitchin (eds.), *Radical Theory/Critical Praxis: Making a Difference Beyond the Academy?* Vernon and Victoria, BC: Praxis (e)Press, pp. 132–46.

14 The United Kingdom

Kye Askins, Kerry Burton, Jo Norcup,
Joe Painter, and James D Sidaway

Introduction

In 1885, incarcerated in a French prison cell, the Russian geographer-anarchist Pyotr Kropotkin, wrote 'What Geography Ought to Be', an impassioned plea for geographers to engage in the work of social justice and critical pedagogy (Kropotkin, 1885). His paper formed part of a report on Geographical Education to the Royal Geographical Society (Keltie, 1885). Following his release from jail the following year, Kropotkin moved to London where he was lauded by the Society. However, 'in spite of a close personal friendship with Scott Keltie (Secretary of the RGS from 1892 to 1915), Kropotkin declined the honour of being elected an official "fellow" of the group. Hostile towards any organization under royal patronage, but committed to the advancement of science, he nevertheless established a close working relationship with its members' (Breitbart, 1981, 143).

This vignette illustrates several themes of this chapter: the task of recovering histories of critical and radical[2] geography in Britain (so that Kropotkin's work has come to be widely read and celebrated, for example); the place of critical pedagogy in those histories; the importance of transnational connections (Kropotkin spent time in France, Canada and the United States, as well as England, Scotland and his native Russia); and the ambiguous and contradictory relationships between dissident and mainstream geography and geographers in Britain.

In what follows we will explore these themes by telling four stories about the development of critical geography in the United Kingdom. We examine the neglected story of a radical journal – *Contemporary Issues in Geography and Education* (*CIGE*) that was published between 1983 and 1990 and focussed on geographical pedagogy. We then return to the topic of the clashes, connections and accommodations between critical geography and the Royal Geographical Society, some 100 years after Kropotkin's involvement. Thirdly, we chart the emergence and development of the on-line Critical Geography Forum mailing list and some of the international networks it helped to engender. Finally, we highlight the lively contribution of activist and participatory geographies and geographers.

DOI: 10.4324/9781315600635-14

It is impossible to circumscribe a uniquely British manifestation of critical geography. Radical ideas, intellectual and political movements and influential individuals all circulate internationally as can be seen from the biographies of many notable British geographers, including David Harvey (Castree and Gregory, 2006), Doreen Massey (Featherstone and Painter, 2013), Derek Gregory and Neil Smith (Slater, 2012a). The radicalism of Smith and Harvey matured in and responded to American conditions and institutions, while Massey's first encounter with the work of French Marxist Louis Althusser also took place in the United States (at the University of Pennsylvania). The circulation and translation of theoretical and political writing from elsewhere has also been important (including the works of Marx, Gramsci, Foucault, Deleuze, Butler and many others). Many other stories could be told and trajectories mapped, such as the early radical geographer Keith Buchanan, who worked in South Africa, Nigeria and, from 1953 to 1975, in Aotearoa/New Zealand (Power and Sidaway, 2004).

Nevertheless, UK institutions, politics and practices have left their imprint on critical geography in various ways, as our stories will show. While recognizing the importance of the contributions of individuals such as Buchanan, Harvey, Massey and Smith, we set the development of critical geography in the United Kingdom in those institutional and political contexts. Our account does not seek to be comprehensive and is thus inevitably partial. Instead, we aim to supplement more conventional textbook narratives of disciplinary development with a collaborative reflection on some inspiring moments, people, debates and publications through which diverse and dynamic geographies and geographers have confronted uneven development, challenged voices of power and sought to enable other voices to be heard. We suggest that the chapter be read in tandem with Avril Maddrell's (2009) account of *Complex Locations: Women's Geographical Work in the UK 1850-1970* as well as Tim Hall's (2014) account of shifting research agendas.

We also seek to strike a balance between distinctively British vantage points and recognition that such positions are constituted out of uneven and sometimes hidden connections (one of the key insights and starting points of radical geography itself). Like Phil Crang, reflecting on the tangled history of British cultural geography, we do not want to write in a way that 'ignores the complex temporalities and geographies of intellectual change in favour of a singular, and singularly located, linearity' or to see trends that do not fit such a narrative as 'necessarily secondary to it (so that, for example, the contemporary emergence of Cultural Geography in Germany or Brazil is rendered as just a product of the time taken for the diffusion of Anglophonic Geographic thought, time-lags that are seen to identify how "behind the times" Geography outside the UK is)' (Crang, 2010, 193).

If the spatial focus of our account is necessarily porous, its temporal focus is also not straightforward. There is a long history of dissident geography in the United Kingdom, as Kropotkin's engagement with the

Royal Geographical Society reveals. We have chosen to focus principally on the histories and geographies of critical geography from the 1980s to the present day. Earlier developments in the 1960s and 1970s, often labelled 'radical' rather than 'critical' geography, are hugely important both in their own terms and for their role in inspiring and enabling subsequent intellectual and political movements. There are continuities, overlaps and affinities as well as tensions and differences between the radical geography of the 1960s and 1970s and the more diverse and diffuse critical geography of the 1980s onwards. However, since the former period has been widely written about elsewhere, our focus here is mainly on the latter, while recognizing that the two cannot (and should not) be sharply separated.

Historical geographies of critical geography

In some respects the trajectory of critical geography in the United Kingdom since the 1970s has been from the margins to the centre. The use and development of critical social theory, a focus on unequal socio-spatial power relations, and vocal commitments to emancipatory politics are now widespread in the discipline's books, journals, conferences, teaching and research. Many radical scholars (albeit disproportionately white and male) have progressed to leading positions in the discipline and in British academia more widely. In other ways, critical geographers, critical pedagogy and radical activism remain on the margins of the subject, especially where the challenges they pose cannot be easily co-opted and domesticated by the structures and practices of mainstream academia.

Reconstructing the historical geographies of critical geography demands attention to those marginal spaces and the recovery of sometimes fragmentary and ephemeral sources. As Maddrell (2009) shows, narratives of persons, publications, moments and movements are often complex, as people navigate their research and teaching across different institutions, temporary contracts, working beyond academia and across multiple forms of media. Many critical geographers transgress academic boundaries, and the impacts and effects of their activities risk being viewed as less significant because of their fractured geographies.

In exploring the less charted histories of critical geography, we are reminded by Wyse (2013, 5) that their archives often exist in the private communications of committee or editorial board members. These archival spaces afford a fertile tapestry of lesser-known accounts which, interwoven with better-known narratives, reveal a broad and more diverse tradition of critical praxis than might otherwise be apparent. Many accounts of pivotal publications, moments, activities and moves exist in the folk memories and informal histories of the subject. They are discussed in the hinterlands of discipline, on the fringes of formal conference spaces, teetering on the edge of footnotes, minutes of research group meetings or occasional newsletters unearthed in private archives of colleagues moving out of their offices and bundling them into recycling bins. Some 'critical' and 'radical' occurrences appear only in the anecdotes of the retired colleagues of those who have died before their accounts could be told in more formal ways.

Similarly, Rowbotham (1999) points out that there is often a threadbare existence of historical archives of radical and critical movements, in part explained by the imperative of presentism: what needs to be done now, what needs to be challenged and campaigned for to achieve desired outcomes? Looking back becomes wrongly equated with being 'backward-looking': politically reactionary, indulged only by those who are able to luxuriate in resources which might be better assigned to contemporary campaigns.

In both its marginal(ized) and more mainstream forms, British critical geography has been shaped by the social, political and institutional contexts in which it has developed. These include the close transatlantic cultural and political-economic relationship in which the United Kingdom was embedded during the twentieth century, the distinctiveness of British capitalism (its long history – the first 'industrial revolution' and long decline vis-à-vis other, later developers) and deep socio-spatial cleavages. Since the Second World War, the United Kingdom has experienced the end of (formal) Empire and the complex territorial structure of a less than stable state. The United Kingdom comprises several nations and regions whose status remains contested and whose relationships with each other, their European neighbours and the world beyond are mediated by imperial legacies.

These features of the British polity have underpinned the strong presence of geography education in schools (it was seen as a vital part of an imperial British education) and in many of the universities. These comprise a hierarchical and class-privileged system. At its apex are Oxford and Cambridge, with a secondary elite axis primarily made-up of members of the Russell Group. The Russell Group was established in 1994 to represent the interests of self-styled 'research intensive' universities. In addition to Oxford and Cambridge the Group includes Bristol, Cardiff, Durham, Edinburgh, Exeter, Glasgow, Queen's University Belfast and several London institutions including University College London and the London School of Economics. Of more than 125 British universities, the 24 Russell Group member institutions currently receive the majority of British research council funding (75%) and account for more than two-thirds of all doctorates awarded in the United Kingdom (Russell Group, 2013). They are also home to many of the best known and, by official measures, most successful geography departments.[3]

The structure has been complicated by successive waves of expansion of higher education and changes in the regulation and governance of universities. In the 1990s a large number of (usually) less research intensive, more teaching-orientated higher education institutions (Institutes of Higher Education and Polytechnics) were permitted to become 'new universities' – a policy move by a Conservative government intent on cheap but large-scale educational restructuring. For a number of years, Human Geography struggled to gain a foothold in the new universities; hampered by a 'lack of effective champions in the highest "corridors of power"' (Johnston, 2004, 57). However, the subject remained strong in Britain's schools, and it is in the schools that our story begins.

'The future is ours to create or to destroy ...'[4]

Contemporary Issues in Geography and Education (*CIGE*) was a radical journal launched in November 1983 by London secondary school geography teacher Dawn Gill. Walford (2000) notes, in a brief account of the journal's origins, that *CIGE* was published under the collective umbrella organization Association for Curriculum Development in Geography (ACDG), which was established to gain funding from the UK's Commission for Racial Equality (CRE). The CRE agreed to fund the distribution of Gill's research into what she found to be inherently racist resource materials used by teachers in London schools. Gill used this funding to reproduce her research in the pages of a new journal where she could promote an emancipatory and critically engaged geography alongside like-minded educators. In March 1983, Gill organized a conference entitled 'Racist Society: Geography Curriculum' at which her growing network of contacts could be seen. Speakers included geographers Derek Gregory, Rex Walford, John Bale, David Wright, Francis Slater, John Huckle and the Chair of the Inner London Education Authority (ILEA), Frances Morrell. Gill worked with and through education and campaigning networks of both ILEA and the Greater London Council (GLC) and was politically active in her local area of Stoke Newington and Hackney in north London. The political activism Gill encountered through her Hackney and GLC connection conjoined with her academic reading. Influenced and inspired by David Harvey's *Social Justice and the City* (1973) and 'radical' Anglo-American geographers involved with *Antipode* and the Union of Socialist Geographers, Gill sought collaborators sympathetic to the critical and radical aspirations covered by the nine main aims of *CIGE* and ACDG. These demanded critical and emancipatory geographical education for school students and called for anti-racist and anti-sexist reconstructed curricula that challenged racism and sexism and highlighted the unequal geographies of power evident in curriculum design and how students experienced and learnt about the world and their place in it.

Gill aimed to produce a campaigning journal written with the academic, educator and student in mind, providing practical resource materials that could be used both within and beyond school classrooms. Each issue was themed with specialist areas for topical debate to inform school teachers who might otherwise feel cut off from critical ideas and engagements. It was not intended to be solely a school teacher's magazine, but a publication space to enable geographers to find a shared voice against right-wing educational policies and proposals and to explore other ways of doing geographical education. Gill's collaborations with geographers in further and higher education, as well as artists and students helped to take the publication to a broader audience. *CIGE* archival papers document Gill seeking out advice from Ian G. Cook (who became co-editor of the series alongside Gill in April 1983), Roger Lee and John Huckle in the run up to the publication of the launch issue. Cook brought with him an invaluable network of academics sympathetic to Gill's ambitions. With Gill's personal and professional network of educators as well as her own political activism, *CIGE* acquired a vibrant and diverse readership and subscription base with international contributors from Australia, Africa and North America.

Eight issues of *CIGE* appeared in as many years. The first two focussed on anti-racism, followed by trade, aid and globalization, apartheid capitalism, ecological crisis, war and peace, gender and geography and anarchism and geography. Each issue contained contributions from academics and educators, as well as artists, activists and school students. The editorial board included subsequently well-known geographers such as David Pepper (who contributed articles to two journal issues and became co-editor alongside Gill and Cook from 1985), Ian G. Cook, Roger Lee, Sarah Whatmore, Peter Jackson, Julian Agyeman, John Huckle, Frances Slater, John Fein, Neil Larkin and Linda Peake. Contributors included Phil O'Keefe, Deborah Potts, Andrew Sayer, Tim O'Riordan, Beverly Naidoo, David Hicks, Barry Munslow, Carol Brickley, Michale Duane, Dennis Hardy, Colin Ward and Myrna Breitbart. Bill Bunge contributed articles and reviews in a number of issues, and *CIGE* was the only place where Gwendolyn Warren, Bunge's collaborator and Director of the Detroit Geographical Expedition and Institute, published in her own words (Norcup, 2015 also see Norcup 2019). The Women and Geography Study Group of the Institute of British Geographers contributed extensively to the 'Geography and Gender' issue. Edited by Jo Little and Sarah Whatmore, the issue contained contributions from Liz Bondi, Linda Peake, Sophie Bowlby, Jo Foord and Rachel Dixey. The artist Peter Kennard allowed his photomontage work to be reproduced gratis, including a 'photomontage essay' that runs throughout the War and Peace issue, and gave the ACDG permission to reproduce his work as a 'poster exhibition' supplementary resource pack for teachers' classrooms. Kennard's art complemented other notable imagery reproduced in the journal series including satirical cartoons by Roddy Megelly and reviews by activists, authors and students.[5] The journal's archive contains records of communications with *Antipode* (and the possibility of a collaborative venture) as well as potential commissions from geographers such as Doreen Massey. The journal's circulation amongst students and teachers across secondary and especially further education and university geography departments was well-known, its resources well used and anticipatory of critical endeavours in later years (Jackson, 1989a).

The *CIGE* correspondence archive underscores Gill's anticipatory vision for a critical geography as all the theme issue that comprise the entire series had been commissioned by the end of 1983. Had publication ambition been successful, the planned thrice-yearly publication would have seen all issues published by the autumn of 1985. *CIGE* was well-organized, with regular Saturday monthly meetings in the staff common room of the Geography Department at Queen Mary College (now QMUL) or at the Institute of Education. However, a lack of resources (personnel, financial) and its reliance on a small volunteer staff led to production time-lags. The *CIGE* archive reveals a broad subscription base spanning numerous university, college and public library institutions. Yet, its unreliable production time added to frustrations from subscribers and other geography publications began to adapt their content to the ideas *CIGE* endorsed. Addressing the role of geography within a pedagogy that responded to the threats and discourses of the Cold War, a 1987 issue on 'War and Peace' echoed the Canadian journal of *Issues in Education and Culture*'s special issue 'On Teaching Peace' (Dalby, 1986). Both these themed issues bought together academics, school teachers and activists to bridge the traditions of radical geography's engagement with peace education and emerging critical engagement with Cold War politics and geopolitical discourses (Burton and Megoran, 2013). The Canadian journal was the first of the two to suffer at the hands of educational reforms that later struck both sides of the north Atlantic.

Specific *CIGE* issues have subsequently taken on an afterlife. The introductory chapter on anarchism in Blunt and Wills' (2000) *Dissident Geographies* cites articles from *CIGE*'s 'Anarchism and Geography' issue, for example. Gill was herself linked into radical and critical geography networks that spanned the globe. Why, then, are Gill and the journal omitted from most recent accounts of the histories of both school geography education and radical/critical geography? It is evident that placing critical geography is itself a political process; temporal and spatial contexts matter. Who and how people write about the past reveals the geographies of power inherent in the realms of critical and radical geography and geographical knowledge production. Just as critical geographers reflect on their own activism in and beyond the academy, so there is a need to consider the historiography of critical and radical geography from a critical perspective: who decides whose stories are included and whose lives and activities remain muted, where are these accounts located, and how are they processed through editorial boards and the spaces of geographical publication.

While radicals of all stripes emphasize education and educational praxis, the geographies of critical geographical education below tertiary level are rarely considered in the historiography of the subject's radical past. *CIGE*'s publishing life, its afterlives and its archives reveal many pioneering school geography teacher- and community-led spaces for engagement with critical geographical education.

UK human geography and critical theory

If the story of *CIGE* reveals one of UK critical geography's hidden histories, a more widely known development was the emergence of the idea of critical geography as a counter to positivism, the quantitative revolution and spatial science. In the United Kingdom, the battle of ideas for and against spatial science was particularly intense at Cambridge University, which had been at the forefront of the quantitative revolution, but became a notable locus for the development of very different approaches to human geography. By the late 1970s 'radical geography' had already established itself in the United States and was increasingly influential in the United Kingdom. An early discussion of the related but specific and distinctive notion of 'critical geography' came in the work of the then-Cambridge-geographer Derek Gregory, whose *Ideology, Science and Human Geography* was a key text in the critique of positivist approaches. Gregory (1978, 170) employed 'critical geography' to refer to his vision of a politically-engaged, post-positivist human geography inspired in part by Frankfurt School critical theory. Gregory cited Bill Bunge and Gwendolyn Warren's Detroit Geographical Expedition and Paolo Freire's radical pedagogy as examples of what such a critical geography might involve.

In the early 1980s Gregory became the co-editor (with his Cambridge colleagues Mark Billinge and Ron Martin) of Macmillan's 'Critical Human Geography' book series. According to its blurb, Critical Human Geography was 'an international series which provides a critical examination and extension of the concepts and consequences of work in human geography and the allied social sciences and humanities'. Some 12 volumes were published in all, starting with *Conceptions of Space in Social Thought* (Sack, 1980), *Geography and the State* (Johnston, 1982), the first edition of Massey's *Spatial Divisions of Labour* (1984) and the influential *Social Relations and Spatial Structures* collection edited by Gregory and John Urry (1985). In these works, and others of the time, 'critical human geography' was marked by a new disciplinary engagement with critical social theory and conceptual critiques of positivism, extending and supplementing the increasingly well-established Marxist approaches. Not all contributors to the Macmillan series were Cambridge- or UK-based, and Cambridge was only one node in an archipelagic geography of the discipline's engagement with critical theory and radical ideas that developed in the 1980s and early 1990s.

One place on the margins of British academia which nurtured more than its fair share of critical geographers was Lampeter in west Wales. When Joe Painter moved to Lampeter's geography department at what was then St David's University College, he joined a small group of human geographers: Paul Cloke, Chris Philo, Phil Crang and Mark Goodwin. Others who later worked in Lampeter included Miles Ogborn, Ian Cook, Ghazi-Walid Falah, Tim Cresswell, David Atkinson, Ulf Strohmayer and Catherine Nash, while earlier members had included Nigel Thrift, Jo Little and

David Sadler. Reinforced by frequent visitors, in that small Welsh town it seemed as if a broadly defined critical geography was virtually hegemonic, a perception reinforced by the Department's undergraduate curriculum.

A volume of essays compiled by Lampeter geographer Chris Philo (1991), *New Words, New Worlds: Reconceptualising Social and Cultural Geography*, disseminated the proceedings of a conference in Edinburgh and did much to renew a critical spirit in British cultural geography. Other sites also strengthened a culturally inflected Marxism in geography, such as University College London, from where Peter Jackson (1989b) authored *Maps of Meaning*, a textbook drawing on the legacies of socialist thinker Raymond Williams and the cultural theorists of the New Left, notably Paul Gilroy and Stuart Hall. The New Left was actively involved with left-leaning social and labour movements during the 1960s–1980s. Academically, it was strong particularly in the polytechnics and the new wave of universities founded after 1960 that were meeting an increased demand for higher education. Whilst Doreen Massey had a long association with the New Left, geography as a discipline was more weakly linked with the movement than disciplines such as sociology and cultural studies. The majority of geographers were in departments located in the older and less politically engaged universities established before the First World War.

Recalling the writing of the *Maps of Meaning*, Jackson notes how it had been informed by a semester in Minnesota and participation in a reading group there. This was co-convened by one of the authors of the North America chapter in *Placing Critical Geographies* and is further testimony to the transatlantic and unbounded influences on 'British' critical geography. Jackson (2005, 746) claims that it also:

> ...benefitted enormously from the intellectual climate at UCL in the mid-1980s where an exceptional group of graduate students worked closely with academic staff, opening up a space for the exploration of feminist, postcolonial and queer theory that was all too rare in other departments at the time.

Radical political economy was another key axis for the development of British critical geography in the 1980s and 1990s. Much of this work was focussed on developing a critique of the uneven and unjust geographies of Thatcherism and its aftermath. James Anderson, Simon Duncan and Ray Hudson were the editors of a landmark set of essays on *Redundant Spaces in Cities and Regions: Studies in Industrial Decline and Social Change* (Hudson et al.,1983), followed soon after by two volumes in the Macmillan Critical Human Geography series: Massey's highly influential *Spatial Divisions of Labour* (1984) and *Geographies of Deindustrialization* (1986) edited by Ron Martin and Bob Rowthorn. These books arose out of a conjuncture of industrial restructuring, financialization and the policies of the Thatcher government that were reworking deeply rooted structural and spatially mediated features of British capitalism, through combinations of force and a fragile hegemony into a neoliberal formula.

These diverse engagements between geography and various strands of critical theory suggest that critical geography has been an intellectual project as much as (or sometimes even more than) a political one. Of course, other sites would reveal different emphases: Loughborough and Nottingham in the 1980s for materialist work on landscape (by Denis Cosgrove, Stephen Daniels and others) is one example. The pioneering distance-learning institution, the Open University (that was established under a Labour government in 1969 and always had a progressive character) is another. The OU's geography department (long a home to Doreen Massey amongst other critical geographers) built its reputation around critical theory and stances (for example, in the 1999 book, *Human Geography Today*) and did so through both teaching (and its particular form of distance learning pedagogy around course books) and research. David Harvey's temporary return to the United Kingdom, as the Halford Mackinder Professor at Oxford University (in the 1980s and early 1990s), saw another critical nucleus emerge through his graduate students and colleagues such as the Belgian Marxist, Eric Sywngedouw. *New Models in Geography* was another landmark. Edited by Bristol-based Nigel Thrift and Richard Peet at Clark University in the United States, on publication in 1989 *New Models* set out the impacts of Marxism and radical geography. Thrift and Peet's collection explicitly contrasted itself with *Models in Geography*, edited by Richard Chorley and Peter Haggett (1967), a book that had helped codify positivism for the discipline 20 years before.

Work on regulation, neoliberalism and locality developed in Manchester in the early 1990s (especially by Jamie Peck and Adam Tickell), as well as the first strands and initially relatively isolated stirrings of queer geography, such as Gill Valentine (then in Reading) or David Bell (then a graduate student in Birmingham). Or harder to pin down to a single site, the immensely productive Danny Dorling's (first at Newcastle, then later at Bristol, Leeds, Sheffield and Oxford) sustained commitment to reinvigorate the study of geographies of wealth, poverty and welfare, that an earlier generation of radicals (notably David Smith at Queen Mary College, University of London) had developed in the 1970s (Smith, 1977).

There may have been diverse currents and a multiplicity of sites, but for many UK geographers, by the early 1990s 'critical geography' and 'human geography since positivism' had come to mean more or less the same thing. Where and when this intellectual project intersected with a variety of forms of activism, it can arguably be regarded as a single tradition of radical and critical geography, albeit one that is broad, loose and highly internally differentiated (and sometimes conflicted). This expansive definition of *actually existing* critical geography may seem like a lazy and/or complacent position and one that is easy for 'insiders' to occupy. It is, though, widespread amongst UK-based geographers and is often motivated by a desire to be inclusive. By the late 1990s, post-Marxist, poststructuralist and feminist research had firmly established critical geography as an intellectual engagement with influence beyond geography. Whilst not wishing to focus only on key texts, the Critical Geographies book series, edited by Tracey Skelton and Gill Valentine (with early input from Sally Lloyd) is worth noting. The series published 20 inter-disciplinary titles between 1999 and 2005, grounded in social theory the titles introduced a number of emerging critical geography strands, including *Entanglements of Power: geographies of domination and resistance* (Sharp et al., 2000), *Mind and Body Spaces: geographies of illness, impairment and disability* (Butler and Parr, 1999), *Children's Geographies: playing, living and learning* (Holloway and Valentine, 2000), each of which retain an influence on the discipline.

The critique of geographical positivism and spatial science that began in the 1970s has given rise to a wealth of geographical scholarship informed by (and contributing to) an enormously diverse body of critical social theory. The rise of post-structuralism in the discipline has led to further diversification, with geographers engaging in a variety of ways with the work of Baudrillard, Butler, Deleuze, Foucault, Haraway, hooks, Irigaray, Lacan, Latour, Spivak and many others. Not all of this work can straightforwardly be identified as critical geography, even on a broad definition of that term, and for a number of self-positioned critical geographers some of this work may be perceived as uncritical and apolitical or as an apology for neoliberalism.

One strand of work that has sometimes formed a lightning rod for such disputes (though it is much more than that) is non-representational theory (Anderson and Harrison, 2010) which draws attention to lived practices, bodily experience, and the affective materialities of human and more-than-human worlds. While many have been sceptical, Woodyer and Geoghegan's (2013) discussion of enchantment suggests ways in which the concerns of non-representational theory might re-animate the critical in geography, in part through an engagement with the 'magical Marxism' espoused by Andy Merrifield (2011). Meanwhile, geographers Ash Amin and Nigel Thrift have explored the implications of affective politics and post-humanist materialism for the Left more generally in *Arts of the Political* (2013). Amin and Thrift have made major contributions to human geography, critical theory and the wider social sciences, many of which resonate with the core concerns of critical geography. However, their interventions have sometimes been greeted with scepticism (and even on occasion with hostility) by some, who see them as conceding too much ground to neoliberalism, writing from positions of power in relatively un-reflexive ways, and/or for using theory to define politics for the Left, rather than drawing on contemporary Left political practice to inform theory and possible future worlds (Barnett, 2013; Featherstone, 2013; see also Amin and Thrift, 2005 and responses).

If 'critical geography' is located at the overlap between geography and critical theory, then it is a very diffuse and loosely defined field, and perhaps too diffuse to be meaningful; if there is nothing much outside the category, then how is the category helpful? For many, activism of some kind (whether in the classroom, the academy or beyond) remains an essential component of critical geography, though this insistence may be tempered with a reluctance to exclude those who share similar political goals but don't consider activism to be their forte. In similar terms, Noel Castree (1999, 2000) documents how 'professionalization' has, to some extent, displaced, or been in lieu of, public activism for many leftist geographers, though attention has been directed to the nature and structure of labour and knowledge in the academy, including the 'scientific' institutions long associated with geography. We return to issues regarding 'scholar activism' later.

Transatlantic circulations

As we have already seen, the growth of critical geography in Britain cannot be separated from developments elsewhere and particularly North America. The complex transatlantic circulation of radical geography in the 1970s merits more reflection than space here permits. However, a good number of UK-based geographers were reading and actively involved in publishing and editing *Antipode*. Others, such as David Smith (1971), were exposed to radical geography by attending the annual Association of American Geographers' meetings and were inspired to report on what they had witnessed as 'the next revolution' [in geographic thought] in the journal *Area* (then still the newsletter of the Institute of British Geographers). From his base at Queen Mary College of the University of London (located in London's East End, long a domain of migrant populations and where relative poverty is in stark contrast to the nearby financial core of the City of London), David Smith (1974, 1977) developed a 'welfare geography' focussed on 'who gets what, where and how?'

As Linda Peake and Eric Sheppard note in the North America chapter in this volume, many British geographers who returned to teach in the United Kingdom after postgraduate training in North American universities brought radical approaches into their research and teaching. This flow was important because, notwithstanding isolated radical presences (mostly outside the universities), such as Kropotkin from 1886 to 1917 and the texts on socialist geography written by the journalist and illustrator J.F. ('Frank') Horrabin in the 1920s (Hepple, 1999), there were few links between the political left and British university geography before the 1970s.

During the 1970s a number of UK-based geographers subscribed to the newsletter of the Union of Socialist Geographers (USG) and a UK affiliate of the USG was established in 1978. The group's founding meetings took place in the evenings of the annual meeting of the Institute of British Geographers in January 1978, where there was lively debate about the extent to which it should be autonomous from the North America-centred original and what kind of structure it should have.[6] One of the corresponding coordinators,

Colm Regan, was based in Dublin and this may have been one reason why the initial report of the meeting in the USG newsletter termed the organization 'a Union of Socialist Geographers...for the British Isles (apologies for the dubious geographical title)' (Regan, 1978, 54). Subsequently, a London group was established and in 1983 published a booklet on socialist perspectives on the relationship between human and physical geography (this was edited by James Mackie, Terry Cannon and Malcolm Forbes). The London School of Economics became one key site of discussion and meetings, but geographers and planners at the polytechnics in London were also involved, and there were some links to the radical secondary school geography teachers whose role we have already sketched.

At the other end of the United Kingdom at the University of St Andrews in Scotland, Joe Doherty (one of *Antipode*'s editors from 1986–1992) had also established a USG node and was a source of inspiration for some in London. The USG's links to activist politics beyond the classroom and academy were modest, however, and after a few years the London group ceased to meet. At about the same time, the labels of socialist geography and radical geography started to be supplanted by references to critical geography. Feminist geography too emerged around the time that the British and London USG groups convened, although as Sophie Bowlby and Jackie Tivers have noted, there was sometimes heated debate between women seeking to establish the relative autonomy of questions about patriarchy and some (mostly, though not exclusively, male) Marxists 'who saw feminism as a distraction from the "true cause of socialism"'. This was quite ironic, considering that feminist geography in the United Kingdom had been born, primarily, out of Marxist geography (Bowlby and Tivers, 2009, 62). In south-east England, Jo Foord, who had been a member of the USG before becoming 'much more involved in the beginnings of the Women and Geography Study Group', notes how 'the existence of a putative USG was instrumental in keeping the idea of socialist feminism alive in the discipline in the late 1970s' (email communication with the authors: 1 November 2012).

Storming the citadel: The RGS-IBG merger and the critical geography forum

While there is more to the history of critical geography in Britain than institutional controversies, the 1995 merger between the Institute of British Geographers (IBG) and the Royal Geographical Society (RGS) was a significant event. Founded in 1830, the RGS was centrally involved in the colonial exploration of Africa, Asia and the Polar Regions. The IBG was formed in 1933 to cater to the growing cadre of university geographers who felt their interests were poorly served by the RGS, which, despite its support for the establishment of university geography departments, remained overwhelmingly concerned with overseas travel and exploration. While scarcely a bastion of radicalism, through its study groups, conferences and journals the IBG provided a set of spaces in which critical approaches to a range of geographical concerns became increasingly prominent during the 1970s and 1980s. The IBG's Women and Geography Study Group was formed in 1980 with explicitly feminist aims, and its 1984 book *Geography and Gender: An Introduction to Feminist Geography* quickly became a key reference point for critical scholarship. The IBG also provided one forum for critical debates about the impact of Thatcherism and what would later come to be labelled as neoliberalism. For example, the 1988 annual conference included conference sessions on Britain's growing north south divide, and papers on that theme and related topics such as privatization and economic restructuring appeared in the pages of the Institute's flagship journal, *Transactions. Transactions* also began occasionally to publish the work of feminist geographers and papers on political ecology, though for many radical and critical geography journals such as *Antipode* (founded 1969), *Environment and Planning D: Society and Space* (founded 1983) and subsequently *Gender, Place and Culture* (founded 1994) provided more productive outlets.

When the IBG put forward proposals in 1993 to merge with the RGS, many critical geographers were strongly opposed to the move. Opponents of the merger raised a range of objections. Many regarded the RGS as not just complicit in British imperialism, but one of its architects and the institutional embodiment of 'geography militant' (Driver, 2001). The organization was also noticeably conservative and masculine. The Society's Royal Charter symbolized its connections to the British establishment, as did its grand, if then somewhat run-down, headquarters next to the Royal Albert Hall in Kensington. Its links to the country's commercial, political and military elites were perhaps no longer what they had been in its Victorian and Edwardian heyday, but remained notable nonetheless. On the other hand, the presence of the more academically credible IBG meant that the RGS was not so well connected to the academic and scientific establishment as other learned societies.

The RGS was, though, a much larger organization than the IBG, with a membership of some 12,000 Fellows compared with the IBG's 1,700. Moreover, a large proportion of RGS Fellows were not academic geographers and many were not involved professionally in geography at all. The perception among many critical geographers in the IBG was that the RGS was less a learned society than an old-fashioned club catering to those with a travelogue view of geography of an often reactionary and distinctly orientalist kind. The difference in size thus raised concerns among some IBG members that academic geographers, and particularly politically committed critical geographers, would be in a small and weak minority in the merged organization. Others argued to the contrary that the merger would turn out to be a reverse take-over, with the RGS having to adapt and change in response to the demands that would be placed on it by the academic membership. As the Director of the merged organization later reflected, 'the memberships of both had concerns about "take-over", probably in equal measure, as well as differing expectations of future development and change'. (Gardner, 2005, 9). Some 15 years after the merger, more conservative and mostly non-academic members mobilized to foreground the Society's traditional 'exploration and discovery' role, but were ultimately outvoted – with the support of most of the professional staff in the Society itself (Maddrell, 2010).

The RGS's corporate links were a particular source of concern to critical geographers. On the same day in June 1993 that an Extraordinary General Meeting of the RGS voted overwhelmingly to approve the merger with the IBG, the Society honoured the mining and minerals company RTZ Corporation (now Rio Tinto) with the Geographical Award 'for the support and encouragement of expeditions'. RTZ provided financial support via the RGS for student fieldwork and expeditions, while being the target of numerous protests by activists concerned about its record on environmental issues and labour rights. Other corporate benefactors included Rolex, Land Rover and most notably the oil company Shell. While some critical geographers were opposed to any form of corporate sponsorship on principle, the main objections were raised to relationships with companies whose commercial activities were thought to be ethically or politically unacceptable. Such objections were twofold: that the money involved was partly attributable to environmental or economic exploitation or political repression, and that the fact of sponsorship would provide both legitimacy for the company and risk reputational damage to the Society. However, it was not until after the merger had taken place that concerns over corporate sponsorship really gained momentum.

There had been a heated but inconclusive debate about the prospective merger with the RGS at the Annual Conference of the IBG at Royal Holloway, University of London in January 1993. In November 1993, following the RGS vote in favour of the merger, the IBG sent out information about the proposals to its own members. After discussion of the issue at the IBG conference in Nottingham in January 1994, the merger proposals were put to a postal vote of the Institute's membership and were approved by a substantial majority of 626 to 261 (Tickell, 1999). On 1 January 1995, the Royal Geographical Society (with the Institute of British Geographers) came into being. A few days later, on Thursday, 5 January, an informal meeting of critical geographers was convened in a bar of Northumbria University during the annual conference of what was now the RGS-IBG. No formal record of the meeting was kept and memories are now hazy, but topics of debate included the formation of a network of critical geographers (and possibly of a separate organization for critical geography) and whether those opposed to the merger should remain within the RGS-IBG.

The still relatively novel technology (even for academics) of the email discussion list offered a seemingly straightforward way to form a network without the administrative burdens of a separate organization, and in August 1995 the Critical Geography Forum (CGF) discussion group began life as crit-geog-forum on mailbase, the UK's national academic mailing list service (now jiscmail). According to the original version of the introduction file:

> 'Critical Geography Forum' is a discussion group which uses electronic mail technology (hence 'e-mail list') to allow members to share ideas, raises questions, provide answers and air opinions. It has been set up in the wake of the merger between the Institute of British Geographers (IBG) and the Royal Geographical Society (RGS) to provide a forum for discussion among critical and radical geographers. While the immediate impetus was opposition to the merger among many critical/radical geographers, the list may be used for the discussion of any topic related to critical/radical geography of a substantive as well as an institutional nature.

With hindsight, the blandly apolitical tone is striking – but at the time it was chosen deliberately to include and encourage a diverse range of participants and to avoid pre-empting the issues for discussion. As there was no pre-constituted community of UK critical geographers to which the discussion group would give voice, the optimistic aim was to allow a network to constitute itself first and then to express its views. However, it was not long before an issue emerged around which many members of crit-geog-forum could unite.

In the two years after the merger, some of the worst fears of critical geographers about the new organization seemed to be realized. Although the activities of the former IBG, including the study groups, journals and conferences continued more or less unchanged under the auspices of the RGS-IBG's new and relatively autonomous Research and Higher Education Division, renewed concerns were raised about the Society's corporate sponsorship. Shortly after the merger in 1995, environmental campaigners and organizations including Sir David Attenborough, Friends of the Earth and the World Wide Fund for Nature wrote an open letter to RTZ in protest against its plans to mine for mineral sands in the forests of Madagascar. According to the *Guardian* newspaper, the RGS's map librarian demanded that a copy of the letter be removed from an exhibition at the Society about endangered forests. Although the Society disputed the accuracy of the *Guardian* story, the response from the Society to one letter of protest from a geographer pointed out that 'RTZ provide a substantial element of funding in support of undergraduate expeditions [...]. Last year this amounted to some 10,000 pounds, and over the years numerous undergraduate expeditions, including some I imagine from [your university], have been beneficiaries'. The unsubtle implication was that by raising complaints against RTZ critical geographers were jeopardizing financial support for their own students. Rio Tinto remains a target for protestors around the world – including at the 2012 London Olympics for which the company provides the winners' medals.

The biggest objections, though, related to the oil company Shell. On 6 November 1995, David Gilbert alerted the CGF to the death sentence passed by the military regime in Nigeria on nine political activists including Ken Saro-Wiwa, author and long-term critic of Shell's links to the Nigerian government and of its activities in the Ogoni region of the Niger delta. Gilbert later described the circumstances in article published in 2009 (in which he argued for a renewed campaign against the relationship between Shell and the RGS-IBG):

> Saro-Wiwa had been a fierce critic both of the Nigerian junta and of Shell, the dominant multinational company in Nigeria [...]. He argued that the company's operations had devastated the environment of the Niger Delta, and that the Ogoni people had received little or no economic benefit from oil extraction. Overnight the Ogoni dispute was transformed from a minor item of African news to the headline story covered by the international press and broadcasting organisations. Shell came under intense pressure from environmental organisations, facing consumer boycotts and protests at shareholder meetings.
>
> (Gilbert, 2009, 522–523)

In the few days before Saro-Wiwa and his fellow activists were killed on 10 November, the CGF discussion group was used to help organize protest letters and petitions against the sentences. After the executions had taken place, attention shifted to Shell's sponsorship of the RGS-IBG, with dozens of posts to the list protesting about the relationship and proposing actions to end it. As Gilbert notes, 'the situation at the RGS-IBG was also transformed. It was not so much that the Nigerian situation provided the opportunity for a concerted campaign on the sponsorship issue, [as] that many geographers reacted with simple disgust at the continuing public relationship with Shell' (Gilbert, 2009, 523).

Following discussions via the CGF, at the RGS-IBG conference at Strathclyde University in Glasgow in January 1996, a proposal to terminate the Society's relationship with Shell was passed by 157 votes to 10. Although not binding on the merged organization, the vote gained considerable international press coverage and put the Ogoni issue back into the media spotlight. In response, the Council of the RGS-IBG asked the prominent environmentalist and former RGS President Crispin Tickell to establish a working party to look into the whole issue of sponsorship and a conference was organized on Petroleum and Nigeria's environment. However, when the matter came before the Council of the RGS-IBG in June 1996, it was decided to continue the sponsorship arrangement. The Working Group met only twice – and it was shut down soon after. A special general meeting of the RGS-IBG followed, the vote favoured continuing the relationship with Shell by 4,309 votes to 1,509 (Wojtas, 1997), a decision widely considered to illustrate the gulf between academic and non-academic membership of the RGS. A number of academic members of the Society resigned in protest, while other critical geographers opted to remain in the RGS-IBG in the hope that change could be promoted from within (Gilbert, 1999, 2009; Tickell, 1999).

As Gilbert points out, although Shell has publicly embraced corporate social responsibility, it is 'both still mired in long term disputes with local groups in many parts of the world, and remains fundamentally a giant machine for turning oil into energy, profit and carbon dioxide' (2009, 525). In August 2007 the ongoing Shell sponsorship was once again raised at the RGS-IBG Annual conference. A session titled 'Corporate involvement in Geography: ethics, power and responsibility in our workplaces' organized by Paul Chatterton and Larch Maxey asked renewed questions on the continued links with Shell. The session led to a leafleting campaign across the conference denouncing the links and subsequent meetings with RGS management. The session also focussed on debates around British based academic publisher Elsevier's role in organizing the largest arms fair in the world, the London based annual Defence Security and Equipment International (DSEI). In 2009, Shell agreed an out of court settlement in a case brought by relatives of Saro-Wiwa and the other campaigners executed by the Nigerian authorities. Elsevier ceased its involvement in DSEI after pressure from academic journal editors and calls for a boycott of journals and other texts published by Elsevier (although it cut across disciplines, for the debate in geography, see Chatterton and Featherstone 2007, Hammett and Newsham 2007 and Kitchin, 2007). By 2017, neither Shell nor RTZ remained on the RGS-IBG's list of corporate supporters.

Whilst some critical geographers have eschewed any direct involvement with the RGS-IBG on principle, the Society's status as the principal collective institutional voice for academic geography has led many others to participate in its activities. The relative autonomy of the Research Groups,[7] the editorial independence of its journals (several of which have been edited by critical geographers) and the lively involvement of critical geographers in the Annual International Conference of the RGS-IBG mean that critical geography has a strong presence in the Society's Research and Higher Education Division. British critical geography has thus been partially professionalized and assimilated into the academic mainstream, though forms of activist and participatory geography (see below), building on significant bodies of critical work outlined earlier in this, and other, chapters, demonstrate continuing radicalism that engages both with and beyond the disputes about internal politics of the RGS-IBG.

In the wake of the Shell controversy, there were some limited efforts to form a separate organization for critical geography. Some argued that this would be a helpful step anyway, regardless of debates about the RGS-IBG. Others doubted whether a formal organization was necessary or even desirable. They suggested that most of the benefits of such a move were realizable through a more informal network, such as that provided by the CGF, and that institutionalizing critical geography could be regressive, leading to new hierarchies and exclusions. Most of the debate took place on the CGF discussion list and all the postings from March 1996 onwards are publicly available at www.jiscmail.ac.uk/crit-geog-forum.

A key strand in the discussions triggered by the RGS-IBG merger was the importance of building a network of critical geographers that was genuinely international in membership and outlook. Although the Critical Geography Forum discussion list had been set-up to respond to UK concerns, it immediately acquired international subscribers. Several subscribers in the United States and Canada, including Nick Blomley, Geraldine Pratt and Neil Smith, offered to act as North American contacts. There had already been discussion among critical geographers in North America about the possibility of organizing a conference of critical geography in Vancouver, Canada. In March 1996 Nick Blomley alerted the CGF to this suggestion and invited discussion about the format and purpose of such an event. A formal Call for Participation was issued in June and the Inaugural International Conference of Critical Geography took place in Vancouver in August 1997. A Steering Committee was formed following the Vancouver meeting to enable further events that were organized in the name of the International Critical Geography Group.

By late 2012 the CGF list had over 3,000 subscribers. Much of the traffic on the list now comprises routine announcements about job vacancies and conferences, though political discussions continue to develop from time to time. For example, it was used to help support Chilean critical geographers (Hirit and Palomino-Schalscha, 2011) who objected to the siting of a conference of the International Geographical Union in the military geographical institute in Chile (where Pinochet's regime once had a core base). The list was also used to publicize the 2010 campaign regarding the balance of research, critical and 'exploration' activities within the RGS-IBG.

Participation, publics and everyday (academic) activisms

Whilst recognizing diverse strands of critical geography, including those we have not considered in depth (feminism, queer theory[8], critical race theory[9] and postcolonial theories, for example), our final section focusses on the debates that emerged since the early 2000s around participatory ways of working. A few geographers had long been using participatory approaches, drawing on a well-established Participatory Action Research (PAR) paradigm grounded in the theory-practice of academic activists and educators such as Paulo Freire, Augusto Boal and Robert Chambers. These approaches gained momentum among UK-based geographers in recent years; especially those disillusioned by the ability of mainstream research projects to effect changes and be inclusive, even where findings are disseminated to policymakers. Key to this participatory impulse is the conviction that those 'beyond' the academy should be more involved in research project design, analysis and dissemination, which (continues to) resonate with renewed concerns about the relevance of the discipline and the 'public university'. Participatory research emphasizes the co-production of knowledge, such that academic knowledge is part of but decentralized in research, in efforts to enact positive social change grounded in communities (Kindon et al., 2007).

The move was internationally driven and supported, inspired by research and collaborations from numerous places, by academics, practitioners and communities. In 2005, steered by Rachel Pain, Duncan Fuller and Paul Chatterton, the RGS-IBG was petitioned to support a Participatory Geographies Working Group (PYGYWG), whose original objectives included:

- increasing the understanding and deployment of participatory principles throughout all aspects of higher education academic geography;
- stimulating and developing critical debate about participatory approaches within and beyond geography;
- encouraging the development of collaborative links within and beyond the academy, and working with non-academic organizations as partners in participatory ways; and
- ensuring that participatory research is firmly linked to debates around public policy, through meaningful collaboration with policy makers, the voluntary sector, activist and interest groups and other vehicles for social action.

(http://www.pygyrg.co.uk/about-us/)

Membership of PYGYWG was open to all with an interest in participatory working, in and beyond the academy.

Participatory approaches, and specifically PAR, have been critiqued as a modernist/instrumentalist paradigm, particularly in relation to their deployment by some development agencies and policymakers. Indeed, when reduced to a set of tools designed to shape public perceptions rather than facilitate more equitable engagement, non-critical processes serve to reiterate rather than subvert hierarchical relations between 'researcher and researched' (see Cooke and Kothari, 2001). Participatory geographers have responded by acknowledging complex relationships to power, emphasizing PAR's performative potential for disrupting such power relations (Cameron and Gibson, 2005). Certainly, PYGYWG offered a forum to open up debates regarding ethical practice, and further question dominant conceptions of research governance via the promotion of a broader vision of collectively negotiated and emergent subjectivities.

In 2007, PYGYWG members voted to apply to the RGS-IBG for Research Group status (Working Groups are 'limited life' entities, and PYGYWG's time was up). The vote was close, as many members struggled with the ethics and politics of being incorporated within a hierarchical institution (an ongoing debate from the time of the Group's formation). Disquiet was further fuelled when the RGS-IBG rejected the initial Constitution put forward, because of its references to activism. The RGS-IBG argued that an explicit commitment to activism risked jeopardizing its charitable status, and insisted that Research Groups should align with the purely educational aims and remit of the Society. The Group ultimately voted to amend its Constitution: 'activism' was replaced with the 'aim to challenge oppressive and unjust social and spatial processes, and pursue social change on these grounds as a goal'. An understanding was reached with the RGS-IBG that, whilst the new Research Group (PYGYRG) could not itself undertake activism, it could (and does!) organize events in which individual members of the Group consider activisms undertaken in their own right as academics, practitioners or communities.

This pragmatic approach has enabled PYGYRG to continue its support for a wide range of members and activities, whilst remaining critical and reflexive. For current members, participatory approaches continue to be implicitly and explicitly linked with activist geographies, engaging directly with communities in efforts towards meaningful social change. Alongside PYGYRG's development there has been an increasing focus in the United Kingdom on what 'activist geographies' might encompass: as multiple and emergent spaces, operating through and co-constructive of diverse geographies. This work highlights the incorporation of 'everyday acts of defiance' within broader action for social change (Chatterton and Pickerill, 2010). Productive critical engagement across participatory and activist geographies is engaging with a range of epistemological perspectives including the phenomenological and poststructural, work on emotions, affect materiality and embodiment, whilst retaining radical and feminist engagement with issues of social and spatial justice (see mrs. c. kinpaisby-hill, 2011).

Simultaneously, there has been resurgence in the United Kingdom of notions of 'relevance' regarding academic endeavour, and facilitating more critical links across universities and communities which pay attention to the struggles of communities within interscaled structures of governance, foregrounding a politics of engagement with and for communities (Autonomous Geographies Collective, 2010). Beyond the dissemination of research via social media, newspapers, websites, radio, where the academia catalyzes public debate primarily through a position of 'expert', more 'organic public geographies' involves engagement of scholars as inter-connected and active with publics (Fuller and Askins, 2010). The latter engagements across university-community are fluid, contextual, complex and demand critical commitment, destabilizing any binary between 'university' and 'outside' or 'academic' and 'public'.

Of increasing concern regarding participatory, activist and public geographies are recent moves to audit the economic and social 'impact' of academic research through the UK's Research Excellence Framework (the government-mandated assessment of university and departmental research capacity and achievements). Those committed to participatory geography joined others in being wary of narrow conceptualizations of 'impact' that re-produce an elite model of power-knowledge relations in which academics-as-the-only-experts are positioned as imparting knowledge to and therefore *having impact upon* passive, non-expert and unknowledgeable publics. Pain et al. (2011) argued that resources and activities within universities should have a vital role to play in progressive social change; calling for an expanded understanding of impact based on the co-production of knowledge between universities and communities, modelled in participatory praxis. One strategy for the REF audit among participatory geographers has been writing collaboratively with participants and specifically making visible the multiplicity of voices present within research processes and knowledge production (see Hawkins et al., 2011). Such writing to disrupt dominant academic scripting is not new, strands of feminist and development geography had actively sought incorporation of the margins, and recognized the imperative of giving voice to those peoples involved in projects, but it has been enlivened through the growth of participatory geographies.

There are also moves in the United Kingdom to re-consider spaces of knowledge production and 'presentation'. There have been increasing attempts to open up seminars and conferences to become spaces of and for public engagement, by making them freely accessible and located within communities that would otherwise have limited or no access to such events. For example, at recent RGS-IBG conferences, PYGYRG has facilitated a series of sessions in local community centres (Manchester, 2009; London, 2013, 2014; Exeter 2015) and organized an art and activism exhibition (Edinburgh 2012). Kenrick and Vinthagen (2008) have called for academics to use the tactic of 'academic seminar blockades', in which research papers are presented in publicly staged seminars that simultaneously constitute direct action. PYGYRG members have staged such seminar blockades at Faslane Nuclear Base, Scotland (2012), the Atomic Weapons Establishment, Reading (2016) and the COP15 Climate Changes talks (2009).

PYGYRG, alongside many other contemporary critical geographers in the United Kingdom and beyond, is also concerned with education issues, bringing our story here full circle and returning to critical perspectives on geographical learning: if critical geography is to continue to problematize power/knowledge relations, classrooms are important sites through which critical praxis is performed. Participatory approaches to pedagogy, recognizing students as active and agentic learners, and resisting growing pressure to treat students primarily as customers, are challenging in a climate of large classes, student fees and competition between universities (for students, funding and status) as well as a heightened emphasis on student employability through putatively 'marketable' skills. In 2010, British government reforms to higher education included stringent funding cuts and undergraduate fee rises, which bought this point to the fore. The reforms became a focal point for a renewed engagement with progressive spatial politics and the resurgence of a student movement (Castree, 2011). Alongside national and localized demonstrations, at least 27 student-led occupations took place in British universities in late 2010. Geography departments were well-represented in these, with staff and students contributing to teach-ins at occupations in Leeds, University College London, Glasgow, Exeter, Newcastle, Manchester and the University of the West of England. The conjuncture in the United Kingdom remains contradictory, however. For some, these protests and the pressure for universities to demonstrate impact and public engagement have offered pathways for critical geographers to confront academic and social injustices (inspired by and building on the work of others – those whose stories are told above as well as the many we are unable to include or do justice to here). Jane Wills (2014, 382) for example judges that 'geography has the potential for subversive influence inside universities, challenging the dominance of market-led competition by developing a civic relationship with the people and communities with whom they share space'. Others are more suspicious of a state agenda of auditing the 'impact' of research (and much of the rhetoric around participation). Tom Slater (2012b, 118) criticizes 'the naïve instrumentalism that afflicts with virulence large sectors of social research in general and human geography in particular'. He goes on to declare that: 'Precious to our scholarship is the ability to ask our own questions drawn from astonishment at the world, from a thirst for intellectual/theoretical discovery, from political outrage and commitment to praxis, and a wish to intervene in ongoing debates' (op cit, 118).

But between such optimism of the will and a certain pessimism of the intellect, diverse strands of critical geography in Britain offer tracks for intellectual contest, engagement, negotiation, reciprocity and departure. Their interactions (in institutional contexts replete with contradictions) promise a future of continued lively debates. We hope that our account in this chapter has shown how 'situated' critical geographies have become central to geographical debate in the United Kingdom, but are also caught in contradictions that evolve unevenly and through contest.

Acknowledgements

We wish to thank Lawrence Berg and Uli Best for commissioning this chapter, for their patience whilst it was written and for their and Henrik Gutzon Larsen's comments on an earlier draft. In addition we are grateful to Richard Baxter, Gavin Brown, Terry Cannon, Noel Castree, Paul Chatterton, Jo Foord, David Gilbert, the late Ron Johnston, Alex Loftus, Avril Maddrell, Clare Madge, Nick Megoran, Claudio Minca, Jon O'Loughlin, Catherine Nash, Miles Ogborn, Chris Philo, Jenny Pickerill, Paul Routledge, Tracey Skelton, Tom Slater and Jane Wills for their comments, critique, recollections and suggestions, although we are responsible for the interpretation of these here and any errors. Likewise we are grateful to Joe Doherty and James Anderson for their challenging questions about the relative significance of radical geography in universities and schools, whilst we remain responsible for the balance of attention each receive here. We thank all others – in our departments and through correspondence and conversation – who have assisted and offered suggestions and apologize to them for not listing even more names here. Part of the chapter is derived from a paper given by Jo Norcup at 'The Geographical Canon?' workshop held by Histories and Philosophies of Geography Research Group of the RGS-IBG at University of Oxford in June 2012. Research into *CIGE* is drawn from doctoral research undertaken by Jo at the University of Glasgow. Although the chapter is a collective effort, James D Sidaway wishes to acknowledge the School of Geography, Queen Mary, University of London, for their hospitality during his visit there in September 2014. The chapter was finalized during that visit to a department where a strong tradition of geographies of social justice endures.

Notes

1 Until the 1990s, the term radical geography was most often used to define work that subsequently has been interpreted as precursors of critical geography in both the United Kingdom and North America. We briefly reflect on the terminological shift later in this chapter.

2 Until the 1990s, the term radical geography was most often used to define work that subsequently has been interpreted as precursors of critical geography in both the United Kingdom and North America. We briefly reflect on the terminological shift later in this chapter.

3 For more on the structure and relative strengths of the discipline in the United Kingdom, see the 2013 report and data for an *International Benchmarking Review of UK Human Geography* that was commissioned by the main research funding agency (Economic and Social Research Council, 2013).

4 *Contemporary Issues in Geography and Education* (1983, 1).

5 Further details about contributors and the journal series can be found in Jo Norcup's University of Glasgow doctoral thesis on 'Awkward Geographies? The historical and cultural geographies of the journal *Contemporary Issues in Geography and Education* (1983–1990)'.

6 These meetings played a similar intellectual and political role to those of the Association of American Geographers (now the American Association of Geographers, AAG) and the Canadian Association of Geographers (CAG). We will have cause to reflect again on the IBG later in this chapter. Suffice to note here that it was established in 1933 as a forum for professional/academic scholarship in geography that broke with the still exploration/establishment orientated RGS just over a century after the later was established (Steel, 1984). The break was, however, by mutual agreement. There was much overlap in membership and the IBG operated out of an office within the larger London headquarters of the RGS. By the 1990s, this continued link would raise the prospect of them merging: with significant consequences for the course of/ debates in critical geography, as we detail below.

7 Research Groups were previously known as Study Groups. The Women and Geography Study Group (WGSG) opted to retain the original name, arguing that the group was one for study as much as research. This was accepted by the RGS-IBG. However, following subsequent debates within the WGSG, it

adopted the name Gender and Feminist Geography Research Group from 2013. A thoughtful discussion of the politics of the name change (including calls to retain the WGSG name) appears in a series of short papers in a special section of *Area* (2013, Volume 45, number 1) on 'Gender or Women? Debating the Future of the Women and Geography Study Group' (Guest edited by Katherine Brickell and Kath Browne).

8 A Space, Sexualities and Queer Research Group was established in the RGS-IBG in 2006. UK-based geographers had played a key role in the development of work on sexuality and space – but as in prior radical and subsequent critical geography, transatlantic (amongst other) connections and influences make it problematic to demarcate a distinctively 'British' queer geography. For a reflection on queer epistemology from a UK-based geographer, see Binnie (1997).

9 Although as we have indicated, anti-racism was a key part of the discussions about pedagogy in *CIGE*, and notwithstanding work from UK-based geographers on whiteness and anti-racism (Abbott 2006, Baldwin 2012, Bonnett 1997, Jackson 1998, Nash 2003), the proportion and positioning of Black and minority ethnic geographers in the United Kingdom was seldom a theme of published discussions. A session led by Black and minority ethnic geographers at the RGS-IBG conference in London in August 2014 was the first event in the history of the conference to foreground the issues. Making reference to literature from North American geographers and initiatives by the AAG, the session was sponsored by the Geographies of Justice Research Group and Higher Education Research Group (subsequently renamed as the Geography and Education Research Group) and convened by Caroline Bressey (University College London), Richard Baxter (Queen Mary, University of London), Tariq Jazeel (University College London) and Lamees Al Mubarak (Queen Mary, University of London). Following this, the Race, Culture and Equality (RACE) Working Group, was established in 2015 (see https://raceingeography.org/about/) and a student-led Black Geographers network was established in 2020 (see https://www.blackgeographers.com/).

References

Abbott, D. 2006. Disrupting the 'whiteness' of fieldwork in geography. *Singapore Journal of Tropical Geography* 27(3), 326–41.

Amin, A. and N. Thrift. 2005. What's left? Just the future. *Antipode* 37(2), 220–38.

Amin, A. and N. Thrift. 2013. *Arts of the Political: New Openings for the Left*. Durham, NC: Duke University Press.

Anderson, B. and P. Harrison. 2010. The promise of non-representational theories. In, Anderson and P. Harrison (eds.), *Taking-Place: Non-Representational Theory and Geography*. Farnham and Burlington, VT: Ashgate, pp. 1–34.

Association for Curriculum Development in Geography. 1983-1990. *Contemporary Issues in Geography and Education* 1(1)-3(2).

Autonomous Geographies Collective. 2010. Beyond scholar activism: Making strategic interventions inside and outside the neoliberal university. *ACME: An International E-Journal for Critical Geographies* 9(2), 245–75.

Baldwin, A. 2012. Whiteness and futurity: Towards a research agenda. *Progress in Human Geography* 36(2), 177–87.

Barnett, C. 2013. Book review essay: Theory as political technology. *Antipode*. http://radicalantipode.files.wordpress.com/2013/07/book-review_barnett-on-amin-and-thrift.ppd (29 December 2020).

Binnie, J. 1997. Coming out of geography: Towards a queer epistemology? *Environment and Planning D: Society and Space* 15(2), 223–37.

Blunt, A. and J. Wills. 2000. *Dissident Geographies: An Introduction to Radical Ideas and Practice*. Harlow: Pearson Education.

Bonnett, A. 1997. Geography, 'race' and Whiteness: Invisible traditions and current challenges. *Area* 29(3), 193–9.

Bowlby, S. and J. Tivers. 2009. Feminist geography, Prehistory of. In, R. Kitchin, N. Thrift et al. (eds.), *The International Encyclopedia of Human Geography*. Oxford: Elsevier, pp. 59–63.

Breitbart, M. 1981. Peter Kropotkin, the anarchist geographer. In, D. Stoddart (ed.), *Geography, Ideology and Social Concern*. Oxford: Blackwell, pp. 134–53.

Butler R. and H. Parr (eds.). 1999. *Mind and Body Spaces: Geographies of Illness, Impairment and Disability*. London: Routledge.

Burton, K. and N. Megoran. 2013. Geographers and peace education from the cold war to the war on terror. Paper presented at Association of American Geographers Annual Meeting, 9-13 April 2013, Los Angeles.

Cameron, J. and K. Gibson. 2005. Participatory action research in a poststructuralist vein. *Geoforum* 36(3), 315–31.

Castree, N. 1999. 'Out there'? In here? Domesticating critical geography. *Area* 31(1), 81–6.

Castree, N. 2000. Professionalisation, activism and the university: Wither critical geography? *Environment and Planning A* 32(6), 955–70.

Castree N. 2011. The future of geography in English universities. *The Geographical Journal* 177(4), 294–9.

Castree, N. and D. Gregory (eds.). 2006. *David Harvey: A Critical Reader*. Oxford: Wiley Blackwell.

Chatterton, P. and D. Featherstone. 2007. Intervention: Elsevier, critical geography and the arms trade. *Political Geography* 26(1), 3–7.

Chatterton, P. and J. Pickerill. 2010. Everyday activism and transitions towards post-capitalist worlds. *Transactions of the Institute of British Geographers* 35, 475–90.

Chorley, E. J. and P. Haggett (eds.). 1967. *New Models in Geography*. London: Methuen.

Cooke, B. and U. Kothari (eds.). 2001. *Participation: The New Tyranny?* London: Zed Books.

Crang, P. 2010. Cultural geography: After a fashion. *Cultural Geographies* 17(2), 191–201.

Dalby, S. (ed.). 1986. *Issues in Education and Culture*, no. 2 (special issue on peace education).

Driver, F. 2001. *Geography Militant: Cultures of Exploration and Empire*. Oxford: Blackwell.

Economic and Social Research Council. 2013. *International Benchmarking Review of UK Human Geography*. https://esrc.ukri.org/files/research/research-and-impact-evaluation/international-benchmbenchm-review-of-uk-human-geography (30 December 2020).

Featherstone, D. 2013. Review of Amin, Ash and Nigel Thrift 2013, Arts of the Political: New Openings for the Left. https://www.societyandspace.org/articles/arts-of-the-political-by-ash-amin-and-nigel-thrthr (30 December 2020).

Featherstone, D. and J. Painter (eds.). 2013. *Spatial Politics: Essays for Doreen Massey*. Oxford: Wiley-Blackwell.

Fuller, D. and K. Askins. 2010. Public geographies II: being organic *Progress in Human Geography* 34(5), 654–67.

Gardner, R. 2005. Recent developments and future prospects. In, D. Popey (ed.), *To the Ends of the Earth. Visions of a Changing World: 175 Years of Exploration and Photography*. London: Bloomsbury.

Gilbert, D. 1999. Sponsorship, academic independence and critical engagement: A forum on Shell, the Ogoni dispute and the Royal Geographical Society (with the Institute of British Geographers). *Ethics, Place and Environment* 2, 219–28.

Gilbert, D. 2009. Time to Shell out? Reflections on the RGS and corporate sponsorship. *ACME: An International E-Journal for Critical Geography* 8(3), 521–9.

Gregory, D. 1978. *Ideology, Science and Human Geography*. London: Hutchinson.

Gregory, D. and J. Urry (eds.). 1985. *Social Relations and Spatial Structures*. London: Macmillan.

Hall, T. 2014. Making their own futures? Research change and diversity amongst contemporary British human geographers. *The Geographical Journal* 180 (1), 39–51.

Hammett D. and A. Newsham. 2007. Intervention: widening the ethical debate. Academia, activism and the arms trade. *Political Geography* 26(1), 10–12.

Harvey, D. 1973. *Social Justice and the City*. London: Edward Arnold.

Hawkins, H., S. Sacks, I. Cook, E. Rawling, H. Griffiths, D. Swift and K. Askins. 2011. Organic public geographies: 'Making the connection'. *Antipode* 43(4), 909–26.

Hepple, L.W. 1999. Socialist Geography in England: J.F. Horrabin and a workers' economic and political geography. *Antipode* 31(1), 80–109.

Hirit, I. and M. Palomino-Schalscha. 2011. Guest Editorial. Geography, the military and critique on the occasion of the 2011 IGU Regional Meeting in Santiago de Chile. *Political Geography* 30(7), 355–7.

Holloway, S.L. and G. Valentine (eds.). 2000. *Children's Geographies: Playing, Living, Learning*. London: Routledge.

Hudson, R., J. Anderson and S. Duncan (eds.). 1983. *Redundant Spaces in Cities and Regions: Studies in Industrial Decline and Social Change*. London: Academic Press.

Jackson, P. 1989a. Challenging racism through geography teaching. *Journal of Geography in Higher Education* 13(1), 5–14.

Jackson, P. 1989b. *Maps of Meaning: An Introduction to Cultural Geography*. London: Unwin Hyman.

Jackson, P. 1998. Constructions of 'whiteness' in the geographical imagination. *Area* 30(2), 99–106.

Jackson, P. 2005. Author's response. Classics in human geography revisited. *Progress in Human Geography* 29(6), 746–7.

Johnston, R.J. 1982. *Geography and the State: An Essay in Political Geography*. London: Macmillan.

Johnston, R. 2004. Institutions and disciplinary fortunes: two moments in the history of UK geography in the 1960s I: geography in the 'plateglass universities'. *Progress in Human Geography* 28 (1), 57–77.

Keltie, J.S. 1885. *Geographical Education. Report to the Council of the Royal Geographical Society*. London: RGS.

Kenrick, J. and S. Vinthagen. 2008. Critique in action: Academic conference blockades. In, A. Zelter (ed.), *Faslane 365: A Year of Anti-nuclear Blockades*. Edinburgh: Luath.

Kindon, S., R. Pain and M. Kesby. 2007. *Participatory Action Research Approaches and Method: Connecting People, Participation and Place*. London: Routledge.

Kitchin, R. 2007. Elsevier, the arms trade, and the forms, means and ends of protest a response to Chatterton and Featherstone. *Political Geography* 26, 499–503.

Kropotkin, P. 1885. What geography ought to be. *The Nineteenth Century* 18, 940–56.

Maddrell, A. 2009. *Complex Locations: Women's Geographical Work in the UK 1850-1970*. Oxford: Wiley-Blackwell.

Maddrell A. 2010. Academic geography as terra incognita: lessons from the 'expedition debate' and another border to cross. *Transactions of The Institute of British Geographers* 35(2), 149–53.

Martin, R. and B. Rowthorn. 1986. *Geographies of Deindustrialization*. London: Macmillan.

Massey, D. 1984. *Spatial Divisions of Labour: Social Structures and the Geography of Production*. London: Macmillan.

Merrifield, A. 2011. *Magical Marxism: Subversive Politics and the Imagination*. London: Pluto Press.

mrs. c. kinpaisby-hill. 2011. Participatory praxis and social justice: towards more fully social geographies. In, V.C. Del Casino, M.E. Thomas, P. Cloke and R. Panelli (eds.), *A Companion to Social Geography*. Oxford: Wiley-Blackwell, pp. 214–34.

Nash, C. 2003. Cultural geography: Anti-racist geographies. *Progress in Human Geography* 27(5), 637–48.

Norcup, J. 2015. Geography education, grey Literature and the geographical canon. *Journal of Historical Geography* 49, pp 64–74.

Norcup, J. 2019. 'Lets here [sic] it for the Brits, you help us here': North American radical geography and British radical geography education. In, T. Barnesand, E. Sheppard, (eds.), *Spatial Histories of Radical Geography: North America and beyond*. Oxford: Wiley-Blackwell, pp. 343–356.

Pain, R., M. Kesby and K. Askins. 2011. Geographies of impact: power, participation and potential. *Area* 43(2), 183–188.

Peet, R. and N. Thrift (eds.). 1989. *New Models in Geography* (two volumes). London: Unwin Hyman.

Philo, C. (compiler). 1991. *New Words, New Worlds – Reconceptualising Social and Cultural Geography*. Aberystwyth: Cambrian Press.

Power, M. and J.D. Sidaway. 2004. The degeneration of tropical geography. *Annals of the Association of American Geographers* 94(3), 585–601.

Regan, C. 1978. Formation of U.S.G. in the British Isles: minutes of Hull meetings January 1978. *Union of Socialist Geographers Newsletter* 3(3), 52–4.

Rowbotham, S. 1999. *Threads Through Time: Writings on History and Autobiography*. London: Penguin.

Russell Group 2013. *Key facts and statistics*. http://www.russellgroup.ac.uk/key-facts-and-statistics (5 October 2013).

Sack, R.D. 1980. *Conceptions of Space in Social Thought: A Geographic Perspective*. London: Macmillan.

Sharp, J., P. Routledge, C. Philo and R. Paddison (eds.). 2000. *Entanglements of Power: Geographies of Domination/Resistance*. London: Routledge.

Slater, T. 2012a. *Rose Street and Revolution: A Tribute to Neil Smith (1954-2012)*. https://blogs.ed.ac.uk/tomslater/neil-smith-tribute (30 December 2020).

Slater, T. 2012b. Commentary. Impacted geographers: a response to Pain, Kesby and Askins. *Area* 44(1), 117–9.

Smith, D.M. 1971. Radical geography: the next revolution? *Area* 3(3), 153–7.

Smith, D.M. 1974. Who gets what where, and how: a welfare focus for human geography. *Geography* 59(4), 289–97.

Smith, D.M. 1977. *Human Geography: A Welfare Approach*. Edward Arnold, London.

Steel, R.W. 1984. *The Institute of British Geographers: The First Fifty Years*. London: Institute of British Geographers.

Tickell, A. 1999. On getting inside the project. *Ethics, Place and Environment* 2(2), 234–238.

Walford, R. 2000. *Geography in British Schools, 1850-2000: Making a World of Difference*. London: Woburn Press.

Wills, J. 2014. Engaging. In, R. Lee, N. Castree, V. Kitchin, A. Paasi, C. Philo, S. Radcliffe, S.M. Roberts and C. Withers (eds.), *The Sage Handbook of Human Geography*. London: Sage, pp. 367–84.

Wojtas, O. 1997. Geographer quits society in Shell protest. *Times Higher Education*. London.

Women and Geography Study Group. 1984. *Geography and Gender: An Introduction to Feminist Geography*. London: Hutchinson.

Woodyer, T. and H. Geoghegan. 2013. (Re)enchanting geography? The nature of being critical and the character of critique in human geography. *Progress in Human Geography* 37(2), 195–214.

Wyse, S. 2013. The founding of the Women and Geography Study Group. *Area* 45(1), 4–6.

15 Placing critical geographies
Australia and Aotearoa New Zealand

*Robyn Dowling, Richard Howitt,
and Robyn Longhurst*

The title of critical geography's oft-identified 'journal of record' – *Antipode* – means not only oppositional, but also it describes Australia and Aotearoa New Zealand and their location at the opposite end of the earth from the western hemisphere home of both Europe's colonial project and the dominant discourses of Anglo-American geography. These two meanings frame the narrative of critical geography we produce in this chapter, in which we suggest that in these Antipodes, critical geography emerged not so much as oppositional to a centre, but as a component of the mainstream. In geographical scholarship, in disciplinary societies like the New Zealand Geographical Society and the Institute of Australian Geographers, in university departments and across sub-disciplinary fields, the radical and engaged activist edges to geographical scholarship have never been far from the centre, nor simply oppositional. The argument of this chapter, then, is that critical geographies in the Antipodes are characterized by both marginalization and centralization, as they address the environmental, social and economic challenges presented in these distinctive places, and their distinctive relationships to material and discursive trends in other places.

Our argument draws inspiration from Raewyn Connell's proposition in *Southern Theory* that: 'Terra nullius, the coloniser's dream, is a sinister presupposition for social science. It is evoked every time we try to theorise the formation of social institutions from scratch, in a blank space' (Connell, 2007, 47). As Connell sees it, Anglo-American theorizing in particular and 'western' and Euro-centric theorizing in general risk imposing a 'grand erasure ... of the experience of the majority of human kind from the foundations of social thought' (2007, 46).

For many Antipodean geographers, their empirical work confronted material realities where the claim of terra nullius was clearly fictitious and delusional. In Australia and Aotearoa New Zealand, the lands and resources claimed by the colonizers were clearly already occupied by peoples whose rights, aspirations and cultures were threatened by the dominant culture of colonial economic and intellectual life. Antipodean environments were increasingly recognised as having unique and complex qualities that were pushed aside by political boosters and economic shysters angling to

DOI: 10.4324/9781315600635-15

whisk commodities and profits to service the voracious appetites of metropolitan markets and global corporations. Even Antipodean rivers failed to comply with the theoretical constructions of metropolitan geographers (Taylor and Stokes, 2005)!

When 'northern' theorists noticed the Antipodes, they often simply got it wrong – as Connell demonstrates in the case of Durkheim's 'radical misunderstanding of Australian Indigenous cultures' (Connell, 2007, 79). In contrast to the sort of social science that 'prefers context-free generalisation' (Connell 2007, 196), 'southern' geography grappled with its simultaneously Antipodean and global contexts in ways that marked what we studied, what we did, what we said and what we understood. In some ways, the 'larrikin culture' of these margins of European intellectual endeavour engaged with a range of underdogs that saw organized labour, colonized peoples, natural heritage and justice as the core business of publicly funded intellectuals.

In economic geography, for example, globalization was long recognized as demanding simultaneous consideration of Europe, North America, East Asia and the global 'south'. Environmental geographers recognised early that science and politics demanded engagement together. Cultural geographers grappled with the suburbs and marginalized places and people – and also turned a critical analytical gaze on the nature and sources of power, privilege and wealth by 'studying up' (Nader, 1974). And urban geographers were confronted with the substantial engagement of organized labour in both government policies and programs through mainstream leftist political parties, and new forms of social movements in Australia's Green Bans and New Zealand's broad-based anti-racism movements that mobilised around global concerns about apartheid and Māori treaty rights.

Box 15.1

Critical geographical collaborations across the Tasman

In November 1980 social scientists from Canada, Australia and New Zealand met in Christchurch at the Three Nations Conference and Workshop on development and underdevelopment. An unprecedented mix of scholars and activists, this meeting challenged many assumptions about scholarly detachment and established close working relationships across the Tasman between radical geographers and other activists. For geographers working with Indigenous groups, the 1992 Christchurch meeting of the Commonwealth Geographical Bureau reinforced these links. Collaborations across the Tasman have been a hallmark of much critical geography in the Antipodes, of which two current ones are noteworthy. The first, explicitly drawing on poststructural political economy (see below) is a group of scholars from the universities of Auckland and Newcastle under the working title of SKCAN (Situated Knowledge Collaborative Auckland and Newcastle). The group aims to progress discussions on situated knowledge in the Australasian context, and in particular its multiple dimensions and directions (SKCAN 2011). A second collaboration also takes the project of situating knowledge as its focus. Building upon a 1998 special issue *Australian Geographer* on 'Gendered Geographies in Australia, Aotearoa/New Zealand and the Asia-Pacific' (Dowling and McGuirk, 1998), was a 2008 special issue on 'Geographies of Sexuality and Gender "Down Under"' (Gorman-Murray, Waitt and Johnston, 2008). The aim of this collaborative effort was also to 'assert the need to speak from an "Australasian" perspective, outlining some contextual differences between Anglo-American and Australasian geographies of sexuality and gender' (Gorman-Murray, Waitt and Johnston, 2008, 236). Geographical knowledge about sexuality, argue the editors, remains centred in the Anglo-American context despite a range of strong contributions on sexuality, queer and gender studies by Australasian scholars. As Lawrence Berg and Robin Kearns (1998, 129) argue: 'Although geographers from the "peripheries" are allowed to participate in such debates [debates about theorizing], they are rarely able to set the agenda or frame the epistemological boundaries' (also see Berg and Kearns, 1996). Consequently, Gorman-Murray, Waitt and Johnston (2008, 239) explain they feel they 'need to speak out from an "Australasian" perspective'.

Emergence of critical and radical geography in the Antipodes was early and influential, at least across the region. There are both significant similarities and profound dissimilarities in the trajectories of critical geography in the two places. Australia and New Zealand are very different: in size, in culture, in history, in political orientation. Yet, they are also strongly connected. Australia and New Zealand share many historical commonalities. Both nations are 'settler colonies' – resident Indigenous peoples were colonized by European settlement in the eighteenth and nineteenth centuries – and both have economies with substantial dependence on resources and/or agriculture. Their urban geographies, especially in terms of metropolitan primacy, are similar (see Dowling and McGuirk, 2008), as are cultural constructions of sex and gender (see Longhurst, 2008). Strong research and personal connections bring geography scholars across the Tasman together, either via the movement of people through universities in both countries, or through shared research interests. Indeed, the two disciplinary groups have combined conferences every four years or so, further fostering intellectual engagement. For these reasons, and also to aid clarity, in this chapter we approach critical geography in these two nations in terms of shared intellectual trajectories and characteristics, whilst pointing out key differences where relevant.

The structure of the chapter is hence as follows. We begin with a brief historical overview of critical geography in the Antipodes. This is followed by three substantive sections that address key issues of critical engagement: postcolonialism, feminism and political economy.

Beginnings

In Aotearoa New Zealand in the late 1980s, there began to emerge in geography a range of diverse theoretical arguments which could broadly be described as 'critical'. At this time, however, in New Zealand the term 'critical' was not used. Instead, the preferred term was 'alternative' geographies and in October 1987, to mark the golden jubilee of geography as a full-fledged university discipline in Aotearoa, the national geography journal *New Zealand Geographer* took a decision to commission a special issue on 'alternative' perspectives in geography. Pip Forer (1987, 113), who at the time was editor, explains: it 'seemed only appropriate to use this [special issue] to encourage publication of work that might break new ground and in some cases might otherwise be slow to appear in printed form'. The guest editors Judith Johnston and Richard Le Heron (1987, 115) state in their introduction:

> The New Zealand geography community is being challenged. At the last two geography conferences the elements of change in New Zealand society and the discipline were both visible and vocal. Questions focus on the relevance of the discipline to political decision-making and state policy formulation, the validity of objective scientific methods for analyzing current issues, the relationship between geography and social and environmental change and the relevance of non-New Zealand models as conceptual frameworks guiding the construction of knowledge and our understanding of society.

The late 1980s was a time when debates around women's rights, the homosexual law reform bill, the pros and cons of nuclear energy, Māori sovereignty and the increasing privatization of public assets were taking place. In the 'Alternative Perspectives' issue of *New Zealand Geographer* (1987), a number of these debates were addressed. For example, Stokes focussed on processes of imperialism in New Zealand paying attention to Māori geography while Eric Pawson addressed 'social processes of urban space' arguing that it is vital that New Zealand geographers engage these themes to examine their own society. Stephen Britton and Richard Le Heron adopted a Marxist approach to analyse regional differentiation and transformation in the New Zealand economy. Also, on the theme of the economy, Ross Barnett analysed the implication of re-privatizing the hospital sector in New Zealand while Warren Moran focussed on rural land use change. In yet another article, Evelyn Stokes, Louise Dooley, Louise Johnson, Jennifer Dixon and Susan Parsons discussed the value of drawing on feminist perspectives in geography. Addressing rapidly changing technologies of the time, Lex Chalmers and Pip Forer critically reviewed what they considered to be overly enthusiastic accounts of the potential of technical advances to teaching and research in geography. Finally, C.R. de Freitas made a case for

New Zealanders to consider carefully the impacts of climate change. These were the local issues that New Zealand geographers felt were pressing and needed to be critically debated in the late 1980s.

Although Marxist and socialist geographers found a voice in North America and western Europe in the early 1970s rallying against poverty and urban segregation based on class, in New Zealand there was relative (but not complete) silence on these issues. Māori society traditionally was based on rank derived from ancestry or whakapapa and yet despite this, until the early 1980s, it was widely claimed that New Zealand was a 'classless society'. Historian Keith Sinclair (1969, 285) wrote that although New Zealand was not a classless society, 'it must be more nearly classless ... than any advanced society in the world'. Given this widely held view, it is perhaps not surprising that Marxist and socialist perspectives did not feature as strongly in geographical work as, say, feminist and anti-racist perspectives.

In Australia, even the founding father of the discipline, Griffith Taylor, constructed a critical geography of sorts as he challenged the dominant settlement policies of the 1930s with arguments about the limited environmental capacities of arid and tropical Australia. While Taylor's views on environmental determinism and race may seem distant from the values of more recent self-avowed critical geography (see, for example, Head's (2000) discussion of Griffith Taylor's 'empty Australia' and also Thom and Mckenzie (2011)), his policy interventions reflect a deep-seated commitment to engagement across the discipline). In development geography, O.H.K. Spate at Australian National University (ANU) challenged the deep-seated racism of national policies towards non-white immigration and development assistance (Spate, 1965), and young women such as Fay Gale (Gale, 1983; Gale and Brookman, 1972) and Janice Monk (1974) turned a similarly critical concern towards Indigenous Australians in urban and rural settings; followed by Elspeth Young (1992, 1995) and Richard Howitt (1993) in more remote areas.

In 1970s Sydney, where the Builders Labourers Federation was transforming the nature of urban planning with Green Bans (Anderson and Jacobs, 1999), geographers formed a branch of the Union of Socialist geographers, led by Ron Horvath, Frank Williamson and Bob Fagan. This collective sponsored the 1981 issue of *Antipode* labelled 'Antipodean Antipode' (edited by Williamson and Fagan joined by urban geographer Vivienne Milligan and the young economic geographer Katherine Gibson) to present papers on class (Horvath and Rogers, 1981), Aboriginal underdevelopment (Drakakis-Smith, 1981), planning in urban and rural settings (Freestone, 1981; Searle, 1981) and housing (Berry, 1981) as well as a review of radical geography in Australia that recognises 'radical' geography was relatively young, and contextualised in a much longer tradition of political radicalism in the organised labour movement (Williamson et al., 1981). Structuralist Marxism and feminism underpinned geographers' strong critiques of Antipodean society throughout the 1980s. The so-called New Cultural Geography drew considerable inspiration from the innovative work of Fay Gale at the University of Adelaide (Anderson and Jacobs, 1997) and nurtured wellsprings of impressive work from Kay Anderson (Anderson, 1991; Anderson and Gale, 1999), Jane Jacobs (Jacobs, 1996) and Richard Baker (Baker, 1999).

Box 15.2

Cultural geography as critical geography in Australia and Aotearoa New Zealand

In critical geography discussions internationally, there have been many words written on the apolitical stance of 'the cultural turn'. To summarize a very broad debate, the argument is that cultural geography's focus on representation, discourse and polyvocality lacks a critical edge and is unable to comprehend and dismantle relations of power, economies and structural inequalities. In an important piece, Anderson and Jacobs (1997) demonstrate how these debates lose much of their resonance in Antipodean critical geographies. These claims, they caution, 'appear to be conditioned by an ambitious, universalising northern disciplinary perspective' and are ignorant of those 'from southern spaces who identify as cultural geographers [who] were implacably imbued as undergraduates with a *political* cultural geography' (13; emphasis in original). Anderson and Jacobs use the example of thinking about Aboriginals and the city to write a history of Australian cultural geography where a critical, politically engaged edge is central. The story of cultural geography in Australia, for Anderson and Jacobs, is deeply political, attuned to the entanglements of power and identity that characterize Australian experience, and able to provide foundations for 'geographic practice that seeks to blur the boundaries between centre and margin' (21).

Australia's place in the changing world economy fuelled not only intellectual engagement with structuralist economics, but also important links to trade union research with powerful analysis of various industries and companies (Fagan, 1981, 1984, 1987, 1988), regions (Gibson, 1991, 1992) as well as an emerging focus on the consequences of economic restructuring for particular social groups, regions and industries (Fagan and Le Heron, 1994; Gibson and Horvath, 1983; Howitt, 1991a, 1991b; Johnson, 1990; O'Neill, 1996) – a theme that brought much discussion and collaboration between geographers in Australian and New Zealand (Britton and Le Heron, 1989; Fagan and Le Heron, 1994). At the same time, feminist geographers were turning attention to both empirical and conceptual concerns in Antipodean geographies (e.g. Gibson-Graham, 1994; Johnson, 1990).

Political economy, poststructuralism and neoliberalism

In the 1980s and 1990s the political and economic landscapes of the Antipodes changed. In New Zealand, neoliberal economic reforms were instituted. These reforms severely weakened the power of unions, cut social welfare benefits and made government state housing less affordable, alongside a growing disparity in wealth in New Zealand (O'Dea, 2000; Te Ara, 2005). More than a decade of Labor Government rule in Australia (1983–1996) saw significant 'structural' adjustments in the Australian economy, including the floating of the Australian dollar, reduction and/or removal of tariffs, and a looser (though still regulated) wage system. These changes were the subject of significant critical geographical scrutiny, with the lens of globalization prominent across the Tasman. Bob Fagan and Michael Webber's (1999) *Global Restructuring: the Australian Experience* brought much of this research together, whilst Fagan and Le Heron's (1994) intervention into the changing role of the state was equally important.

As the 1990s drew to a close, the political climate drifted further to the right and poststructuralist approaches to political economy emerged. Critical social scientists such as Wendy Larner began to reflect, drawing on poststructuralist theory, on the processes of neoliberalism examining trade liberalization, deregulation, commodification and marketization (see Larner 1996, 1997, 1998; Larner and Le Heron, 2002). Antipodean critical geographers have been at the forefront of critical debates on the concept of neoliberalism itself, arguing that its proliferation in academic discourse has had the consequence of giving it greater credence in policy realms (Larner, 2000) and, more recently, that it is an open assemblage constituted by both state and market actors (McGuirk and Dowling, 2009). Gibson-Graham's (1996) *The End of Capitalism (as we knew it)* draws on both poststructuralism and feminist theory to discursively dismantle the economy; to bring to the fore non-capitalist economic practices, processes and identities. McManus and Pritchard (2000) offer a response to structural change across rural Australia.

Box 15.3

Suburban geographies

Suburban landscapes of detached houses dotted across the edges of cities are prominent in Australia and Aotearoa New Zealand and feminist, political economy, postcolonial and poststructuralist perspectives are flecked through geographic understandings of these landscapes. Moreover, suburban scholarship also highlights the ways in which Antipodean critical geographies produce 'southern theory'. A strong thread of political economy runs through the emphasis placed on suburban employment and unemployment. In the 1980s, Fagan highlighted the uneven and negative impacts of restructuring on suburban and inner urban employment centres. Ten years later, suburban employment patterns were used to challenge the notion of a global city, especially its strong connection to the CBD and office employment. The global connectivity of suburban employment centres, whether they be in manufacturing or business services, Fagan argued, is in evidence but takes a different form to that of the stereotypical CBD office tower (see Fagan and Dowling, 2005). Feminist perspectives on suburbia have a long heritage and contemporary currency. Well-known here is the work of Johnson (2000), as well as Fincher and Jacobs' *Cities of Difference* (1998). The postcolonial elements of suburban landscapes, and a critique of planning, are also reflected in suburban scholarship (Johnson 1994).

Postcolonialism

Antipodean geographies are deeply inscribed by the impact of British colonization and the socio-spatial legacies of colonial intrusion into Indigenous geographies and contested identities, polities and sovereignties. For critical geographers, the need to respond to the dominance of colonial cultures and persistent Indigenous presences has held significant challenges, leading to thoughtful mobilizations of critical discourses of race, culture and identity. Working to advance Indigenous rights has led to geographical practices on both sides of the Tasman Sea that have pushed geographers to address political and ethical implications of methodological and conceptual tools.

Two particular areas of scholarship worth discussing in more depth as a way of 'placing critical geographies' in the context of Aotearoa are feminist geography and Māori geography, both of which took root in New Zealand in the mid to late 1980s especially at Waikato University. Dame Evelyn Stokes played a major role in both these critical developments. Stokes was initially appointed in 1964 to University of Auckland, although she taught in both Auckland and Hamilton, before being appointed a year later as a foundation staff member to the newly created University of Waikato. Stokes remained at Waikato until her death in 2005. Australian geographer Johnson (2005, 92) in a special issue of *New Zealand Geographer* devoted to Stokes' work describes her as 'a wise but stroppy woman, someone who would have no truck with fools but who opened her heart, her mind and her home to those whose intellect and politics she respected' (also see Bedford, 2005; Cant, 2005). Stokes was known as the 'kuia' (Māori word for a wise old woman) of the geography department. For decades she promoted Māori studies and worked hard to integrate local Māori communities into the university's activities. It was in the aforementioned 1987 'Alternative Perspectives' issue that Stokes published her now often cited article 'Maori Geography or Geography of Maoris' in which she challenges Eurocentric thinking (Stokes, 1987a). This article was an important attempt to 'assess New Zealand social and economic geography within the context of colonialism and the effects of the relationship between immigrant ruling culture and indigenous people' (Stokes, 1987a, 119). At this time, many Māori were attempting to chart a more autonomous path of development. Questions were being posed about unconscious ethnocentric presumptions that continually positioned Māori as Other to Pākehā. There were calls for Māori sovereignty. Māori were demanding to be able to determine their own destiny and were calling for land to be returned to its rightful Māori owners. Stokes (1987a, 121) argues:

> Maori geography is another way of viewing the world, another dimension, another perspective on New Zealand geography. But – kia tupato, Pakeha. Be careful Pakeha. Tread warily. This is not your history or geography. Do not expect all to be revealed to you. You must be prepared to serve a long apprenticeship of learning on the marae. You must know the language and understand the culture ... Do not expect that because you are an academic or experienced researcher in the Pakeha world that all this will come easily to you.

Stokes paved the way for critical geographical and historical scholarship on colonial relationships, Māori, whiteness, indigeneity and related topics. For example, Brian Murton (2006), focusing on the North Kaipara beach between about 1900 and 1971, shows how the cultural politics of resource management were enacted both from the top-down (the Marine Department's implementation of policy) and the bottom-up (resistance from Māori) in relation to protecting *toheroa*, the giant surf clam. Wikitoria August (2005, 117) examines 'Māori women and the ways in which their bodies are constituted within particular cultural spaces, namely at *urupā* [cemeteries] and sites where food is gathered'. August explores the impacts of colonization on bodily rituals associated with these cultural spaces. These are just two examples of a range of critical work both by local and international scholars and activists who have built on Stokes' contribution over the past two decades.

In Australia, the pioneering work of Gale and Monk was followed by that of Young (1995) in reshaping links between Indigenous rights and geography. Drawing on postcolonial theory and engagement with both community scale analysis and struggles, geographers in Australia were actively reorienting the discipline's response to issues of deeply entrenched racism (Howitt and Jackson, 1998). Cross-Tasman engagement on these issue (Davies, 2003) predated the 1980 Three Nations Conference in Christchurch, which brought together radical scholars and activists from the Antipodes and Canada, but were greatly strengthened in a New Zealand-hosted meeting of the Commonwealth Geographical Bureau in 1990 (Cant et al., 1993). Work on Native title (Agius et al., 2004; Davies, 2003; Gooder and Jacobs, 2000; Jackson, 1997; Mercer, 1993; 1997) and environmental management (Jackson et al., 2005; Muller, 2008; Muller et al., 2009; Suchet, 1996) and Indigenous geographies continues to build on this work.

Box 15.4

Postcolonial geographies – Antipodean specificities

The particularities of postcolonial politics in Australia and New Zealand has underpinned some specific contributions to postcolonial geographies from southern theorists. In Aotearoa, the political imperative to reconfigure the legal standing of the Treaty of Waitangi and the implications of treaty rights for governance of resources, property and intercultural relations (McHugh, 1996) focussed attention on Māori geographies in innovative ways that also intruded into geographical theories of place, identity and research practice in New Zealand (Barclay-Kerr, 1991; Cant, 1990; Coombes, 2007; McClean et al., 1997; Murton, 2006; Pawson and Cant, 1992; Smith, 1999; Stokes, 1987a; 1987b; 1992; 1993; Tipa and Nelson, 2008).

In Australia, Gale's seminal contributions as researcher and teacher demanded recognition that the spatial margins and marginalized Indigenous peoples were inescapably central to the construction of the colonial project. From this genealogy, Jacob's influential account in *Edge of Empire* unsettled many assumptions about the nature of postcolonial geographies. Amongst a range of contributions, Baker's work on oral geographies (e.g. Baker, 1999), Howitt on remote resource projects (Howitt, 1989, 1991a, 1991b, 1992a, 1992b) and Shaw (2000, 2006) on Australian whiteness all helped to open discursive spaces to reconsider key state interventions in Indigenous programs such as reconciliation, Native title and social inclusion in ways that have methodologically and ethically emphasized Indigenous rights as demanding consideration in narrating Australian geographies.

The centrality of these concerns to the disciplinary discourse in Australia was reflected in 2002 when the Institute of Australian Geographers discussed Indigenous rights as a disciplinary concern at its annual conference and established a new study group with a mandate not simply to study Indigenous issues, but to advocate Indigenous rights and advance links between geography and the Indigenous peoples' movement.

The internationally influential work of Anderson (Anderson, 1991, 1993a, 1993b,1998, 2000, 2003, 2005, 2007; Anderson and Gale, 1992) and Jacobs (Gelder and Jacobs, 1995; Gooder and Jacobs, 2000; Huggins et al., 1995; Jacobs, 1988, 1990, 1996, 1997) has drawn deeply from the particularities of Antipodean perspectives on colonial processes to emphasize significant links between ideology, politics and power across space and time in innovative ways.

Gendered geographies and gendered geography departments

Feminist geography has had a strong, early and pervasive influence across Australia and Aotearoa New Zealand. The University of Waikato has been central here (see Stokes et al., 1987), beginning with an 'Alternative Perspectives' issue of *New Zealand Geographer* explaining:

Inspired by the pioneering work of Ann Magee in the establishment of a programme in Women's Studies at the University of Waikato [including courses in geography with a feminist perspective] there has been a deliberate effort to maintain a feminist perspective in the geography taught at Waikato (Stokes, 1987b, 139).

Over the intervening years, Waikato has continued to embrace feminism as a politically useful perspective (see Longhurst and Johnston, 2005) and scholars such as Wendy Larner, Louise Johnson, Lawrence Berg, Lynda Johnston, Margaret Begg, Yvonne Underhill-Sem and more recently Carey-Ann Morrison have all taught gender issues and/or completed doctoral theses that highlight gender issues at Waikato. Other university geography departments in Aotearoa have, generally speaking, been less keen on taking up feminist perspectives (although see Law et al., 1999; Panelli, 2003), and there is continuing concern about the place of women in Australian geography and the presence of feminist perspectives in undergraduate education (Johnson, 2000). Notwithstanding, the intellectual contributions of Antipodean feminist geography to critical geographies have been at least threefold.

First, feminist frameworks and perspectives have been important in informing a range of critical geographies. For example, work on labour markets and economic restructuring has included consideration of their gendered impacts (e.g. Johnson, 1990), which continues through poststructural political economy (e.g. Gibson-Graham, 1996; Larner, 2000). Feminist thought is similarly inflected through postcolonial geographies, geographies of sexuality and critical geographies of ethnicity and race. Feminist geographies of embodiment (e.g. Longhurst, 2001; Teather, 1999) have established critical new research pathways. There remains debate, nonetheless, on the relative invisibility of feminism as a key driver of critical geographies in Australia.

Second, and where feminist frameworks are most apparent in Antipodean critical geographies, is the elaboration of the gendered nature of urban spaces. As in North America, suburban neighbourhoods, houses and lives became the focus of feminist critiques, pointing out the narrow perceptions of gender assumed to operate in these spaces, and their connections to heterosexism (Johnson, 1990; Johnson and Valentine, 1995). More recently, the gendered narratives of processes of property development have been highlighted. Fincher (2004), for example, demonstrates the limited conceptions of family and gender articulated by high-rise property developers in Melbourne. Law (1999) similarly scrutinizes gendered assumptions, though in this case the assumptions of the women and transport literature. Rather than see women simply as a demographic dimension of transport studies, Law argues, geographers need to think about gendered patterns of mobility, and, as a consequence, new conceptualizations of the city.

Third, the necessary and intricate connections between gender and other axes of identity have been at the forefront of these feminist geographies. Gendered bodies, sexualities and racial identities, to list just a few, remain key foci, in relation to home (Gorman-Murray, 2008; Pulvirenti, 2000), urban structure (Fincher and Jacobs, 1998) and urban politics (Anderson and Jacobs, 1997).

Absences and ways forward in Antipodean critical geographies

Over the past decade, specific avenues of critical geography have continued to be developed in Australia and Aotearoa New Zealand, not just poststructuralist geographies of the economy, Māori or (post)colonial geographies, or feminist and sexuality geographies but also others. Some key examples within the New Zealand context are participatory geographies (Kindon et al., 2007); critiques of 'mainstream' population geographies (Underhill-Sem, 2001); rural geographies (Panelli, 2003); queer geographies (Johnston, 2012); geographies of the body (Longhurst, 2001); critical cultural studies (Cupples and Glynn, 2009); and work that links culture, health and place (Kearns and Moon, 2002). In Australia, the geographically rich 'community economies' collective is working towards transforming future economic and political possibilities in a wide range of places (see www.communityeconomies.org), complemented by activist work conceptualizing new models of housing provision (Crabtree 2008). Radical geographical critiques have emerged in a range of policy areas, including urban planning (Dowling and McGuirk, 2008; Forster, 2004; Hillier, 1993; Sandercock, 1995), social inclusion and engagement (Fincher, 1995; Fincher & Jacobs 1998) and the interplay of justice and sustainability across a range of settings (Kirkpatrick, 1988, 1998, 2011). This has focussed much attention of critical geographers, as has the persistence of developmentalism in national development assistance, disaster relief and resource management settings across the Asia-Pacific (Hirsch, 1988; Koczberski and Curry, 2004). While this has seen Antipodean geography integrated into many 'northern' discourses, the southern critique continues to challenge many of the absences in the mainstreams of critical discourse in Anglo-American geography. There are of course also many other examples too numerous to mention here – projects that highlight 'down under' spaces, places and issues, span diverse theoretical territory and speak to local and global politics.

However, there are also some noticeable gaps in the critical geographies literature that are worth reflecting upon. While proud traditions of political engagement amongst diverse geographers across the Antipodes have developed since the emergence of radical and critical geography as an approach to disciplinary theory and practice, there are also important absences in Antipodean critical geographies. For example, fostering participation of geographers from diverse ethnic backgrounds has lagged behind the discipline's examination of ethnic diversity, immigration and demography. Much of the best efforts of Antipodean radicals remain less accessible than desirable because it is published in reports and community outlets rather than peer-reviewed outlets. While significant theoretical work across

Antipodean geography has been highly political in its impact and orientation, it has been less explicitly engaged with critical political geographies than the political geographies developing elsewhere.

As bodies of work evolve, it is worth paying attention not to that which has emerged but also to what is missing and what might be addressed in the future. We think an important direction for future research continues to be indigenous people, spaces and places. As we have highlighted, there is undoubtedly some excellent work that has long been carried out in this area, but there are still many opportunities for geographers and others to engage politically with issues of race, ethnicity, indigeneity and colonization. Living 'down under' one quickly becomes aware that issues of biculturalism, multiculturalism, racism, dispossession and (post)colonisalism are part of the everyday. They are part of local, regional and national politics and cannot, nor should not be, disentangled from a range of other issues. It is our hope that geographers and others working in Australasia will continue to draw on and develop a rich array of theories and methods not just to develop interesting and varied academic treatises but to prompt political action, action that seeks to understand, question, challenge and change the conditions under which people live their lives.

References

Agius, P., J. Davies, R. Howitt, S. Jarvis and R. Williams. 2004. Comprehensive native title negotiations in South Australia. In, M. Langton, M. Teehan, L. Palmer and K. Shain (eds.). *Honour Among Nations? Treaties and Agreements with Indigenous People*. Melbourne: Melbourne University Press, pp. 203–19.

Anderson, K. 1991. *Vancouver's Chinatown: Racial Discourse in Canada, 1875-1980*. Montreal: McGill-Queen's University Press.

Anderson, K. 1993a. Place narratives and the origins of Sydney's Aboriginal settlement, 1972-73. *Journal of Historical Geography* 19, 314–55.

Anderson, K. 1993b. Otherness, culture and capital: 'Chinatown's' transformation under Australian multiculturalism. In, G. Clarke, D. Forbes and R. Francis (eds.), *Multiculturalism, Difference and Postmodernism*. Melbourne: Longmans Cheshire, pp. 68–89.

Anderson, K. 1998. Science and the savage: The Linnean Society of NSW, 1874-1900. *Ecumene* 5, 125–43.

Anderson, K. 2000. Thinking 'postnationally': Dialogue across multicultural, indigenous and settler spaces. *Annals of the Association of American Geographers* 90, 381–91.

Anderson, K. 2003. White natures: Sydney's Royal Agricultural Show in post-humanist perspective. *Transactions of the Institute of British Geographers* 28, 422–41.

Anderson, K. 2005. Australia and the 'State of Nature/Native': Griffith Taylor Lecture, Geographical Society of New South Wales, 2004. *Australian Geographer* 36, 267–82.

Anderson, K. 2007. *Race and the Crisis of Humanism*. London: Routledge.

Anderson, K. and F. Gale. 1992. *Inventing Places: Studies in Cultural Geography*. Melbourne: Longman Cheshire.

Anderson, K. and F. Gale. 1999. *Cultural Geographies*. Sydney: Longman.

Anderson, K. and J.M. Jacobs. 1997. From urban Aborigines to Aboriginality and the city: One path through the history of Australian cultural geography. *Australian Geographical Studies* 35, 12–22.

Anderson, K. and J.M. Jacobs. 1999: Geographies of publicity and privacy: Residential activism in Sydney in the 1970s. *Environment and Planning A* 31, 1017–30.

August, W. 2005. Maori women: Bodies, spaces, sacredness and mana. *New Zealand Geographer* 61, 117–23.

Baker, R. 1999. *Land Is Life: From Bush to Town - the Story of the Yanyuwa People.* Sydney: Allen & Unwin.

Barclay-Kerr, K. 1991. Conflict over Waikato Coal: Maori land rights. In, J. Connell and R. Howitt (eds.), *Mining and Indigenous Peoples in Australasia.* Sydney: Sydney University Press, pp. 183–95.

Bedford, R. 2005. Obituary – Evelyn Mary Stokes (nee Dinsdale) DNZM, MA (NZ), PhD (Syr) 5 December 1936 – 11 August 2005. *New Zealand Geographer* 61, 242–9.

Berg, L.D. and R.A. Kearns. 1996. Naming as norming: "Race", gender, and the identity politics of naming places in Aotearoa/New Zealand. *Environment and Planning D: Society and Space* 14, 99–122.

Berg, L.D. and R.A. Kearns. 1998. America unlimited. *Environment and Planning D: Society and Space* 16, 128–32.

Berry, M. 1981. Posing the housing question in Australia: Elements of a theoretical framework for a Marxist analysis of housing. *Antipode* 13, 3–14.

Britton, S. and R. Le Heron. 1989. Capitalist relations and the geography of industrial restructuring: a debate. Workshop on 'Enterprises and Restructuring', 23-24 September 1988. Canberra: Department of Geography and Oceanography, University College, University of New South Wales.

Cant, G. 1990. Waitangi: Treaty and tribunal. *New Zealand Journal of Geography* 89, 7–12.

Cant, G. 2005. Windows into the work of a geographical educator: Evelyn Stokes from the University of Waikato. *New Zealand Geographer* 61, 89–91.

Cant, G., J. Overton and E. Pawson. 1993. *Indigenous Land Rights in Commonwealth Countries: Dispossession, Negotiation and Community Action.* Christchurch: Department of Geography, University of Canterbury and the Ngai Tahu Maori Trust Board for the Commonwealth Geographical Bureau.

Connell, R. 2007. *Southern Theory: The Global Dynamics of Knowledge in Social Science.* Sydney: Allen & Unwin.

Coombes, B. 2007. Defending community? Indigeneity, self-determination and institutional ambivalence in the restoration of Lake Whakaki. *Geoforum* 38, 60–72.

Crabtree, L. 2008. The role of tenure, work and cooperativism in sustainable urban livelihoods. *Acme* 7(2), 260–82.

Cupples, J. and K. Glynn. (eds.). 2009. *New Zealand Geographer* 65(1) (Special Issue: Countercartographies: New (Zealand) Cultural Studies/Geographies and the City).

Davies, J. 2003. Contemporary geographies of Indigenous rights and interests in rural Australia. *Australian Geographer* 34, 19–45.

Dowling, R. and P.M. McGuirk. 1998. Gendered geographies in Australia, Aotearoa/ New Zealand and the Asia-Pacific. *Australian Geographer* 29(3), 279–91.

Dowling R. and P.M. McGuirk. 2008. Cities of Australia and the Pacific Islands. In, S. Brunn, M. Hays-Mitchell and D. Ziegler (eds.), *Cities of the World*, 4th edition. Rowman and Littlefield.

Drakakis-Smith, D. 1981. Aboriginal underdevelopment in Australia. *Antipode* 13, 35–44.

Fagan, B. 1984. Corporate strategy and regional uneven development in Australia: the case of BHP Ltd. In, M. Taylor (ed.), *The Geography of Australian Corporate Power*. Sydney: Croom Helm, pp. 91–123.

Fagan, B. 1987. Australia's BHP Ltd – an emerging transnational resources corporation. *Raw Materials Report* 4, 46–55.

Fagan, B. and R. Dowling. 2005. Neoliberalism and Suburban Employment: Western Sydney in the 1990s. *Geographical Research* 43(1), 71–81.

Fagan, R.H. 1981. Geographically uneven development: Restructuring of the Australian aluminium industry. *Australian Geographic* 19, 141–60.

Fagan, R.H. 1988. Corporate structure and Australian industry: A geography of industrial restructuring. In, R.L. Heathcote (ed.), *The Australian Experience: Essays in Australian Land Settlement and Resource Management*. Melbourne: Longman Cheshire, pp. 25–36.

Fagan, R.H. and R.D. Le Heron. 1994. Reinterpreting the geography of accumulation: The global shift and local restructuring. *Environment and Planning D: Society and Space* 12, 265–85.

Fagan, R.H. and M. Webber. 1999. *Global Restructuring: the Australian Experience*. Melbourne: Oxford University Press.

Fincher, R. 1995. Women, immigration and the state: Issues of difference and social justice. In, A. Edwards and S. Magarey (eds.), *Women in a Restructuring Australia: Work and Welfare*. Sydney: Allen and Unwin, pp. 203–22, 280–303.

Fincher, R. 2004. Gender and life course in the narratives of Melbourne's high-rise housing developers. *Australian Geographical Studies* 42, 325–38.

Fincher, R. and J. Jacobs. 1998. *Cities of Difference*. New York: Guildford Press.

Forer, P. 1987. Alternative perspectives. *New Zealand Geographer* 4(3) (Special Issue: Alternative Perspectives).

Forster, C. 2004. *Australian Cities: Continuity and Change*. Melbourne: Oxford University Press.

Freestone, R. 1981. Planning for profit in urban Australia 1900-1930: A descriptive prolegomenon. *Antipode* 13, 15–26.

Gale, F. 1983. *We Are Bosses Ourselves: The Status and Role of Aboriginal Women Today*. Canberra: Australian Institute of Aboriginal Studies.

Gale, F. and A. Brookman. 1972. *Urban Aborigines*. Canberra: Australian National University Press.

Gelder, K. and J.M. Jacobs. 1995. Uncanny Australia. *Ecumene* 2, 171–83.

Gibson-Graham, J.K. 1994. 'Stuffed if I know!': Reflections on post-modern feminist social research. *Gender, Place and Culture* 1, 205–24.

Gibson-Graham, J.K. 1996. *The End of Capitalism (As We Knew It): A Feminist Critique of Political Economy*. Oxford: Blackwell.

Gibson, K. 1991. Company towns and class processes: A study of the coal towns of Central Queensland. *Environment and Planning D: Society and Space* 9, 285–308.

Gibson, K. 1992. Hewers of cake and drawers of tea: Women and restructuring in the coalfields of the Bowen Basin. *Rethinking Marxism* 5, 29–56.

Gibson, K.D. and R.J. Horvath. 1983. Global capital and the restructuring crisis in Australian manufacturing. *Economic Geography* 59, 178–94.

Gooder, H. and J. Jacobs. 2000. 'On the border of the unsayable': the apology in post-colonizing Australia. *Interventions: International Journal of Postcolonial Studies* 2, 229–47.

Gorman-Murray, A. 2008. Queering the family home: Narratives from gay, lesbian and bisexual youth coming out in supportive family homes in Australia. *Gender, Place and Culture* 15, 31–44.

Gorman-Murray, A., G. Waitt and L. Johnston (guest eds.). 2008. *Australian Geographer* 39(3) (Special Issue: 'Geographies of Sexuality and Gender 'Down Under').

Head, L. 2000: *Second Nature: The History and Implications of Australia as Aboriginal Landscape*. Syracuse: Syracuse University Press.

Hillier, J. 1993. To boldly go where no planners have ever. *Environment and Planning D: Society and Space* 11, 89–113.

Hirsch, P. 1988. Dammed or damned? Hydropower versus people's power. *Bulletin of Concerned Asian Scholars* 20(1), 2–10.

Horvath, R.J. and P. Rogers. 1981. Class structure in Australian history: A review article. *Antipode* 13, 45–9.

Howitt, R. 1989. Resource development and Aborigines: The case of Roebourne. *Australian Geographic Studies* 27, 155–69.

Howitt, R. 1991a. Aborigines and gold mining in Central Australia. In, J. Connell and R. Howitt (eds.), *Mining and Indigenous Peoples in Australasia*. Sydney: University of Sydney Press, pp. 119–37.

Howitt, R. 1991b. Aborigines and restructuring in the mining sector: Vested and representative interests. *Australian Geographer* 22, 117–9.

Howitt, R. 1992a. The political relevance of locality studies: A remote Antipodean viewpoint. *Area* 24, 73–81.

Howitt, R. 1992b. Weipa: Industrialisation and indigenous rights in a remote Australian mining area. *Geography* 77, 223–35.

Howitt, R. (1993). Aborigines, mining and regional restructuring: applied peoples' geography? In, G. Cant, J. Overton, E. Pawson and R. Baker (eds.), *Indigenous Land Rights in Commonwealth Countries: Dispossession, Negotiation and Community Action*. Christchurch: Department of Geography, University of Canterbury and the Ngai Tahu Maori Trust Board for the Commonwealth Geographical Bureau, pp. 144–53.

Howitt, R. and S. Jackson. 1998. Some things do change: Indigenous rights, geographers and geography in Australia. *Australian Geographer* 29, 155–73.

Huggins, J., R. Huggins and J.M. Jacobs. 1995. Kooramindanjie: Place and the post-colonial. *History Workshop Journal* 39, 164–81.

Jackson, S. 1997. A disturbing story: The fiction of rationality in land use planning in Aboriginal Australia. *Australian Planner* 34, 221–26.

Jackson, S., M. Storrs and J. Morrison. 2005. Recognition of Aboriginal rights, interests and values in river research and management: Perspectives from northern Australia. *Ecological Management & Restoration* 6, 105–10.

Jacobs, J.M. 1988. Politics and the cultural landscape: The case of Aboriginal land rights. *Australian Geographic* 26, 249–63.

Jacobs, J.M. 1990: *The Politics of the Past: Redevelopment in London*. London: University of London.

Jacobs, J.M. 1996: *Edge of Empire: Postcolonialism and the City*. London and New York: Routledge.

Jacobs, J.M. 1997. Resisting reconciliation: The social geographies of (post)colonial Australia. In, S. Pile and M. Keith (eds.), *Geographies of Resistance*. London and New York: Routledge, pp. 203–18.

Johnston, J. and R. Le Heron. 1987. A deepening stream. *New Zealand Geographer* 43(3), 115–7.

Johnson, L. 1990. New patriarchal economies in the Australian textile industry. *Antipode* 22, 1–32.

Johnson, L. 1994. Occupying the suburban frontier: Accommodating difference on Melbourne's urban fringe. In, A. Blunt and G. Rose (eds.), *Writing Women and Space: Colonial and Post-colonial Perspectives*. New York: Guilford, pp. 141–68.

Johnson, L. 2005. Stoking the fires of enquiry – Evelyn Stokes. Contribution to Waikato and New Zealand geography, 1988–90. *New Zealand Geographer* 61, 92–3.

Johnson, L. and G. Valentine. 1995. Wherever I lay my girlfriend, that's my home: The performance and surveillance of lesbian identities in domestic environments. In, D. Bell and G. Valentine (eds.), *Mapping Desire: Geographies of Sexualities*. London: Routledge, pp. 99–113.

Johnson, L. (with J. Huggins and J. Jacobs). 2000. *Placebound: Australian Feminist Geographies*. Sydney: Oxford University Press.

Johnston, L. 2012. Sites of excess: The spatial politics of touch for drag queens in Aotearoa New Zealand. *Emotion, Space and Society* 5(1), 1–9.

Kearns, R. and G. Moon. 2002. From medical to health geography: Novelty, place and theory after a decade of change. *Progress in Human Geography* 26(5), 605–25.

Kindon, S., R. Pain and M. Kesby (eds.). 2007. *Participatory Action Research: Connecting People, Participation and Place*. London: Routledge.

Kirkpatrick, J.B. 1988. Heritage and development in Tasmania. *Australian Geographer* 19(1), 46–63.

Kirkpatrick, J.B. 2011. The political ecology of soil and species conservation in a 'Big Australia'. *Geographical Research* 49(3), 276–85.

Kirkpatrick, J.B. 1998. Nature conservation and the regional forest agreement process. *Australian Journal of Environmental Management* 5(1), 31–37.

Koczberski, G. and G. Curry. 2004. Divided communities and contested landscapes: Mobility, development and shifting identities in migrant destination sites in Papua New Guinea. *Asia Pacific Viewpoint* 45(3), 357–71.

Larner, W. 1996. The "new boys": Restructuring in New Zealand 1984-1994. *Social Politics* 3, 32–56.

Larner, W. 1997. "A means to an end": Neo-liberalism and state processes in New Zealand. *Studies in Political Economy* 52, 7–38.

Larner, W. 1998. Hitching a ride on a tiger's back: Globalisation and spatial imaginaries in New Zealand. *Environment and Planning D: Society and Space* 16, 599–614.

Larner, W. 2000. Neo-liberalism: Policy, ideology, governmentality. *Studies in Political Economy* 63, 5–26.

Larner, W. and R. Le Heron. 2002. From economic globalisation to globalising economic processes: Towards post-structural political economies. *Geoforum* 33, 415–9.

Law, R. 1999. Beyond 'women and transport': Towards new geographies of gender and daily mobility. *Progress in Human Geography* 23(4), 567–88.

Law, R., H. Campbell and J. Dolan (eds.). 1999. *Masculinities in Aotearoa/New Zealand*, Palmerston North: Dunmore Press.

Longhurst, R. 2001. *Bodies: Exploring Fluid Boundaries*. London: Routledge.

Longhurst, R. 2008. Afterword: Geographies of sexuality and gender 'down under', *Australian Geographer* 39(3), 381–7.

Longhurst, R. and J. Johnston. 2005. Changing bodies, spaces, places and politics: Feminist geography at the University of Waikato. *New Zealand Geographer* 61(2), 94–101.

McClean, R., L.D. Berg and M.M. Roche. 1997. Responsible geographies: Co-creating knowledge in Aotearoa. *New Zealand Geographer* 53, 9–15.

McGuirk P.M. and R. Dowling. 2009. Master-planned residential developments: Beyond iconic spaces of neoliberalism? *Asia Pacific Viewpoint* 50, 120–34.

McHugh, P.G. 1996. The legal and constitutional position of the Crown in resource management. In, R. Howitt, J. Connell and P. Hirsch (eds.), *Resources, Nations and Indigenous Peoples: Case Studies from Australasia, Melanesia and Southeast Asia.* Melbourne: Oxford University Press, pp. 300–16.

McManus, P. and B. Pritchard (eds.). 2000. *Land of Discontent: The Dynamics of Change in Rural and Regional Australia.* Sydney: UNSWPress.

Mercer, D. 1993. Terra nullius, Aboriginal sovereignty and land rights in Australia: The debate continues. *Political Geography* 12, 299–318.

Mercer, D. 1997. Aboriginal self-determination and indigenous land title in post-Mabo Australia. *Political Geography* 16, 189–212.

Monk, J. 1974. Australian Aboriginal social and economic life: Some community differences and their causes. In, L.J. Evenden and F.F. Cunningham (eds.), *Cultural Discord in the Modern World; Geographical Themes.* Vancouver: Tantalus Research, pp. 157–74.

Muller, S. 2008. Community-based management of saltwater country, Northern Australia. *Development* 51, 139–43.

Muller, S., E.R. Power, S. Suchet-Pearson, S. Wright and K. Lloyd. 2009. "Quarantine matters!": quotidian relationships around quarantine in Australia's northern borderlands. *Environment and Planning A* 41, 780–95.

Murton, B. 2006. 'Toheroa Wars': Cultural politics and everyday resistance on a northern New Zealand beach. *New Zealand Geographer* 62, 25–38.

Nader, L. 1974. Up the anthropologist – perspective gained from studying up. In, D. Hymes (ed.), *Reinventing Anthropology.* New York: Vintage Books, pp. 284–311.

O'Dea, D. 2000. *The Changes in New Zealand's Income Distribution.* Treasury Working Papers Series 00/13. New Zealand Treasury http://ideas.repec.org/p/nzt/nztwps/00-13.html (27 December 2020).

O'Neill, P.M. 1996. In what sense a region's problem? The place of redistribution in Australia's internationalization strategy. *Regional Studies* 30(4), 401–411.

Panelli, R. 2003. Gender research in rural geography. *Gender, Place and Culture* 10, 281–89.

Pawson, E. and G. Cant. 1992. Land rights in historical and contemporary context. *Applied Geography* 12, 95–108.

Pulvirenti, M. 2000. The morality of immigrant home ownership: gender, work and Italian-Australian *sistemazione. Australian Geographer* 31, 237–49.

Sandercock, L. 1995. Voices from the borderlands: A meditation on a metaphor. *Journal of Planning Education and Research* 14, 77–88.

Searle, G.H. 1981. The role of the state in regional development: The example of non-metropolitan New South Wales. *Antipode* 13, 27–34.

Shaw, W.S. 2000. Ways of whiteness: Harlemising Sydney's Aboriginal Redfern. *Australian Geographical Studies* 38, 291–305.

Shaw, W.S. 2006. Decolonizing geographies of whiteness. *Antipode* 38, 851–69.

Sinclair, K. 1969. *A History of New Zealand*, 2nd edition. Harmondsworth: Penguin.

SKCAN (Situated Knowledge Collaborative Auckland and Newcastle). 2011. Kitchen stories: an introduction to the situated knowledges sessions. Paper presented to the Institute of Australian Geographers' Conference, Wollongong, July 2011.

Smith, L.T. 1999. *Decolonizing Methodologies: Research and Indigenous Peoples.* Dunedin and London: University of Otago Press and Zed Books.

Spate, O.H.K. 1965. *Let Me Enjoy: Essays, Partly Geographical.* Canberra: Australian National University.

Stokes, E. 1987a. Maori geography or geography of Maoris. *New Zealand Geographer* 43, 118–23.

Stokes, E. 1987b. By way of introduction. *New Zealand Geographer* 43, 139.

Stokes, E. 1992. The Treaty of Waitangi and the Waitangi Tibunal: Maori claims in New Zealand. *Applied Geography* 12, 176–91.

Stokes, E. 1993. The Treaty of Waitangi and the Waitangi Tribunal: Maori Claims in New Zealand. In, G. Cant, J. Overton and E. Pawson (eds.), *Indigenous Land Rights in Commonwealth Countries: Dispossession, Negotiation and Community Action.* Christchurch: Department of Geography, University of Canterbury and Ngai Tahu Maori Trust Board, pp. 66–80.

Stokes, E., L. Dooley, L. Johnson, J. Dixon and S. Parsons. 1987. Feminist perspectives in geography: A collective statement. *New Zealand Geographer* 43, 139–49.

Suchet, S. 1996. Nurturing culture through country: Resource management strategies and aspirations of local landowning families in Napranum. *Australian Geographical Studies* 34, 200–15.

Taylor, M.P. and R. Stokes. 2005. When is a river not a river? Consideration of the legal definition of a river for geomorphologists practising in New South Wales, Australia. *Australian Geographer* 36, 183–200.

Te Ara. 2005. Distribution of disposable income, 1982 and 1996. *The Encyclopedia of New Zealand.* https://teara.govt.nz/en/diagram/3731/distribution-of-disposable-income-1982-and-1996 (27 December 2020).

Teather, E. (ed.). 1999. *Embodied Geographies: Spaces, Bodies and Rites of Passage.* London and New York: Routledge.

Thom, B. and F. Mckenzie. 2011. The population policy debate from a natural resource perspective: reflections from the Wentworth Group. *Geographical Research* 49(3), 348–62.

Tipa, G. and Nelson, K. 2008. Introducing cultural opportunities: A framework for incorporating cultural perspectives in contemporary resource management. *Journal of Environmental Policy & Planning* 10, 313–37.

Underhill-Sem, Y. 2001. Maternities in "out-of-the-way places": Epistemological possibilities for retheorising population geography. *International Journal of Population Geography* 7, 447–60.

Williamson, F., V. Milligan, K. Gibson and B. Fagan. 1981. Antipodean Antipode: An introduction. *Antipode* 13, 1–3.

Young, E. 1992. Aboriginal land rights in Australia: Expectations, achievements and implications. *Applied Geography* 12, 146–61.

Young, E. 1995. *Third World in the First: Development and Indigenous Peoples.* London: Routledge.

Index

Note: Italicized, bold and bold italics page numbers refer to figures and tables; Page numbers followed by "n" refer to notes.

Printed in the United States
by Baker & Taylor Publisher Services

Printed in the United States
by Baker & Taylor Publisher Services